DIRECT METHANE TO METHANOL

DIRECT METHANE TO METHANOL

Foundations and Prospects of the Process

VLADIMIR ARUTYUNOV

ELSEVIER

AMSTERDAM • BOSTON • HEIDELBERG • LONDON • NEW YORK • OXFORD
PARIS • SAN DIEGO • SAN FRANCISCO • SINGAPORE • SYDNEY • TOKYO

Elsevier
225, Wyman Street, Waltham, MA 02451, USA
The Boulevard, Langford Lane, Kidlington, Oxford OX5 1GB, UK
Radarweg 29, PO Box 211, 1000 AE Amsterdam, The Netherlands

British Library Cataloguing in Publication Data
A catalogue record for this book is available from the British Library

Library of Congress Cataloging-in-Publication Data
A catalog record for this book is available from the Library of Congress

ISBN: 978-0-444-63253-1

Printed and bound in Great Britain
14 15 16 17 18 10 9 8 7 6 5 4 3 2 1

Contents

Direct Methane To Methanol: Foundations and Prospects of the Process

Vladimir Arutyunov

Semenov Institute of Chemical Physics, Russian Academy of Sciences, Moscow, Russia

The direct conversion of methane and other light hydrocarbons to methanol (DMTM) and other oxygenates has been a long-sought goal of the gas industry for decades. In spite of some experience of industrial applications in the 1930–1940s, nowadays this process remains in the sphere of academic rather than practical interest. However, some new tendencies in the world's gas industry rekindled interest in this process. The monograph reviews the modern state of knowledge in this area and discusses some promising practical applications.

INTRODUCTION

Intense scientific research and technological development in the field of direct conversion of natural gas to chemicals has been going on for a century. The reason for the unflagging interest in this problem is twofold. On the one hand, it has been found that, under certain conditions, the oxidation of even simple hydrocarbons, such as methane, is accompanied by a rapidly increasing complexity of the system, with the formation of a multitude of different products, from water and carbon oxides to various carbon compounds, including such complex as fullerenes and carbon nanotubes. Therefore, even though the oxidation of methane and its closest homologues gave, in due time, the main body of data underlying the modern concepts of the kinetics of gas-phase reactions, these processes even now seem sophisticated and inexhaustible, like the nature itself. On the other hand, the chemical conversion of light alkanes, the main components of natural and associated gas, has become one of the most important trends in modern petro-chemistry and in a full-fledged new industrial sector: gas chemistry.

The wide variety of products of the oxidative conversion of natural gas includes hydrogen, syngas, ethylene, other olefins, oxygenates, and aromatic compounds. The scope of the present monograph is much narrower, limited to the direct conversion of natural gas to oxygen-containing compounds (oxygenates), mainly methanol and formaldehyde, under predominantly homogeneous gas-phase conditions. Nowadays, this direction revealed potentiality for new practical applications, in particular concerning low resources and remote gas fields, shale gas, and other unconventional gas resources.

Since the predominant and most hardly convertible component of natural and associated gas is methane, with the main product of the direct oxidative conversion of methane and other light alkanes C_2–C_4 being methanol, the overall process of natural gas oxidative conversion to oxygenates has traditionally been referred to as Direct Methane To Methanol (DMTM), a term that will be used throughout the monograph. It covers the totality of processes of conversion of natural gas light alkanes to oxygenates. In addition, the partial oxidation of methane to methanol is the subject of the overwhelming majority of published academic and technological investigations on the direct gas-phase conversion of alkanes to oxygenates. On the other hand, there are only limited number publications concerning heavier hydrocarbons; therefore, the focus will be on analysis of methane partial oxidation to oxygenates, with generalization to other alkanes given separately.

As an industrial technology, DMTM has seen a sweeping rise and subsequent decline, mainly due to the lack of understanding of the underlying complex chemistry. However, scientific research on the process have never stopped, as evidenced by a large number of publications, including very recent, cited in the monograph. Of the most attractive features of DMTM, apart of its technological simplicity, are a very efficient internal energy self-supply by the partial oxidation of the hydrocarbon itself and the possibility of direct formation of nonequilibrium products that cannot be produced in significant amounts under thermodynamically equilibrium conditions.

Current interest in DMTM is prompted by significant changes in the gas industry, which strongly needs low-scale technologies capable of operating on low resources, remote and unconventional gas fields. In the technological niche of relatively simple small-capacity processes, the direct partial oxidation of natural and associated gas to methanol is still one of the most promising and elaborated directions. Along with fundamental advances in studying the mechanism of the oxidation of hydrocarbons at moderate temperatures and high pressures, it gives a good chance to revive DMTM as an industrial technology.

In this connection, a good deal of the monograph is devoted to the prospects of practical applications of DMTM and its possible future role in the gas industry. There is a vast room for further improvements of the process given that the interest from the industry will promote necessary efforts. Thus, the need in alternative simple and flexible low-scale Gas-to-Liquid (GTL) technologies may become a real driving force for a DMTM refinement.

The monograph reviews the modern state of knowledge on DMTM with emphasis on its most evident practical applications. It summarizes and supplements our previous reviews [1–4] and monographs [5, 6] on the subject, including the problem of modeling the homogeneous–heterogeneous oxidation of light alkanes [7], as well as a number of recent interesting but somewhat contradictory experimental results. The monograph covers more than 300 publications on the high-pressure low-temperature oxidation of methane and other gaseous hydrocarbons over a period from the beginning of the last century up to now and offers a unique comprehensive analysis of practically all relevant experimental and theoretical data on the DMTM process. As well it discusses the real practical prospects of DMTM.

We believe that it will provide researchers and engineers working in the field of gas production and processing with necessary theoretical knowledge and experimental data for development and application of new processes involving the direct oxidation of natural gas. We hope that it will be useful for all those who work in the field of gas processing and gas chemistry by facilitating their efforts in practical development and utilization of new effective gas chemical process.

The author feels obliged to express his deep gratitude to late Professor V.I. Vedeneev for many years of joint work on the DMTM process and especially its mechanism and to late Professor O.V. Krylov for very fruitful cooperation and discussions of the whole field of hydrocarbon oxidation. The author is also grateful to Professor V.I. Savchenko and Dr. V.M. Rudakov for long-standing collaboration, which has produced practically all of our recent results on DMTM. Author thanks Dr. V.N. Smirnov for his big help in preparation of English version of this monograph.

CHAPTER

1

Historical Review on the DMTM

As far back as the beginning of the last century, Bone revealed the principal possibility of producing valuable oxygenates by direct oxidation of methane [8–10]. However, a real interest in the problem arose in the 1920–30s in connection with the emergence of DMTM-based industrial processes [11,12]. Another stimulus was to verify some of the then theoretical concepts concerning the mechanism of the gas-phase oxidation of hydrocarbons [13]. At the beginning of the 1930s, several different groups of researchers [14–18] have almost simultaneously demonstrated the possibility of achieving a high selectivity of methanol formation in the high-pressure gas-phase oxidation of methane. This gave an impetus to subsequent efforts to increase the yield of oxygenates and to develop industrial processes for their production by the direct oxidation of methane and heavier alkanes. Along with some early patents on the catalytic oxidation of methane to oxygenates [19], there have appeared the first patents on the high-pressure gas-phase partial oxidation of methane [20,21].

In the 1930s, thorough investigations of DMTM were performed by Newitt and coworkers [16–18,22,23]. They studied gas-phase oxidation of methane and other hydrocarbons at high pressures to elucidate a number of fundamental issues concerning the role of alcohols in the oxidation of hydrocarbons. These works were first to demonstrate the possibility of achieving high yields of alcohols and aldehydes in the direct oxidation of hydrocarbons. Later, it was shown that methanol is also formed at atmospheric pressure, even during the induction period [24]. Although the selectivity of methanol formation for some of the stages of methane oxidation at atmospheric pressure attains ~20%, the integrated selectivity does not exceed 5% [25]. Thus, high pressure is a key factor in providing significant methanol yields.

Already in the first works in the 1930s, the optimal conditions for DMTM were identified: high pressure (~100 atm) [16,17,26], moderate temperature (400–500 °C) [27], and low oxygen concentration [15,27]. The attainable selectivity of methanol was demonstrated to be as high as 60% [16,17]. A number of kinetic features of the process were also established: the methanol selectivity decreased sharply with increasing oxygen concentration [15,27], and a high CO/CO_2 ratio in the products was observed [10]. Since then up to now, the main objective of most studies has been to determine conditions that would provide high and stable yields of the target products, so as to make DMTM competitive with other technologies.

By the middle of the last century, the partial oxidation of methane and, later, propane, butane, and mixtures thereof extracted from oil-associated, crude stabilization, and processing gases became a widespread petrochemical process in the United States [28].

Direct Methane to Methanol
http://dx.doi.org/10.1016/B978-0-444-63253-1.00001-5

The subsequent decline of this technology in the late 1950s was associated with the rapid progress of competing technologies based on the preconversion of hydrocarbons into syngas and subsequent catalytic synthesis of target products and with the development of the market of propane and butane as domestic fuel and raw material for many petrochemical processes. Another factor that contributed to this was the difficulty of extracting individual components from a complex mixture of products of the nonselective gas-phase oxidation of C_3-C_4 hydrocarbons, which were used in the then technologies. Important, in our view, is the fact that the partial oxidation technology of that period was based solely on empirical knowledge. Small scope of fundamental research and the lack of clear understanding of the mechanism became a serious obstacle to the improvement of the process.

Although studies on methane partial oxidation have never stopped and continued beyond this period, and even attempts to introduce new industrial processes have been made (see, e.g., [29]), none of them found practical implementation. At the same time, under the influence of the theory of branched-chain reactions, developed by Semenov and his co-workers [30], a new understanding of mechanism of the oxidation of hydrocarbons began to take shape. Results of more than half a century of research in the oxidation of hydrocarbons and new concepts of hydrocarbon oxidation mechanism were summarized in a fundamental monograph by Shtern [13].

In the mid-1980s, interest in the direct production of oxygenates from methane rekindled due to the rapid growth of the role of natural gas in the global energy mix, the oil crisis of the 1970s, and the acute need in a clean motor fuel in major industrial countries. Methanol was considered as a convenient raw material for production of components of ecologically clean motor fuels and even as a potential fuel, features that boosted interest in technologies of its production from nonoil rough materials. To some extent, this interest was excited by several experimental studies that reported very high yields of methanol in the DMTM process [31,32].

This interest induced a number of reviews on the subject [31,33–38]. However, these reviews primarily dealt with the catalytic oxidation of methane, reflecting the traditional focus on catalytic technologies and the lack, at that time, of a clear understanding of the real mechanism of the DMTM process. With the exception of [31], they contained no new original data or conclusions, being based predominantly on compilations of previous results.

Since then, dozens of experimental works have been published. The most evident and widespread drawbacks of some of these papers, especially in the early period, which significantly complicate their analysis, are an incomplete presentation of the experimental conditions and the practice of simultaneously changing several experimental parameters. It may be supposed that, in the absence of clear ideas about the reaction mechanism, some researches practiced "random" (rather than time-consuming systematic) search for optimal conditions of methanol production. Such approach significantly depreciates the results and, in some, cases makes their analysis and inclusion in the common pool of DMTM data practically impossible.

Another common drawback of some of the works consists in attempts to make promising technological predictions based solely on a limited set of their own experimental results, without serious analysis, comparison with available data, or treatment within the framework of theoretical concepts. However, the claimed high parameters of the process are usually poorly reproducible, if at all. In addition, discrepancies between the results of experiments performed under very close conditions in the absence of a serious analysis of the underlying reasons give rise to the skeptical attitude regarding the practical applicability of DMTM.

A vast body of interesting but contradictory data published in recent years and a new level of theoretical understanding of the DMTM mechanism motivated us to perform a new comprehensive analysis of the process. Such analysis showed that the potentialities of the DMTM process is high enough but needs serious elaboration.

According to cost assessments, made some time ago, for the DMTM process to be competitive with traditional technologies, it should provide a selectivity of methanol formation (or the sum of valuable organic products) in excess of 77% at a reasonable conversion of methane, no less than 5% [39]. However, at present, for most of the experiments that can be reproduced at the industrial level, at a methane conversion higher than 5%, the methanol selectivity S_{CH3OH} does not exceed 50% (Appendix I). Nevertheless, even these results open possibilities for some interesting technological applications of the processes. Although several works reported significantly higher values of S_{CH3OH}, up to 60% [40,41], 80% [42,43], and even >90% [44] (see Appendix I), the authors of [45] and others failed to reproduce them. Possible reasons for such significant discrepancies will be discussed below.

By now, dozens of experimental studies on the gas-phase partial oxidation of methane, as well as ethane and heavier homologues, to oxygenates has been published. Facilities with various types of reactors, including static reactors, well-stirred flow reactors, and plug flow reactors of various sizes with a diameter of 5–30 mm operating at a flow rate from a few liters to 1000 m^3/h, have been used. The pressure was varied up to thousands of atmospheres. Along with the main parameters of the process, such as pressure, temperature, flow rate, and mixture composition, the effects of reactor surface material, heterogeneous catalysts, homogeneous promoters, and various physical methods of initiation of the process have been examined. The experiments were carried out using both premixed and individual supply of oxidant and hydrocarbon into the reactor, either cold or preheated. The oxidant was oxygen or air.

The most reliable experimental data on DMTM used in this analysis are summarized in Appendixes I and II. Note that many of the figures in the monograph, while plotted based on data from the cited original papers, are not necessarily present in these papers.

Since no detailed description of the experimental techniques used in the cited works will be given in the main text, we would like to note that most experiments in the 1930s, as well as experiments at very high (thousands of atmospheres) pressures [46–48] were performed in static reactors. In later studies, predominantly flow reactors have been used, which are more suited to the conditions of the practical implementation of the process. Experiments in flow reactors have been carried out over a wide pressure range (1–300 bar) and initial temperatures of 300–600 °C and above, with the reaction time ranging from a second to tens of minutes. In addition, more exotic reactors, such a cylinder of a high-compression internal combustion engine [49] or a rapid compression machine [50] were used.

Appendix I contains a summary of the experimental conditions and the results of the most informative (in our opinion) works on DMTM, an analysis of which enabled to draw most of the subsequent conclusions. Looking ahead and anticipating the results of the analysis, it is worthwhile to note that flow-reactor experiments on the DMTM, prototypes of real industrial processes, cannot be viewed as studying one and the same process. It is necessary to consider three fundamentally different groups of experiments, investigating, strictly speaking, and different reactions. The first group of experiments deals with fast-flow gas-phase oxidation, with the residence time of the reactants in the reactor being not much longer than the time of their diffusion to the surface, typically a few seconds. In the second-group of experiments, the oxidation is slow, with a residence time of the reactants above 1 min, which is much longer than the time of their diffusion to the surface. Finally, the oxidation in the presence of a heterogeneous catalyst, a type of processes beyond the scope of this monograph, but which cannot be left unattended at all. That is why one of the most important parameters to be considered while characterizing the experimental condition is the residence time. The discussion of purposes of such division and the influence of this difference in experimental conditions on obtaining results is one of the important points of the monograph.

Appendix 1, with the rare exception, includes only experiments with full or almost full conversion of oxygen which allows their comparison. As a parameter that characterizes the composition of reagents, where it is possible to determine, not absolute concentration of oxygen but the ratio of its concentration to the sum of methane and oxygen concentrations $[O_2]/([CH_4] + [O_2])$ are used. A part of data in Appendix 1 was calculated or estimated from data or figures presented in original papers. These estimated data are highlighted by italics.

In works focused on practical applications, the yield of liquid DMTM products is often characterized not by the selectivity of their formation or their concentration, but by the mass per unit volume of gas mixture passed through the reactor. The results of such works are summarized in Appendix II. It should be borne in mind that, the liquid phase is usually separated from the postreaction gas–liquid mixture by cooling it with ambient-temperature water. In this case, according to phase equilibrium calculations, the exhaust gas can carry away 5–15% of the methanol produced and an even a larger fraction of formaldehyde [51]. Therefore, the actual yield of oxygenates may exceed the corresponding experimental values. A more complete methanol extraction, with a carried-away amount of less than 1%, requires cooling the exhaust gas to a temperature close to 0 °C.

To provide the reader with the entire scope of data and facts about this process alongside with their comprehensive analysis, the chapters 2 through 4 discuss the products of DMTM, the main parameters of the process and the influence of gas composition on its behavior. These chapters present some unique data in this respect, some from sources difficult to access or published only in Russian.

Chapters 5 through 9 are devoted to the peculiarities of the DMTM mechanism and the most important consequences ensuing from them. These chapters provide a basic theoretical understanding of the main features of this complex branched-chain oxidation process. The reader will find explanations of some unusual and unobvious features of the process and an in-depth analysis of its characteristics and possibilities.

Chapter 10 examines the partial oxidation of heavier homologues of methane. It is the most rapidly developing direction in the modern studies of the DMTM process, motivated by the need to effectively process associated oil gases and heavier components of natural gas.

The last two chapters discuss the technological prospects of the process and its place among the other gas chemical technologies. These chapters describe the most recent progress in the field, including the results obtained by author and his colleagues on the optimal organization and possible practical applications of the process.

CHAPTER

2

Oxidation Products

MAIN PRODUCTS OF THE DMTM PROCESS

A high selectivity of methanol formation is the main feature of the oxidation of rich methane–oxygen mixtures ($[CH_4]/[O_2] = 10–30$) at high pressures and moderate temperatures. The other two main liquid products of this process are water and formaldehyde. Although only trace amounts of formaldehyde, if any, have been detected in a number of studies, it can be argued that this is most likely a result of its subsequent removal in secondary processes, including those on the surface of the reactor or catalyst.

Both methanol and formaldehyde are equally abundant in primary products of the gas-phase reaction, being formed at similar rates at low oxygen concentrations, but via different pathways. This is especially evident in the initial stage of the process (at low methane conversion) in experiments at relatively low pressures, 1.5–3 atm [52] (Fig. 2.1). At 425–500 °C and low conversions of methane in a 10-mm-diameter quartz flow reactor, the yield of formaldehyde is even greater than that of methanol. However, with increasing temperature and, correspondingly, increasing conversion, the integral selectivity of formaldehyde formation decreases faster, which makes methanol the predominant product.

The ratio of the main products in the initial stage of the process at $T = 300–500$ °C, $P = 21–41$ atm, and $[O_2] = 1.2–6.5\%$ (Fig. 2.2) was obtained in [53]. At the initial stage, formaldehyde and methanol are virtually the only oxidation products, with the selectivity of formaldehyde formation being more than two times that of methanol. Note, however, that, while the selectivity of formation of relatively stable methanol remains almost constant in time, that of chemically less stable formaldehyde decreases rapidly, apparently due to its removal in secondary reactions yielding carbon oxides.

Importantly, no carbon oxides were detected in the primary products at very short reaction times under rapid compression machine conditions at maximum degrees of compression of $\alpha_m = 237$ to $\alpha_m = 420$ [50]. A complete set of partial oxidation products detected in these experiments includes CH_3OH, CH_2O, CO, H_2, C_2H_6, C_2H_4, C_2H_2, C_3H_6, and C_2H_5OH. Figure 2.3 displays the dependence of the conversion of the reagents and the yields of oxygenates on the maximum degree of compression. The maximum yield of methanol is ~3 times higher than that of formaldehyde. The appearance of H_2, CO, and methane coupling products (C_2H_6, C_2H_4, C_2H_2, and C_3H_6) is observed starting from $\alpha_m = 297$, i.e., under conditions where the maximum formaldehyde yield is achieved. In this case, the yields of CO, H_2, and C_2H_6

Direct Methane to Methanol
http://dx.doi.org/10.1016/B978-0-444-63253-1.00002-7

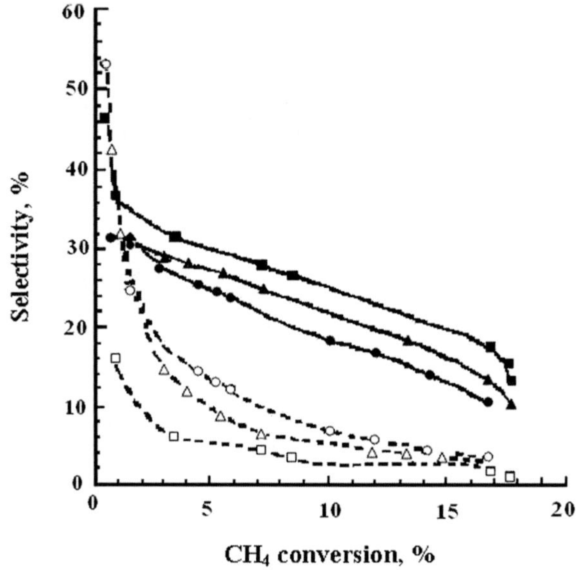

FIGURE 2.1 Selectivity of methanol (closed symbols, solid lines) and formaldehyde (open symbols, dotted lines) formation as a function of the methane conversion at $P = 1.5$ (●), 2.0 (▲), and 3.0 (■) atm, $[O_2] = 20\%$, $t_r = 96$ s [52].

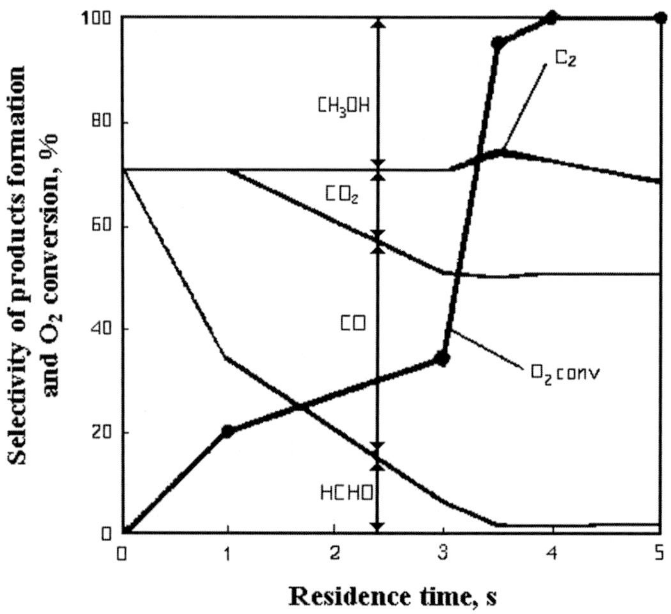

FIGURE 2.2 Product selectivity as a function of the residence time at $T = 400\,°C$, $P = 41$ atm, and $CH_4/O_2/N_2 = 30{:}1{:}4$ [53].

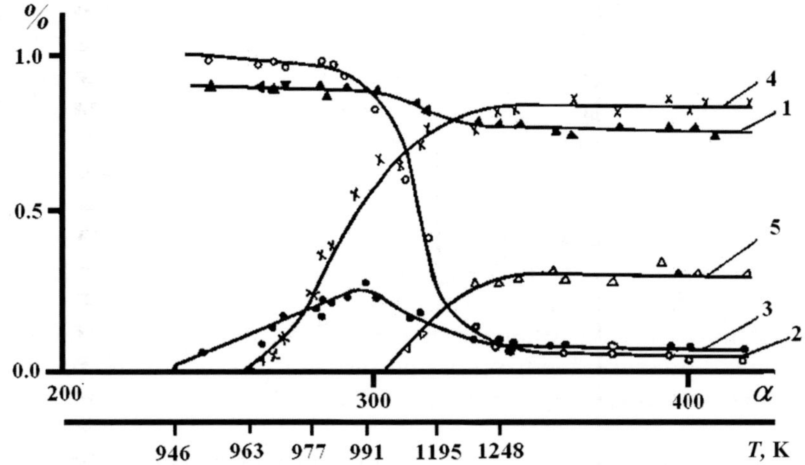

FIGURE 2.3 Conversion of reagents and the yield of oxygenates as a function of the maximum degree of compression α_m: (1) CH_4 ($\times10^{-2}$), (2) O_2 ($\times10^{-1}$), (3) CH_2O, (4) CH_3OH, and (5) C_2H_5OH ($\times10$) [50].

were ~7%, ~3%, and ~1.4%, respectively. The yields of the other products were substantially lower.

The major gas-phase products of the DMTM process are carbon oxides and hydrogen, with the yield of carbon monoxide at high pressures being several times that of carbon dioxide. The main pathways of formation of carbon dioxide in this process are apparently not directly related to carbon monoxide, since according to a number of studies, carbon dioxide is formed before carbon monoxide in the induction period [54]. With rising temperature, the yields of ethane and ethylene increase rapidly.

Although methane oxidation is a complex branched-chain reaction, the ratio of the major DMTM products for typical conditions can be illustrated by the following gross reaction scheme, which reflects the fundamental nonselectivity of the gas-phase process:

$$2CH_4 + 2O_2 \rightarrow \begin{cases} 0.9CH_3OH + 0.1(CH_2O + H_2) \\ 0.9(CO + 2H_2O) + 0.1(CO_2 + H_2O + H_2) \end{cases} \qquad (2.1)$$

This scheme shows that the process is principally unselective with respect to oxygenates with two principal carbon products, CH_3OH and CO, which are formed in approximately equal amounts. Therefore, it is difficult to expect a very high selectivity of CH_3OH formation. The consumption of both reactants is approximately equal, but only about 25% of the initial oxygen passes into oxygenates. Water is a principle liquid product that dilutes methanol formed. When the conditions of the process are changed, the ratios between the contributions of these reaction branches and between the products in each can vary severalfold; in addition, the scheme takes into account, by far, not all the products.

THE RATIO CH_3OH/CH_2O

The fact that formaldehyde, along with methanol, is the principal product of the DMTM process has been demonstrated in many studies, but then it is quickly removed by secondary reactions. Formaldehyde, dominating at low pressures, probably due to some specific features of the oxidation mechanism, remains present in significant quantities at high pressures as well, in both gas-phase and catalytic processes.

The concurrent formation of methanol and formaldehyde during the oxidation of methane at atmospheric pressure in a static quartz reactor ($T = 440-450\,°C$, $t_r = 1.5-4.3$ min) was observed in [24]. Under these conditions, the methanol-to-formaldehyde ratio increased monotonically with decreasing residence time (Fig. 2.4), which led the authors to suggest that, at atmospheric pressure, methanol is the primary product of the process. However, even at the shortest reaction time ($t_r = 1.5$ min), the selectivity of methanol formation was only 4.6%.

Based on thermodynamic considerations, the authors of [55] assumed that, at high temperatures and low pressures, formaldehyde is the predominant product, while methanol is the principal product at high pressures and low temperatures. Although the yield of products for the gas-phase DMTM is determined not by the thermodynamics but by the kinetics of the process, this assumption is generally true. Formaldehyde is the main product of partial methane oxidation, including catalytic, at atmospheric pressure, with methanol being the predominant product at high pressures. However, according to [56], even at a pressure of 50 atm and a high space flow rate, 10^5/h ($T = 375\,°C$), the process conducted in the presence of a molybdenum or chromium catalyst showed almost the same selectivity of methanol and formaldehyde formation, which indicates that both the products are primary. With decreasing space flow rate, the formaldehyde yield decreased rapidly, while the methanol yield passed through a maximum.

The catalytic partial oxidation of methane to formaldehyde, considered in many comprehensive reviews – see, e.g., [57] – is not the subject of the present analysis, limited only to

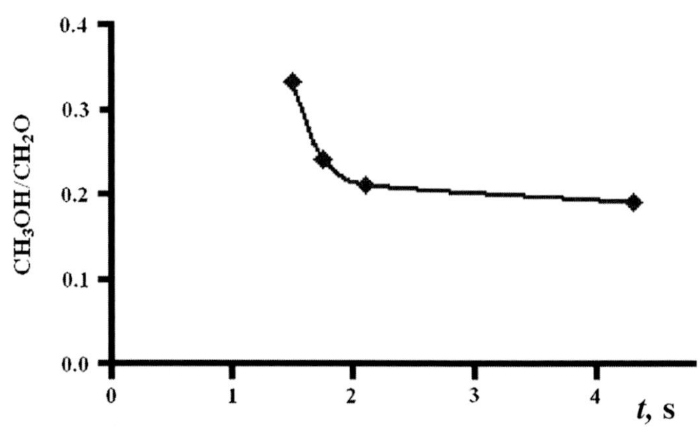

FIGURE 2.4 The CH_3OH/CH_2O ratio as a function of the reaction time. $P = 1$ atm, $T = 440-450\,°C$. *According to data from [24].* (For colour version of this figure, the reader is referred to the online version of this book.)

gas-phase homogeneous processes in a void reactor under conditions in which significant amounts of methanol are formed. We would like only to point out that one of the most active catalytic systems, vanadium oxide (5%V_2O_5–SiO_2), operating at 650 °C and atmospheric pressure provided a methane conversion per pass of 13.5% and a formaldehyde yield of 4.73% [58]. A higher yield could be reached only by recycling the mixture and concurrently separating the accumulated product. The authors of [59] demonstrated the possibility of achieving a formaldehyde yield of up to 50%, as compared to 2% per pass, on a V_2O_5–SiO_2 supported catalyst ($T = 840$–900 K, $P = 1$ atm), while maintaining high selectivity of formaldehyde formation (up 56%) and a high degree (89%) of methane conversion.

One of the most thorough comparisons of formaldehyde formation in the presence of oxide catalysts and in a void reactor was performed in [57]. Figure 2.5 displays the dependences of the methane conversion, selectivity of CH_2O formation, and the yield of CH_2O in an empty quartz reactor on the reaction temperature for a 1:5 methane–air mixture at a pressure of $P = 5$ atm and a reaction time of $t_r = 2.3$ s.

As can be seen from Fig. 2.5, the maximum CH_2O yield in the empty quartz reactor reaches 3.46%, a result only slightly inferior to the best results obtained in the presence of oxide catalysts. The pressure dependence of the CH_2O yield at 625 °C is shown in Fig. 2.6. As can be seen, the optimum pressure for the noncatalytic conversion of methane to formaldehyde is ~5 atm. However, the reproducibility of the results obtained in the absence of a catalyst was reported to be poor.

According to the estimates of the authors of this study, the secondary thermal reactions of formaldehyde make it unlikely that, under their experimental conditions, its yield would exceed 8%, expectedly much lower, even at very short residence times of the reactants in the reaction zone. Thus, according to [57], the results obtained in a void reactor at slightly elevated pressures (~5 atm) can compete with the best results acquired in the presence of a catalyst. This is also confirmed by the data from [60], according to which the catalysis or promotion of the process by NO additives do not enable to surpass the results reported in [57].

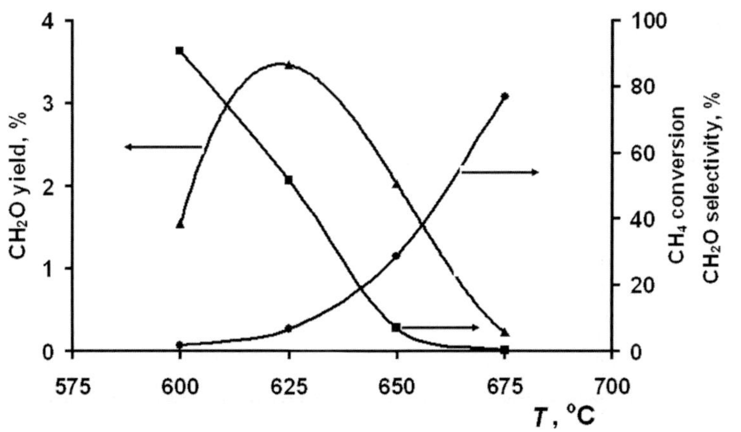

FIGURE 2.5 Effect of temperature on the formaldehyde formation in an empty quartz reactor. Methane–air ratio 1:5, $P = 5$ atm, reaction time = 2.3 s (●) CH_4 conversion, (■) CH_2O selectivity, (▲) CH_2O yield. *According to data from [57].*

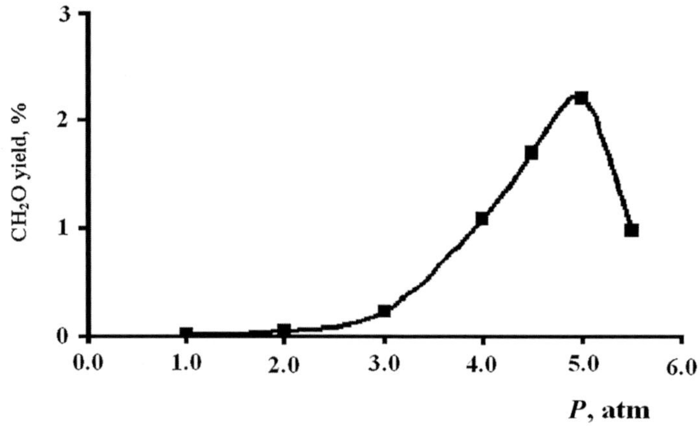

FIGURE 2.6 Effect of pressure on the formaldehyde yield in a void quartz reactor at a methane-to-air ratio of 1:5, $T = 625\ °C$, and reaction time of 2.3 s [57].

In this case, the filling of the reactor with an 'inert' solid phase, as well as an increase in the reaction time, significantly reduce the yield of formaldehyde.

As another demonstration of the possibility of achieving practically the same results in the partial gas-phase oxidation of methane at a sufficiently high temperature as in the catalytic oxidation, it is worthwhile to mention a study of the gas-phase oxidation of methane at very short residence times and high temperatures [61]. Experiments in a flow quartz reactor at $T = 1173-1273$ K, $P = 2$ atm, a ratio of reactants of $X_{O2}/X_{CH4} = 0.5-15.0$ (at 90% dilution with Ar), and an interaction time of 20–60 ms gave a highest formaldehyde yield of 2.4%. Under these conditions, an appreciable formation of the products was observed only at temperatures above 1198 K. Appearing first, CH_2O and C_2H_6 seem to be the primary reaction products. Carbon oxides are formed only at temperatures above 1223 K; the selectivity of CO formation passes through a maximum (35% at 1248 K), while the CO_2 yield increases rapidly with temperature. The selectivity of formation of methane coupling products, C_2H_6 and C_2H_4, also features a maximum (52%) at 1223 K, and then, at $T = 1233-1273$ K, remains at a level of 30%. According to [61], in the temperature range covered, the temperature has little effect on the ratio of the products of partial oxidation and oxidative coupling, since the precursor of both types of products is the methyl radical. Increasing the reactant ratio X_{O2}/X_{CH4}, already rather high for normal conditions, still further, from 2 (a value at which the major products are CH_2O and C_2H_6) to 5 results in a very rapid decrease in the selectivity of their formation. At lower values of X_{O2}/X_{CH4}, the methane-coupling products are predominant, but this is accompanied by a sharp decrease in the methane conversion.

Kinetic simulations closely describe the results and show that, at such high temperatures, the main pathway of CH_2O formation is the direct oxidation of methyl radicals $CH_3^{\bullet}$ by oxygen. It is also formed in part by the dehydrogenation or oxidation of methoxy radicals $CH_3O^{\bullet}$.

An analysis of the NOx-promoted homogeneous partial oxidation of methane at atmospheric pressure performed within the framework of a detailed kinetic model [62] confirmed the existence of fundamental limitations on the CH_2O yield. Under all realistic conditions, the calculated maximum yield was below 10%. Adding NOx, along with increasing the reaction rate, promotes the conversion of $HO_2^{\bullet}$ and $H^{\bullet}$, relatively slow-reacting radicals, into $OH^{\bullet}$, a more reactive radical, a factor that increases the CH_2O yield. However, in any case, the CH_2O yield is limited because $OH^{\bullet}$ radicals react with CH_2O faster than with CH_4. The effect of pressure on the current yield of formaldehyde was minimal.

At the same time, a pressure rise greatly increases the methanol yield, thereby increasing the CH_3OH/CH_2O ratio. According to the DMTM mechanism [63—66], which will be discussed below, at high pressures, the ratio of the rates of formation of methanol and formaldehyde and, consequently, the yield thereof, is primarily determined by the competition between the reactions of the methoxy radical $CH_3O^{\bullet}$, an oxidation intermediate, with methane and oxygen. This ratio depends, among other things, on the ratio between the concentrations of CH_4 and O_2 in the mixture and the temperature of the process.

$$CH_3O^{\bullet} + CH_4 \rightarrow CH_3OH + CH_3^{\bullet} \tag{2.2}$$

$$CH_3O^{\bullet} + O_2 \rightarrow CH_2O + HO_2^{\bullet} \tag{2.3}$$

At typical temperatures of the DMTM process, the rate constant of reaction (2.3) is ~5 times that of reaction (2.2); however, a significantly higher concentration of methane (by a factor of 20—30) makes methanol a predominant product. For example, in experiments at $P = 40$ and 60 atm, $T = 450\,°C$, $[O_2] = 2.8\%$ and short reaction times ($t_r \approx 1.2$ s), the CH_3OH/CH_2O ratio, especially at low conversions, was typically 3—5 [67].

Measurements [68] in a quartz flow reactor at an oxygen concentration of 7% with respect to methane, reaction times from 2.5 to ~20 s, and pressures of up 30 atm enabled to obtain the pressure dependences of the selectivity of formation of the sum of C_1-oxygenates and of the CH_3OH/(sum of C_1-oxygenates) ratio (Fig. 2.7). At moderate pressures, intermediate between those optimal for the formation of formaldehyde and methanol, the selectivity of

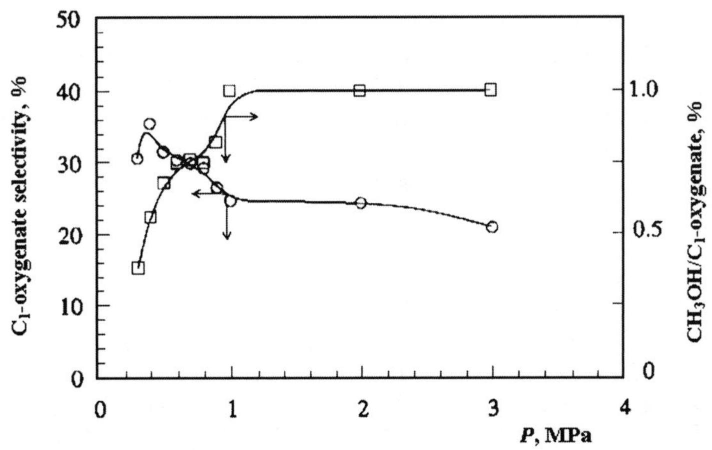

FIGURE 2.7 Pressure dependence of the (○) C_1-oxygenate selectivity and (□) CH_3OH-to-C_1-oxygenates ratio: $[O_2] = 0.07[CH_4]$; $T = 625\,°C$ (0.3 MPa), 575 °C (0.4 MPa), 550 °C (0.5 MPa), 525 °C (0.6—1.0 MPa), and 450 °C (2 and 3 MPa) [68].

formation of the sum of C_1-oxygenates first increases with pressure, but then declines. This may be due to a more rapid decrease in the selectivity of formaldehyde formation compared to the increase in the selectivity of methanol formation under these conditions. In any case, the decrease of the overall selectivity of C_1-oxygenates is accompanied by a rapid growth of the CH_3OH/(sum of C_1-oxygenates) ratio until it reaches a constant value of 1.0 at $P \geq 10$ atm, pressures at which formaldehyde was no longer detected in the reaction products. Note, however, that, in these experiments, the pressure was increased concurrently with decreasing the reaction temperature, from 625 °C at $P = 3$ atm to 450 °C at $P = 20$ and 30 atm.

These data make it possible to plot the pressure dependence of the CH_3OH/CH_2O ratio (Fig. 2.8). This figure also displays the result from [24] for $P = 1$ atm and a minimal reaction time of 1.5 min.

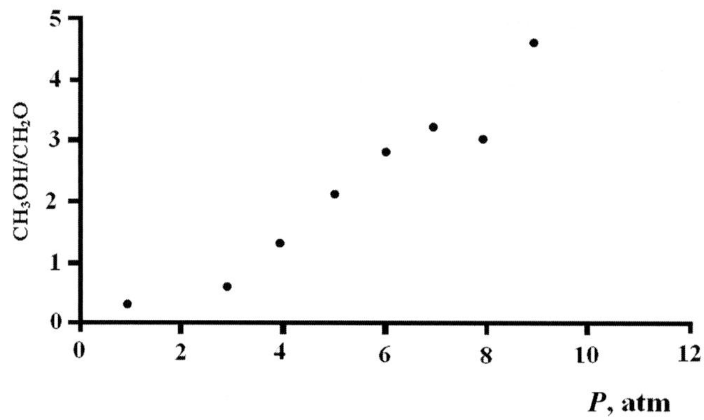

FIGURE 2.8 Pressure dependence of the CH_3OH/CH_2O ratio. *According to the data from [68] and [24].*

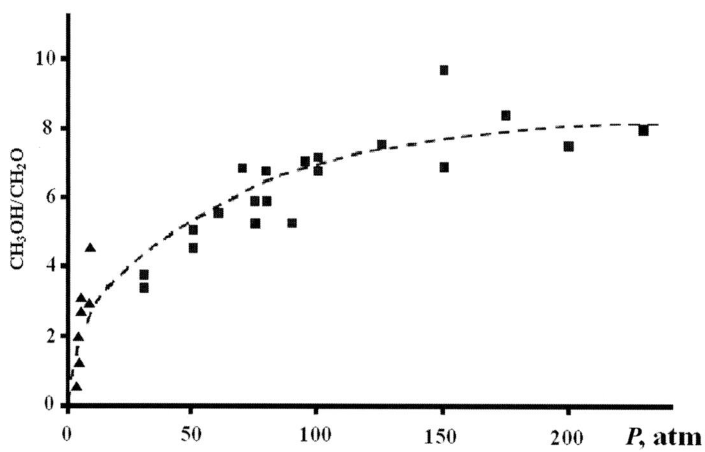

FIGURE 2.9 Pressure dependence of the CH_3OH/CH_2O ratio: (■) [69] and (▲) [68]. (For colour version of this figure, the reader is referred to the online version of this book.)

At high pressures, the pressure dependence of the CH_3OH/CH_2O ratio can be obtained based on the data from [69]. In Fig. 2.9, this dependence is presented together with the low-pressure data from [68], which appear to be in qualitative agreement. That the dependence from [68] is steeper and that the authors of this work detected no measurable formaldehyde concentrations at pressures above 10 atm can be explained by a different probability of CH_2O loss at the reactor surface in these two works. This dependence shows that the methanol/formaldehyde ratio in the oxidation products can be varied over a wide range by selecting suitable conditions of the process, primarily the pressure, as well as the temperature and initial mixture composition.

THE RATIO CO/CO_2

Although both carbon oxides are present as the main products of the partial oxidation of methane under conditions typical of DMTM, in the overwhelming majority of the studies, the carbon monoxide yield was substantially higher than that of carbon dioxide. For example, in [67], at $P = 40$ and 60 atm, $T = 450\,°C$, $[O_2] = 2.8\%$, and $t_r = 1.2$ s, the CO/CO_2 ratio was ≥ 10. The predominance of carbon dioxide in some experiments with a long residence time of the reactants in a static reactor [16,54] is most likely associated with the contributions from the heterogeneous oxidation of methane and heterogeneous conversion of primary oxidation products. This also probably explains why CO_2 was observed to appear before CO during the ignition delay period [54]. On the contrary, the kinetic modeling of the gas-phase process [70] predicts an earlier appearance of CO with respect to CO_2.

According to experimental data [69], at an initial oxygen concentration of ~2.8%, the concentration of carbon monoxide, the main gas-phase reaction product at the reactor outlet, is ~1.5%, remaining practically constant over a pressure range from 30 to 230 atm (Fig. 2.10). The carbon dioxide concentration is several times lower and increases slightly with pressure,

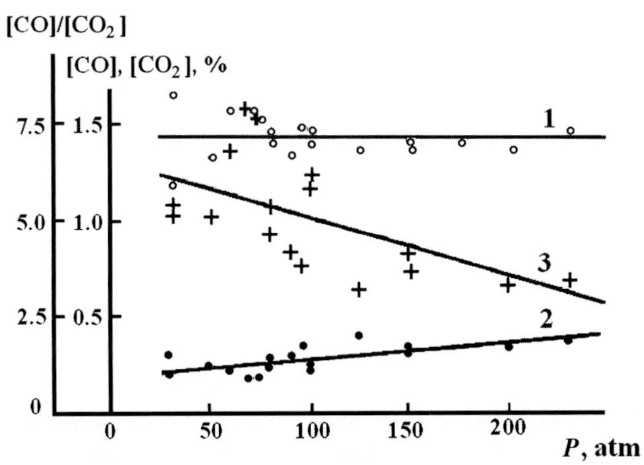

FIGURE 2.10 Pressure dependence of the concentrations of carbon oxides and the [CO]/[CO_2] ratio at $T = 400\,°C$ and $[O_2] \approx 2.8\%$: (1) CO, (2) CO_2, and (3) [CO]/[CO_2] [69].

resulting in a corresponding decrease in the CO/CO_2 ratio. That the CO_2 concentration increases with pressure may be partly due to the formation and subsequent decomposition of formic acid according to the mechanism proposed in [71].

BYPRODUCTS

In early studies of the DMTM process, the composition of the products was carefully analyzed, and a large number of various oxygen-containing and oxygen-free hydrocarbon oxidation products were separated and identified. In the later works, the focus was largely on the yield and kinetics of formation of the main products. Therefore, the scarcity of information on byproducts in contemporary works seems to reflect a decline in interest in their formation in this process. However, in part, it may also be associated with the changeover to more rigid conditions and shorter reaction times, which is less favourable for their formation.

Along with the aforementioned basic DMTM products, many works report the formation of small amounts of various liquid products, such as ethanol (1.5–2%), acetone (up to ~1%), and formic acid (up to ~0.8%). The formation of small amounts of higher alcohols, aldehydes, and organic acids has also been observed. For the oxidation of gases with a high concentration of ethane and heavier alkanes, the yield of C_2–C_3 alcohols, aldehydes, and acids can be significantly higher. Although hydrogen peroxide and organic peroxides are undoubtedly formed during the oxidation as intermediates, even special attempts to detect them in the reaction products failed [16]. Despite every effort, hydrogen peroxide was not observed during the induction period of the ignition of rich (from 1.9 to 8.8% O_2 with respect to $CH_4 + O_2$) methane–air mixtures at pressures from 58 to 110 atm [54]. Trace amounts of these compounds were detected only at atmospheric pressure [72].

Below are listed the organic compounds observed in various works concerning the DMTM process:

In [47], along with gaseous ethane, various condensable products, such as methyl formate, formaldehyde, formic acid, and acetic acid, were observed.

Measurements [46] in a static reactor at pressures from 150 to 1000 atm, temperatures of ~250 °C, and reaction times of tens of minutes showed that, along with the traditional set of products, methyl formate formed as the formaldehyde yield increased, presumably as a result of a secondary reaction between formaldehyde and methanol.

In [48], at pressures of 1700 and 3400 atm, along with methanol and formaldehyde, trace amounts of other oxygenates, such as $HCOOH$, C_2H_5OH, and CH_3COCH_3, were also detected.

Dimethyl ether (DME) is rarely detected in the DMTM products; however, in [73], it was reported to be formed on a reduced copper surface, along with other products of methanol decomposition.

In the patent [74], it was claimed that the oxidation of natural gas containing ~4% ethane and 1% heavier alkanes C_3+ by air in a pilot plant at a gas flow rate of 760 m^3/h, $P = 100$ atm, initial temperature of $T = 430$ °C, total oxygen content of 5%, and a reaction time of less than 1 s produced a gas mixture composed of a wide variety of oxygen-containing products (vol %): CH_3OH, 1.65; C_2H_5OH, 0.03; C_3H_7OH, 0.01; i-C_4H_9OH, 0.001; CH_2O, 0.050; CH_3COOH, 0.006; $(CH_3)_2O$, 0.050; $(CH_3)_2CO$, 0.017; $CH_3C_2H_5CO$, 0.006; and H_2O, 2.020. The collected liquid contained, along with water, methanol, and formaldehyde, appreciable amounts of DME $(CH_3)_2O$ (2.35–2.90 wt%), ethanol (1.24–1.40 wt%), and acetone $(CH_3)_2CO$ (0.96–1.2 wt%). In addition, C_3H_7OH, i-C_4H_9OH, and other organic compounds were detected.

These same products were reported in the subsequent works carried out on the same pilot plant. The patent [75] states that, at $P = 100$ atm, $T = 330–370\,°C$, $[O_2] \sim 3\%$, and residence time of the gas mixture in the reactor of 0.5 s, the products contained the following compounds (vol %): CH_3OH, 1.65; C_2H_5OH, 0.028; C_3H_7OH, 0.002; i-C_4H_9OH, 0.0009; CH_2O, 0.23; CH_3COOH, 0.0008; $(CH_3)_2O$, 0.03; $CH_3C_2H_5O$, 0.04; $(CH_3)_2CO$, 0.19; $CH_3C_2H_5CO$, 0.005; and H_2O, 1.28. In [76], at $P = 5.0–100.0$ atm, $T = 370–450\,°C$, and $[O_2] = 2–4\%$, the same liquid-phase products as in [74] were identified. Surprisingly, formic acid, found in many other studies, was not among the listed products, although acetic acid and other more complex compounds were present. In addition, given short reaction times (0.2–0.5 s) and the lack of a description of the methods used to analyze such low concentrations of complex organic products in the gas phase, it cannot be excluded that the reported compounds, at least in part, were formed directly in the liquid phase due to the interaction between the primary oxidation products.

Experiments [77] in a stainless steel reactor (Alonized A106 steel) revealed the formation of a significant amount of DME (with a selectivity of ~20%) due to a corresponding decrease in the selectivity of methanol formation (up to ~30–40%), apparently as a result of its dehydration. The formation of DME on other surfaces (glass, Teflon, stainless steel, copper, silver, tin, and other metals) was not observed. Purging the reactor with a KOH solution decreased the DME yield, a result indicative of a heterogeneous mechanism of its formation.

In [53], at $T = 300–500\,°C$, $P = 21–41$ atm, $[O_2] = 1.2–6.5\%$, and $t_r = 3–15$ s, along with methanol, formaldehyde, and carbon oxides, the formation of ethanol, acetaldehyde, and C_2 hydrocarbons was observed.

In [69], at $T = 400\,°C$, $[O_2] = 2.7–2.9\%$, and $P = 30–230$ atm, the liquid oxidation products consistently contained 0.5–1.5% ethanol, acetone, and organic acids.

The authors of [67] detected trace amounts of acetaldehyde, methyl formate, ethanol, and formic acid.

In experiments on a pilot plant with a flow stainless steel reactor at a gas flow rate of ~1000 m^3/h [78], among the oxidation byproducts, normal $C_2–C_4$ alcohols and acetone were identified, the yield of which relative to the methanol yield ranged from ~2% for ethanol to less than 0.2% for butanol. The formation of $C_2–C_4$ products in this work is most likely associated with the use of natural gas, although its composition was not specified.

Experiments [79] in a Pyrex reactor with a preheating section at a long residence time of the reactants demonstrated record-high values of the yield and selectivity of methanol, as well as the formation of a wide variety of complex organic products, including propane, acetaldehyde, methyl formate, dimethoxymethane (methylal), acetone, and DME. Moreover, at high pressures and low flow rates of the reagents (i.e. long residence times in the reactor), dimethoxymethane was one of the main products, with a selectivity of formation up to 16%. The selectivity of methyl formate formation reached nearly 1%. The explanation within the framework of mechanisms involving the cage effect, suggested by the authors to account for the formation of such complex products in the gas phase, seems too exotic and less persuasive than a trivial involvement of heterogeneous processes. The yields of these products increase with the residence time, indicative of their formation in secondary reactions.

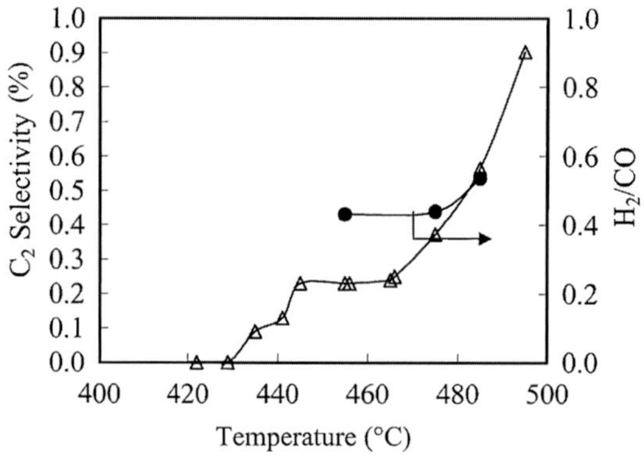

FIGURE 2.11 Effect of temperatures on the (△) C_2 selectivity and (●) H_2 production: $P = 5.0$ MPa, $CH_4/O_2/N_2 = 100/10/10$ (ml/min) [86].

It is likely that some complex organic compounds, such as methyl formate and methylal, may be formed by reactions in the separated liquid prior to its analysis. This conclusion is supported by the fact that, in experiments on a rapid compression machine at very short reaction times [50], no any complex organic products were detected.

The mechanism of the formation of the observed gaseous byproducts is more obvious:

At temperatures above 500 °C, appreciable amounts of methane-coupling products appear, notably ethane and ethylene, whereas above 600 °C, their heavier homologues are formed as well [80].

At temperatures above 550 °C in [81, 82], acetylene was detected.

The formation of C_2H_6 at $T > 440$ °C ($P = 50$ atm, $t_r \approx 2$) was observed in [83]. At $T = 470$ °C, the selectivity of ethane formation reached ~8%.

In [84] ($P = 50$ atm, $[O_2] = 1.5–12.0\%$, $t_r \approx 15$ s), the C_2 products appeared at $T > 450$ °C, and then their selectivity increased rapidly with the temperature.

The formation of hydrogen was mentioned in [41,67,85,86]; the H_2/CO ratio was found to be ~0.4–0.5 [86] (Fig. 2.11). The source of molecular hydrogen was presumably the fast dissociation of CH_2O to H_2 and CO.

RESIDUAL OXYGEN

In some of the earlier studies, problems with oxygen balance were faced, which was drawn up considerably more poorly than the balance in carbon and hydrogen, with the uncertainty only occasionally being less than 10%. For example, the authors of [87] pointed out that, with a good balance in carbon and a somewhat less good in hydrogen, the balance in oxygen always showed a deficit from 50 to 10%, which clearly exceeded the measurement error. In contrast, in experiments with natural gas, an excess of oxygen was observed. The authors suggested that oxygen is consumed to oxidize the copper catalyst present in the system, but immediately pointed out that, when glass was used as the catalyst, oxygen loss was equally high. The authors also refer to [26], which showed that only 70% of the initial O_2 is

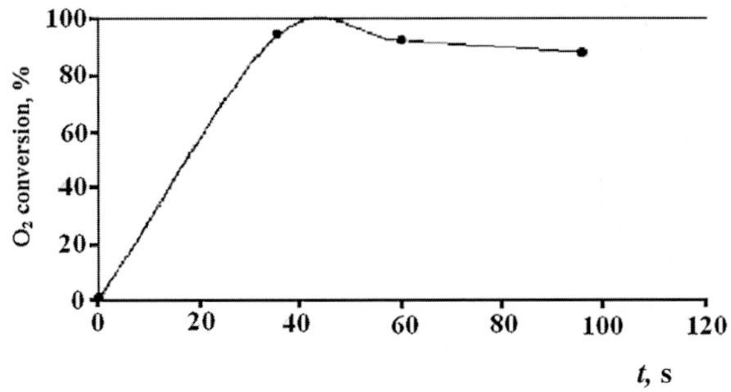

FIGURE 2.12 Oxygen conversion of as a function of the reaction time; $P = 34$ atm, $T = 430$ °C, $CH_4/O_2 = 16$ [85].

accounted for in experiments with natural gas and only 40% in experiments with propane; no explanation was given.

Another important observation is that, in this and many other studies, oxygen was detected in the gaseous reaction products, a fact inexplicable in terms of the branched-chain mechanism of gas-phase methane oxidation, the simulation of which invariably leads to the complete consumption of oxygen. The presence of oxygen in the gaseous reaction products was tested in a series of studies with great care, but with the same result: some oxygen remained unreacted. In experiments [48] at pressures of 1700 and 3400 atm and an initial oxygen concentration of 8%, the reaction clearly stopped after 30 min, leaving ~20% unreacted O_2. The consistent presence of oxygen in the gaseous reaction products, on average ~5% of its initial concentration, was reported in [69]. In this work, special measures were undertaken to exclude such trivial factors as oxygen penetration into the system downstream from the reactor and the unreliability of the analysis, which was carried out by various methods, including chromatographic analysis, and chemical titration. In [68], the O_2 conversion was always less than 100% (99.0–99.7%), as it was in [88], ranging within 92–98%.

We believe that the most feasible explanation for the presence of oxygen in the gaseous reaction products is the formation of secondary oxygen. This is most clearly illustrated by the data from [85], which show that, after reaching its maximum, the oxygen conversion then decreases with time (Fig. 2.12).

In our opinion, based on the results of a kinetic modeling of the process, secondary oxygen is formed by the decomposition of hydrogen peroxide, a species that, at the time of complete conversion of oxygen and actual stop of the branched-chain process of methane oxidation, may contain up to 20% of the oxygen originally present in the mixture.

$$2H_2O_2 \rightarrow 2H_2O + O_2 \tag{2.4}$$

In fact, this overall reaction does not reflect the radical nature of hydrogen peroxide decomposition, which is accompanied by the formation of significant concentrations of hydroxyl radicals. It is the secondary reactions of the hydroxyl radical,

$$H_2O_2 \rightarrow 2OH^{\bullet} \tag{2.5}$$

$$OH^{\bullet} + CH_3OH \rightarrow H_2O + CH_3O^{\bullet} \tag{2.6}$$

$$OH^{\bullet} + CH_3OH \rightarrow H_2O + CH_2OH^{\bullet} \tag{2.7}$$

$$OH^{\bullet} + CH_2O \rightarrow H_2O + CHO^{\bullet} \tag{2.8}$$

that cause, as observed in many studies, a decrease in the postmaximum concentration of methanol and, especially, formaldehyde after the complete conversion of the initial oxygen and termination of branched-chain oxidation. In this case, in the temperature range characteristic of DMTM, 400–600 °C, the rate constant of gas-phase reaction (2.8) is more than twice those of reactions (2.6) and (2.7), which may be one of the reasons for the faster removal of formaldehyde as compared to methanol.

THE $\Delta CH_4/\Delta O_2$ RATIO

Depending on the target products, the stoichiometric ratio of the reaction of methane with oxygen can vary widely, from $\Delta CH_4/\Delta O_2 = 1{:}2$ for complete oxidation,

$$CH_4 + 2O_2 \rightarrow CO_2 + 2H_2O \tag{2.9}$$

to $\Delta CH_4/\Delta O_2 = 2{:}1$ for the hypothetical overall reactions

$$CH_4 + 1/2O_2 \rightarrow CO + 2H_2 \tag{2.10}$$

$$CH_4 + 1/2O_2 \rightarrow CH_3OH \tag{2.11}$$

$$CH_4 + 1/2O_2 \rightarrow CH_2O + H_2 \tag{2.12}$$

and even to higher values if the oxidation yields considerable amounts of oxygen-free products, such as C_2H_6 and C_2H_4. In reality, according to scheme (2.1), the gas-phase DMTM process produces two main groups of products,

$$\begin{aligned} 2CH_4 + 2O_2 &\rightarrow CH_3OH \\ &\rightarrow CO + 2H_2O, \end{aligned} \tag{2.13}$$

characterized by a reactant conversion ratio of $\Delta CH_4/\Delta O_2 \approx 1$. Lower values of this ratio are indicative of either a redistribution of these pathways in favour of the formation of CO and H_2O or the formation of considerable amounts of CO_2, most likely in deep-oxidation reactions on the surface of the reactor. The $\Delta CH_4/\Delta O_2 > 1$ ratio signifies the predominance of oxygenates (methanol and formaldehyde), or C_2+ and other methane coupling products.

Analysis of [51] (Fig. 2.13) shows that the $\Delta CH_4/\Delta O_2$ ratio increases with pressure, which corresponds to an increase in the yield of C_1-oxygenates. The increase in the residence time from 1.8 to 9.0 s significantly reduces the $\Delta CH_4/\Delta O_2$ ratio, probably because of the heterogeneous decomposition of oxygenates to carbon oxides and water. The reaction temperature has little effect on this ratio, whereas increasing the oxygen concentration naturally reduces the $\Delta CH_4/\Delta O_2$ ratio because of a deeper oxidation of the primary reaction products.

As claimed in the patent [44], a very high value of $\Delta CH_4/\Delta O_2 = 2.67$, corresponding to a methane conversion of 13.36%, was achieved at an initial oxygen concentration of only 5%. Note, however, that, at the claimed selectivity of methanol formation of over 92%, values of $[CH_4]/[O_2] > 2$ are unrealistic, a contradiction implying a serious error.

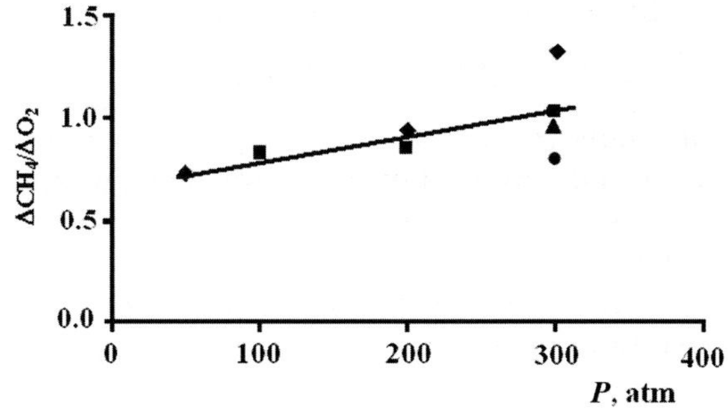

FIGURE 2.13 Dependence of the $\Delta CH_4/\Delta O_2$ ratio on the pressure. (◆) $T = 475\,°C$, $O_2 = 3\%$, $t_r = 1.8$ s, (■) $T = 475\,°C$, $[O_2] = 3\%$, $t_r = 9.0$ s, (▲) $T = 375\,°C$, $[O_2] = 3\%$, $t_r = 9.0$ s, (●) $T = 475\,°C$, $[O_2] = 6\%$, $t_r = 9.0$ s. *According to data from [51].*

YIELD OF METHANOL AND OXYGENATES

Since a low oxygen concentration is a fundamental prerequisite for a high selectivity of methanol formation, which decreases rapidly with increasing initial concentration of oxygen and, consequently, the methane conversion (Fig. 2.14), the methanol yield in a single pass through the reactor is small, rarely exceeding 2–2.5%.

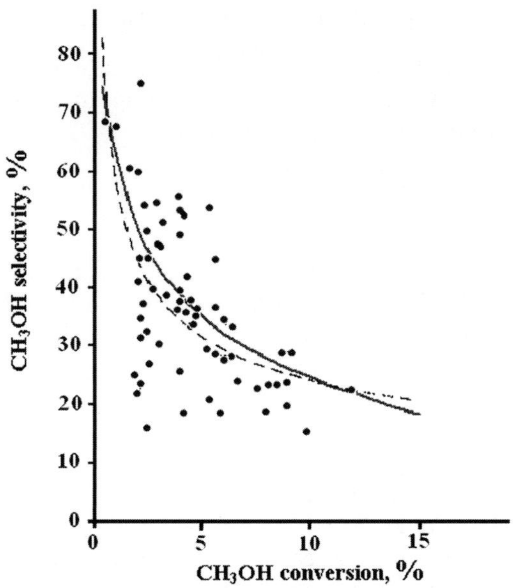

FIGURE 2.14 Dependence of the CH_3OH selectivity on the CH_4 conversion obtained from fast-flow experiments with $t_r < 20$ s and $P \geq 30$ atm (based on the data from Appendix I). The dashed line is the trend of experimental data, whereas the solid line represents the results of kinetic simulations of the gas-phase DMTM process at $P = 100$ atm and $T = 420\,°C$ [89]. (For colour version of this figure, the reader is referred to the online version of this book.)

Figure 2.14 shows the results of the most reliable fast-flow experiments at a residence time of less than 20 s and pressure of above 30 atm (Appendix I). Excluded are experiments with a spread oxygen supply, at several points along the reactor, and unreliable data from patents. Despite a large scatter of the experimental points, a clear trend of methanol selectivity quickly reducing with increasing methane conversion manifests itself, with the overall trend line of the data being almost identical with the results of kinetic modeling of the gas-phase DMTM process at $P = 100$ atm and $T = 420$ °C from [89].

Figure 2.15 displays the dependence of the methanol yield on the methane conversion, which shows a clear tendency to be limited by ~2.5%. As can be seen, like in the case of the selectivity, the experimental trend line is in close agreement with the predictions of the kinetic simulations of the gas-phase DMTM process.

Since a high selectivity of methanol formation or the sum of oxygenates is critical for the economic efficiency of the process, studies in which high values of this parameter were claimed have always attracted a lot of attention. Cost evaluations in the mid-1980s [39] set specific target values of these indicators that make the process cost-effective in the economic conditions of that time: at a methane conversion of no less then 5%, the selectivity of methanol formation should be over 77%, which is significantly higher than the values obtained in the vast majority of fast-flow experiments (Fig. 2.14). However, almost simultaneously with the publication of this assessment, several experimental results that satisfy this criterion were reported. In [42,44,79] and a number of other works by the same group of authors, as well as in [43], the methanol yield per pass of reaction mixture through the reactor was claimed to be more than 6–7%, with the selectivity of its formation reaching 80 or even 90% (Appendix I). However, attempts of other teams [45] and later attempts of the authors themselves to reproduce such promising results proved unsuccessful.

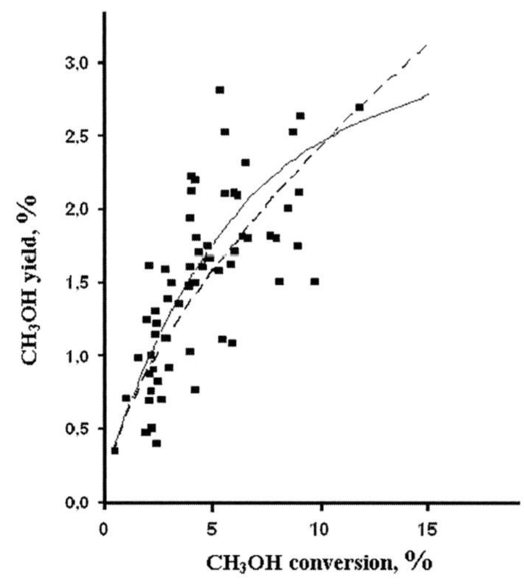

FIGURE 2.15 Dependence of the CH_3OH yield on the CH_4 conversion obtained from fast-flow experiments at $t_r < 20$ s and $P \geq 30$ atm (based on the data from Appendix I). The dashed line is the trend of experimental data, and the solid line represent the results of kinetic simulations of the gas-phase DMTM process at $P = 100$ atm and $T = 420$ °C [89]. (For colour version of this figure, the reader is referred to the online version of this book.)

Along with the studies of Gesser and co-workers [42,44,79], a very high selectivity (>70%) and yield (>10%) of methanol formation were reported only by one other group [43]. As in Gesser et al.'s studies, the experiments were performed at long (a few minutes) residence times of the reactants in a monocrystalline sapphire reactor and relatively low (less than 400 °C) temperatures. Only trace amounts of formaldehyde were detected, which is quite expected at such long residence times of the products in the hot zone of the reactor. In addition, traces of ethane and ethylene were detected.

Although an increase in the reactor surface area in these experiments by introducing quartz wool, MgO, Al_2O_3, and zeolite caused a decrease in the degree of methane conversion and in the selectivity of methanol formation, it is difficult to suggest any other explanation of the results, especially a nearly 80% selectivity of methanol formation, except for an essentially heterogeneous nature of the process under these conditions. This assumption is supported by the observation of thermokinetic oscillations with a characteristic period of about a minute [90]. A further argument in favour of the heterogeneous nature of the process is a very weak pressure dependence of the selectivity of methanol formation at a complete conversion of oxygen and a constant temperature of 340–350 °C over a wide pressure range, 30–70 atm (Fig. 2.16), which is untypical of gas-phase processes. As regards the claim of the authors that the process is insensitive to temperature, the range of its variation (less than 30 °C at fixed other conditions) was too small to draw such a conclusion.

At complete oxygen conversion, the selectivity of methanol formation in these experiments was influenced neither by changing the residence time of the mixture in the reactor nor by quenching the products at the reactor outlet. In contrast to [90], in [43], there were no signs of any oscillations: as claimed by the authors, the process was stable and reproducible. In general, the heterogeneous nature of the process in this study remains the most likely explanation of its high characteristics, substantially different from the results of most other studies.

It should be noted that all the experiments in which a very high methanol yield was obtained were carried out at a very low flow rate of the reaction mixture and, accordingly, a

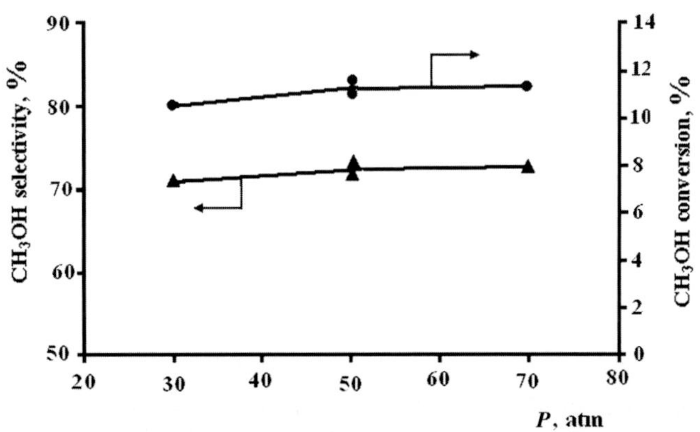

FIGURE 2.16 Pressure dependence of CH_4 conversion (●) and CH_3OH selectivity (▲) at complete oxygen conversion. $T = 340–350$ °C, and $[O_2] = 6.5\%$. *According to data from [43].*

long time of its residence in the reactor, ~100 s or more. At such a long residence time in the reactor and temperatures of 300 °C and above, quartz or any other of the used materials can no longer be considered inert with respect to methane–oxygen mixtures. Thus, under these conditions, the reaction is strongly influenced by the surface and cannot be regarded as a purely gas-phase process.

The assumption of the essentially heterogeneous nature of the reaction in experiments with a high declared methanol yield is consistent with their characteristic manifestations: a poor reproducibility and a wide scatter of experimental points even in one series of experiments. For example, in [42], experiments performed under similar conditions resulted in methanol yields of more than 7% and, on the other hand, as low as 1–2%. The data presented in the paper provide no clue as to how so significant differences are related to various parameters of the process, such as the pressure, temperature, or initial oxygen concentration. Unfortunately, in many experiments, more than one parameter was changed simultaneously, which makes it impossible to establish definite dependences, including that on the residence time of the mixture in the reactor. The absence or low concentration of formaldehyde in these works is most likely due to its decomposition on the reactor walls for such a long time. Difficult to explain is why CO_2 was not detected in a number of experiments while being abundant in other experiments under very similar conditions in the same study.

Experiments [91] show that, in typical laboratory reactors with a diameter of about 10 mm, a considerable percentage of conversion of the reactants occurs on the reactor's quartz surface, even at pressures of up to 80 atm and reaction times of only a few seconds. A reaction time of a few minutes not only manifold exceeds the typical time of radial diffusion of reactants to the surface of the reactor, but also is comparable with the time of axial diffusion over the characteristic length of the reactor. For example, under typical conditions realized in [42] ($P = 35$ atm, $T \approx 450$ °C), axial diffusion over a distance of 10 cm takes ~400 s, as compared to the reaction time of ~100 s. In fact, kinetic simulations of the gas-phase process [63–66] demonstrated that, because of a significant induction period, the overwhelming fraction of the reactants is converted within a much shorter time (and, hence, space) interval, constituting only a small part of the total duration of the process. Therefore, at such low flow rates, the back diffusion of the products from the region of active transformation of the reactants can have a significant impact on the processes during the induction period, specifically through unpredictable and uncontrollable changes in the state of reactor surface.

An argument in favour of a significant influence of back axial diffusion and thermal or chemical feedback through the reactor surface at reaction times longer than 100 s is that, at pressures of 25–35 atm, temperatures of 410–475 °C, and oxygen concentrations of 3–9%, methane oxidation exhibits temperature oscillations with a period of ~10–100 s [90]. Because of a long induction period, most of the conversion of methane and, accordingly, of the heat release occur within a time interval much shorter than the total residence time of the reactants in the reactor, so the oscillation period is well commensurate with the characteristic time of diffusion over the width of the reagent conversion zone. Therefore, back axial diffusion can have a substantial influence on the processes occurring during the induction period and, consequently, the process as a whole.

It is very instructive to compare the results of [42] with those of the experiments from [45], with the latter being intentionally performed under nearly identical conditions to check the findings of [42]. The selectivity and yield of methanol formation in [45] turned out to be more than two times lower than the values reported in [42] and consistent with the results of most of the other studies. Among the possible reasons for the discrepancy, the authors of [45] suggest the influence of minor difficult-to-identify structural features of the two experimental facilities, as well as the possible influence of the back diffusion of products. Note that [45] also reports a very small effect of the pressure and oxygen concentration on the selectivity of methanol formation, which, as discussed above, is most likely indicative of a significant contribution from heterogeneous processes to methane conversion, which is quite natural for reaction times of 100–300 s.

Another possible reason for the differences may be that, in the studies where a high yield of oxygenates was observed, either a special section for preheating the reactants [43,79] or a special heater for maintaining varying temperature along the reactor length [44] was used. Preheating to a sufficiently high (250–285 °C) temperature allows the reagents intensely interact with the surface and with each other on the surface for a long time. This introduces uncertainties in the composition of the reaction mixture supplied into the main part of the reactor. Therefore, it is likely that the decrease in the reaction temperature and increase in the selectivity of methanol formation, observed in these works upon preheating the mixture entering the reactor, are in fact associated with such a change in its composition during this preheating. In [79], the authors pointed out that, at a high oxygen concentration, preheating is critical for obtaining a high selectivity of methanol formation.

The conclusion that heterogeneous processes play a significant role in experiments with a declared high methanol yield is supported by observation of a wide variety of complex organic products, such as methyl formate, dimethoxymethane (methylal), acetone, DME [79], which are not detected for the short-time DMTM process. Moreover, under these conditions, dimethoxymethane (methylal) becomes one of the most abundant products (with a selectivity of up to 16%). Interpretation of the formation of such complex products within the framework of a gas-phase mechanism involving the 'cage effect' [79] seems to be very exotic and less persuasive than a trivial occurrence of heterogeneous processes. The assumption on the heterogeneous formation of complex organic products is supported by the fact that their yield increases with the residence time of the reactants in the reactor, pointing to their secondary nature.

Note also that, in [79], at relatively high concentrations of oxygen (over 4%), a methanol-formation selectivity of over 60% was observed only at residence times of a few minutes.

Another indication of the reaction deviating from the gas-phase mode at large residence times is the difference in the $\Delta[CH_4]/\Delta[O_2]$ ratio, which is generally much greater than 1 or even 2 at these times. Note that, for the slow ($\sim$10 h) low-temperature ($\sim$200 °C) oxidation of methane to methanol in a static reactor at a pressure of 40 atm (initiated by an electrically heated wire spiral), i.e., under conditions where it is difficult to deny a significant role of the surface in the initial period of the reaction, the selectivity of methanol formation reached 90%, but then decreased to 22% [92].

Among the other factors, in addition, surface catalysis and back diffusion, that contribute to the high selectivity of methanol formation in the aforementioned papers, it is worthwhile to single out a low temperature, since a reaction occurring mainly on the surface is characterized by an intense heat transfer to the reactor wall and, consequently, only a slight heating of the reaction mixture.

To illustrate the hypothesis of a substantial changeover of the oxidation mechanism depending on the residence time of the reactants in the reactor, in addition to experiments with $t_r < 20$ s (Figs 2.14 and 2.15), we selected the second group of experiments from those presented in Appendix I. Figures 2.17 and 2.18 show the dependences of the selectivity

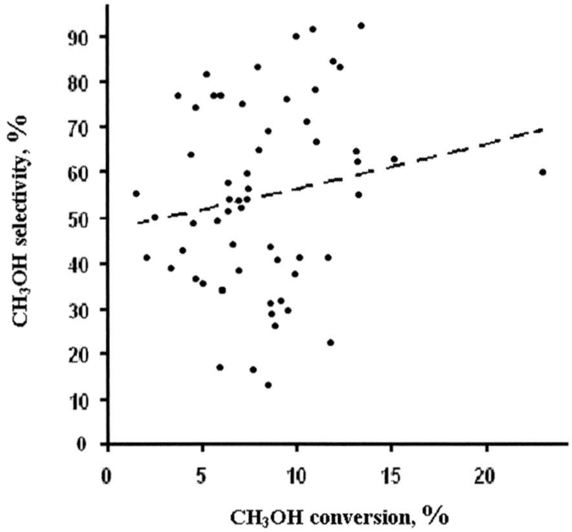

FIGURE 2.17 Dependence of the CH_3OH selectivity on the CH_4 conversion from experiments with a slow flow ($t_r > 20$ s). $P \geq 25$ atm (plotted based on the data from Appendix I). Dashed line – the trend of experimental data.

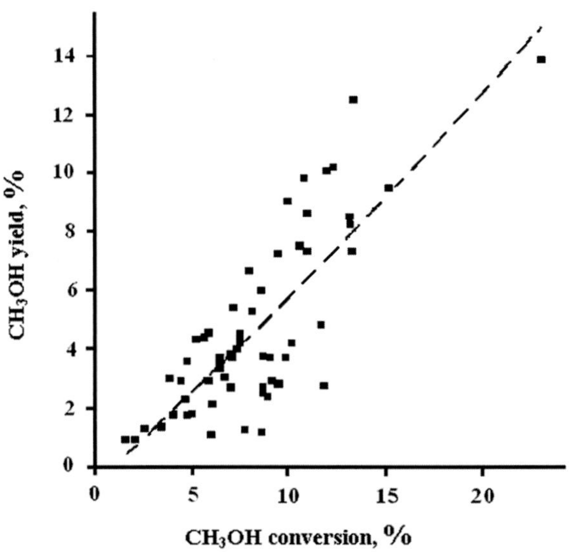

FIGURE 2.18 Dependence of the CH_3OH yield on the CH_4 conversion from slow-flow experiments ($t_r > 20$ s); $P \geq 25$ atm (based on the data from Appendix I). The dashed line shows the trend of experimental data.

and yield of methanol formation obtained in experiments (Appendix I) in which the reaction time exceeded 20 s ($P > 25$ atm). Figure 2.18 demonstrates the absence of any relation between the methane conversion and the selectivity of methanol formation for this group of experiments. Formally, the trend line even shows a growth of the selectivity with increasing conversion, a behaviour unrealistic for a gas-phase process. Accordingly, the methanol yield shows a tendency of unlimited growth with increasing conversion, unfeasible for a gas-phase process.

Based on this comparison, we can argue that the DMTM proceeds as a gas-phase process only at sufficiently short reaction times, not much longer than the time of diffusion of the reactants to the reactor walls. The actual yield of liquid products under these conditions, obtained from the experimental data presented in Appendix II, is shown in Fig. 2.19. According to these data, the output of methanol per pass through the reactor at a single-point injection of oxygen into the reactor can reach just over 20 g/m^3 of passed gas. This value is reached at an initial concentration of oxygen in the mixture of ~4%. At higher oxygen concentrations, the methanol yield practically ceases to grow, and its concentration in the liquid oxidation products decreases rapidly in favour of water.

The assumption that a slow DMTM process is of heterogeneous nature readily accounts for a significant influence of the reactor design on the parameters of the process [41,86]. In this study, a quartz insert was firmly planted into a stainless steel reactor with the help of O-ring fluorinated rubber gaskets, a design that eliminates the contact of the hot reactants and products with the metal surface. The products quickly flowed out of the hot zone, thereby quenching radical reactions and preventing further oxidation. The presented data suggest that the residence time was ~50–100 s. In this case, the selectivity of methanol

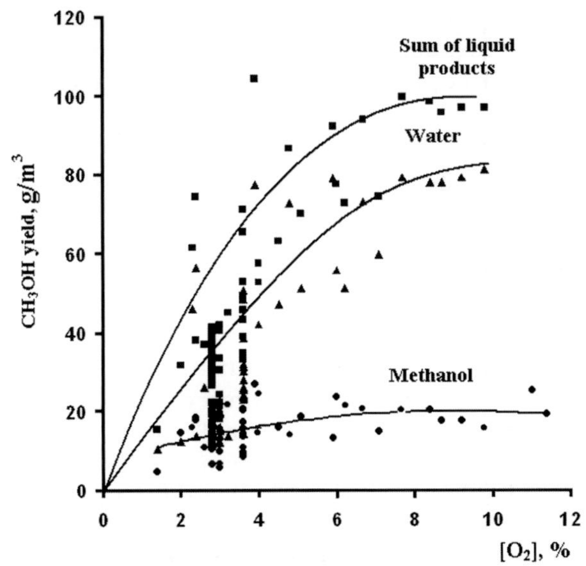

FIGURE 2.19 Dependence of the yield of liquid oxidation products on the oxygen concentration (based on the data from Appendix II).

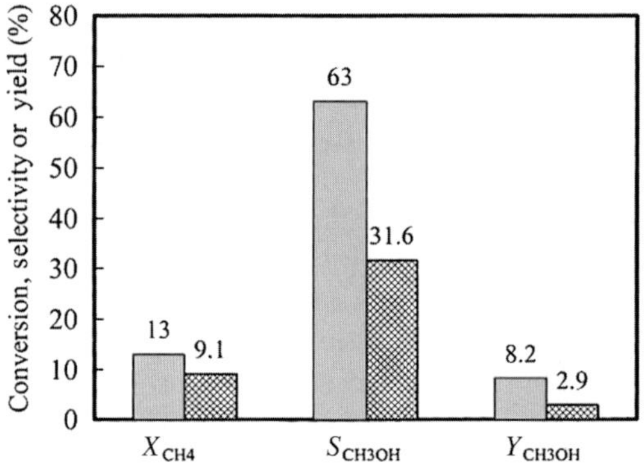

FIGURE 2.20 Effect of a complete (grey bars) and incomplete (dotted bars) isolation of the hot reaction zone from the metal surface on the methane conversion X_{CH4}, methanol selectivity S_{CH3OH}, and methanol yield Y_{CH3OH}, at $P = 50$ atm, $T = 450\,°C$, and $CH_4/O_2/N_2 = 100{:}10{:}10$ [86].

formation at $P = 50$ atm, $T = 450\,°C$, and $CH_4/O_2/N_2 = 100{:}10{:}10$ was as high as 62.3%, with the yield reaching 8.2%. When the O-gaskets were removed, the selectivity and yield decreased to values typical of these conditions, 31.6 and 2.9%, respectively (Fig. 2.20). Among the other factors affecting the selectivity and yield of methanol formation were the replacement of the quartz insert by a silicate glass one, placement or removal of a quartz sheath for thermocouple coaxially positioned along the centre of the reactor, and temperature profile in the reactor.

Thus, it seems that all the known cases of very high (over 60%) selectivity of methanol formation at a methane conversion of over 5% are associated with slow heterogeneous—homogeneous processes. At long residence times of the reactants in laboratory reactors, the influence of a surface, including a quartz one, on the process is so great that, under these conditions, in contrast to high-temperature fast-flow conditions, the process can no longer be considered homogeneous. Accordingly, the results obtained under these conditions can hardly be described in the framework of gas-phase kinetic models developed for qualitative and even quantitative description of the short-time DMTM process. A quantitative description of such slow gas-surface processes is impossible without a fairly full account of not only heterogeneous stages, but also of thermokinetic phenomena, including the back diffusion of products, variations of the temperature and state of the reactor surface under the influence of the reaction, heat transfer through the reactor walls along the flow, etc. In all probability, it is these phenomena that manifest themselves through oscillatory modes and poor reproducible differences in the yield of the products.

Summarizing the experimental results on the effect of the conditions on the yield of oxygenates in the DMTM process, it is necessary to point to the inadequacy of the traditional division of this process into two types: gas-phase homogeneous and heterogeneous-catalytic. The influence of the solid phase, even a relatively 'inert' reactor surface, on this radical process, has been clearly underestimated. The entire process, at any rate on the laboratory scale, should be viewed as fundamentally heterogeneous—homogeneous. In fact, it is such a heterogeneous—homogeneous process that has been actually studied in almost all experimental works on the subject. Only a few studies, in reactors with a relatively inert surface, at pressures above 70 atm and residence times of no more than a few seconds, can be considered as dealing with a nearly homogeneous process amenable to simulation within the framework of homogeneous models. With a high degree of confidence, one can expect that, in industrial-scale reactors, with a typical diameter of a few tens of centimetres, at pressures of tens of atmospheres and reaction times of about a second, the process can be qualitatively and even quantitatively described by gas-phase models and interpreted in terms of the underlying concepts. A practical benchmark for the expected yield of methanol on such facilities can be based on the best results presented in Appendix II and in Fig. 2.19 or even can somewhat surpass them.

As for sensational results, periodically appearing in the literature, it seems that all of them are related to a complex nonlinear kinetics of the interaction of gas-phase and gas-surface processes, being poorly reproducible and difficult to describe, even qualitatively. This interaction gives rise to a situation where a substantial part of the conversion occurs immediately on the surface or the composition of the gas phase is changed by heterogeneous processes so that the subsequent gas-phase reaction proceeds in a substantially modified mixture. What is certain, however, is that, even if such intricate heterogeneous–homogeneous phenomena could be understood and kinetically described, they are unlikely to have a chance to be a suitable basis for industrial-scale processes due to their long duration and instability of the conditions at the surface. A further refinement of technologies designed for practical applications should be based on a deeper analysis of the mechanism of the homogeneous reaction, its stepwise nature, and the ability to control this nonlinear chain-branched process.

OXIDATION PRODUCTS OF METHANE HOMOLOGUES

The number of papers devoted to the partial oxidation of methane homologues is noticeably smaller than those dealing with a methane oxidation. A large variety of products formed in this case makes it extremely difficult to perform any quantitative assessment of the real fractions of individual products and their dependence on the specific parameters of the process. In general, while the basic classes of compounds formed remain essentially the same (alcohols, aldehydes, and organic acids), the transition to methane homologues expectedly increases the yield of compounds with a large number of carbon atoms, up to that present in the parent alkane.

One of the first studies of ethane oxidation under static conditions at pressures of 15–100 atm [22] showed that the oxidation of 9:1 C_2H_6–O_2 mixtures yields, along with carbon oxides, methane, and water, various oxygenates, such as methanol, ethanol, formaldehyde, acetaldehyde, and formic and acetic acids. At higher pressures, the yield of C_2 products (ethanol, acetaldehyde, and acetic acid) increases, whereas the yield of methanol and formaldehyde decreases.

Practically, the same products (ethanol, acetaldehyde, methanol, formaldehyde, and acetic acid) were observed during the partial oxidation of ethane in a static quartz reactor at $T = 440–450\,°C$, $t_r = 1.5–4.3$ min, and atmospheric pressure [24].

For the oxidation of propane at high pressure in static reactors [18,23,26], aldehydes were represented by propionic and acetic aldehydes, normal alcohols, by methanol, ethanol, and propanol, and acids, mainly by acetic acid with small amounts of propionic and formic acids.

The growth of the pressure leads to a significant increase in the yields of isopropanol, acetone, and organic acids, a decrease in the yields of aldehydes and alcohols (with a change in the ratio between the normal alcohols in favour of propanol), an increase in the CO_2 yield, and a decrease in the CO yield. At pressures above 20 atm, propylene, abundant at lower pressures, completely disappeared from the reaction products.

As for the products of the gas-phase oxidation of butane, pentane, and heavier hydrocarbons at high pressures, only disembodied data can be found. The authors of [26] studied the oxidation of normal butane, pentane, and heptane. Among the products of the interaction of *n*-butane with oxygen at pressures from 33 to 160 atm, along with the products characteristic of propane oxidation, butyl alcohols were found, their fraction increasing with the pressure. In [93], the oxidation of isobutane was reported to produce acetone, *tert*-butanol, and *tert*-butyl hydroperoxide.

CHAPTER

3

The Main Parameters of the Process

EFFECT OF PRESSURE ON THE TEMPERATURE AND RATE OF THE PROCESS

As established in the earlier works on the direct oxidation of hydrocarbon gases, one of the two basic conditions for producing a significant yield of alcohols, along with a low concentration of oxygen, is high pressure. High pressure not only increases the selectivity of formation of alcohols, but also reduces the temperature of the process, a factor favourable for preserving the formed oxygen-containing products.

The pressure dependence of the temperatures of onset of conversion and of complete oxygen conversion in a quartz reactor can be plotted (Fig. 3.1) based on the data from [45] and [68]. In addition, the figure shows the data from [91] at $P = 80$ atm. Although these works differ somewhat in experimental conditions, particularly, in the oxygen concentration and the

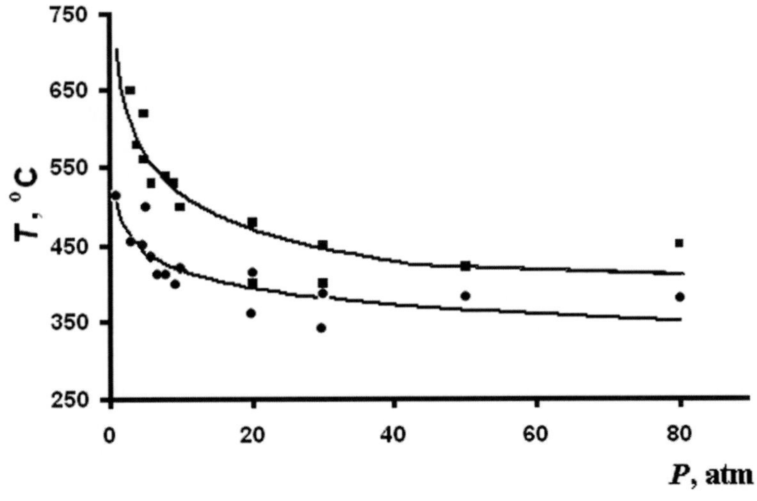

FIGURE 3.1 Dependence of the temperatures of (■) onset of conversion and (●) complete oxygen conversion on the pressure plotted based on the data from [45], [68], and [91].

http://dx.doi.org/10.1016/B978-0-444-63253-1.00003-9

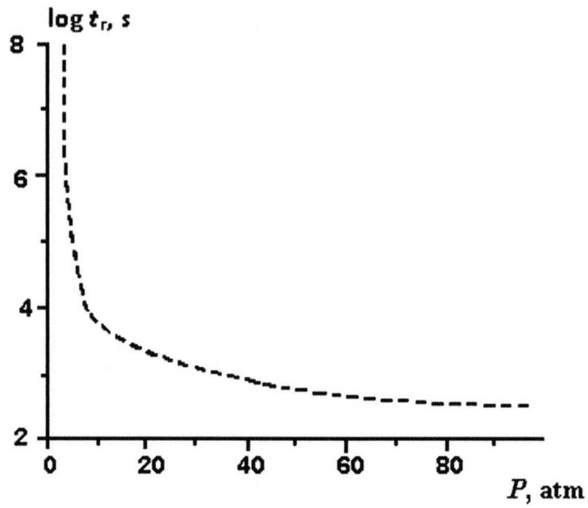

FIGURE 3.2 Calculated pressure dependence of the reaction time t_r at $T = 650$ K [94].

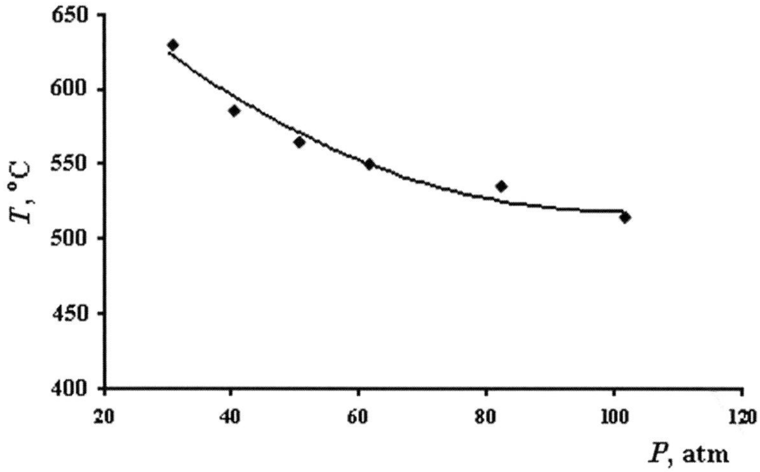

FIGURE 3.3 Pressure dependence of the temperature of complete oxygen conversion at $[O_2]_0 = 14\%$ and $t_r = 0.6-0.8$ s [81].

residence time of the mixture in the reactor (Appendix I), Fig. 3.1 clearly demonstrates how the temperature of the process depends on the pressure. As can be seen, the temperature of the process decreases most rapidly with increasing pressure at low pressures, with the dependence becoming weak above 30 atm and seemingly disappearing at pressures above 80 atm.

Note that this experimental dependence is quite consistent with the pressure dependence of the reaction time (Fig. 3.2) obtained in a kinetic modeling of the process [94].

A similar pressure dependence of the temperature of complete oxygen conversion in a quartz flow reactor filled with 30–80- and 80–120-mesh inert sand packings, at a residence time of the mixture in the reactor of 0.6–0.8 s and an oxygen concentration of $[O_2]_0 = 14\%$ (Fig. 3.3), was obtained in [81].

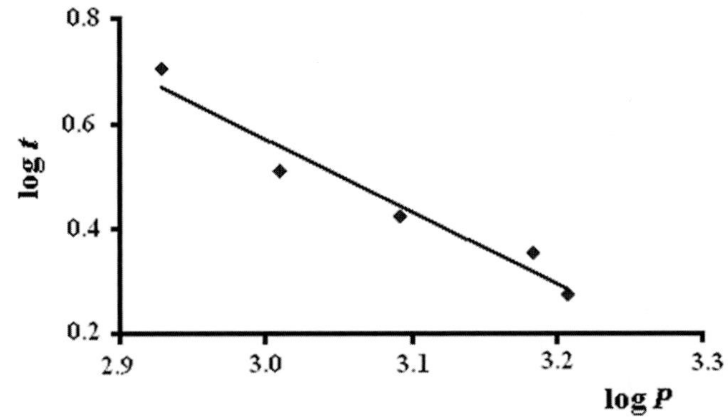

FIGURE 3.4 Pressure dependence of the ignition delay time for an 88% $CH_4 + 12\%$ air mixture at $T = 406\ °C$ [54]. (For colour version of this figure, the reader is referred to the online version of this book.)

A detailed study of the slow reaction in very rich methane–air mixtures during the ignition delay at pressures of 58–110 atm [54] showed that autoignition occurs at $T \approx 350\ °C$, well below the ignition temperature at low pressures. Increasing the pressure in an 88% $CH_4 + 12\%$ air mixture ($[O_2]_0 = 2.8\%$ of $[CH_4] + [O_2]$) from 58 to 110 atm at $T = 406\ °C$ caused a decrease in the ignition delay (Fig. 3.4), i.e., an increase in the conversion rate, which obeyed the power dependence

$$\tau_{ig} \sim P^{-1.4} \tag{3.1}$$

The observed change in the temperature and reaction time reflects changes in the mechanism of the process caused by increasing pressure. Therefore, it is not surprising that, in about the same pressure range, the methanol yield grows most rapidly with increasing pressure.

EFFECT OF PRESSURE ON THE YIELD OF THE PARTIAL METHANE OXIDATION PRODUCTS

That the CH_3OH yield increases rapidly with the pressure (Fig. 3.5) was shown in one of the early studies in a static reactor [15].

A significant increase in the yield of methanol (and formaldehyde) with increasing pressure in a copper reactor was observed in [51,95] (Fig. 3.6). The authors emphasized the role of pressure in reducing the initial temperature and in suppressing the side reactions on the surface of the reactor. As the pressure was raised, the growth rate of the methanol yield gradually slowed down and virtually stopped occurring at pressures of 300–500 atm. As the residence time of the reactants in the reactor was increased from 1.8 to 9.0 s, the methanol yield declined significantly.

Although the formaldehyde yield in [51] also increased with the pressure up to ~200 atm, it was very small. The CH_3OH/CH_2O ratio increased almost linearly with the pressure (Fig. 3.7); at the same pressure and residence time, it was about an order of magnitude higher than that reported in [69] (Fig. 2.9). Most likely, this is due to a higher rate of CH_2O loss on the

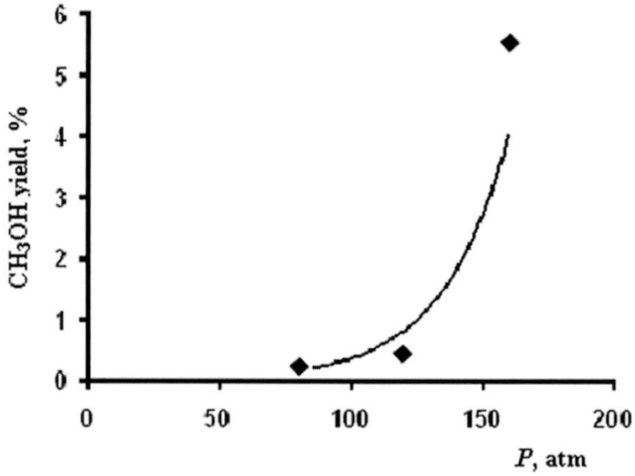

FIGURE 3.5 Yield of CH_3OH as a function of the pressure in the static reactor at $T = 350\ °C$ and $CH_4/O_2 = 9:1$ [15].

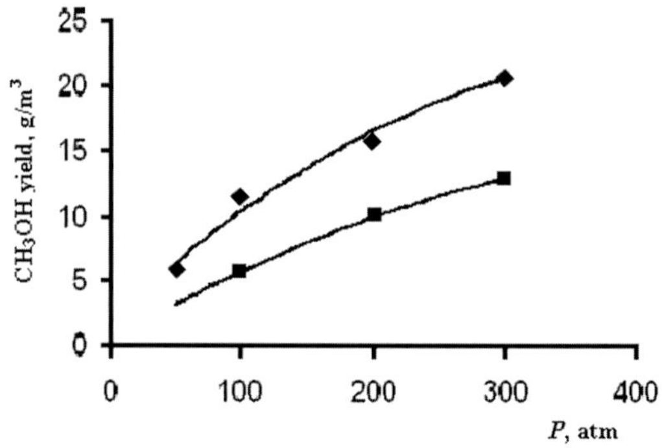

FIGURE 3.6 Pressure dependence of the CH_3OH yield in a copper-surface reactor at $T = 475\ °C$, $[O_2]_0 = 3\%$: (◆) $t_r = 1.8$ s and (■) $t_r = 9.0$ s. *Based on the data from [51].*

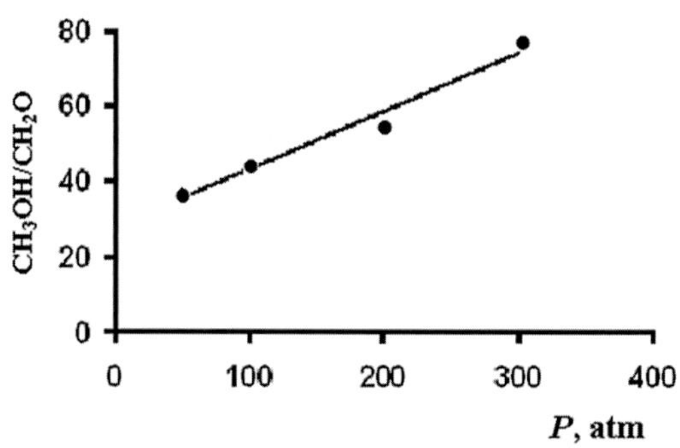

FIGURE 3.7 Pressure dependence of the CH_3OH/CH_2O ratio for the conversion in a copper-surface reactor at $T = 475\ °C$, $[O_2]_0 = 3\%$, and $t_r = 1.8$ s. *Based on the data from [51].*

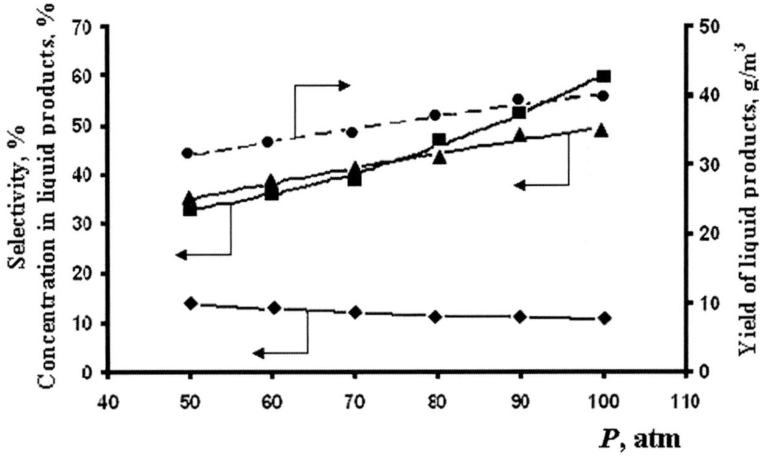

FIGURE 3.8 Pressure dependence of the (●) total yield of liquid organic products, (■) total selectivity of formation of liquid organic products, (▲) concentration of methanol in liquid product, and (◆) concentration of the sum of aldehydes. *Based on the data from [76].*

surface of the copper reactor in [51] as compared to that on the surface of the stainless steel reactor used in [69].

Experiments [76] in a stainless steel flow reactor at a short residence time (~0.2 s), $T = 410$ °C, and $[O_2]_0 = 2.8\%$ showed (Fig. 3.8) that the total selectivity of formation of liquid organic products, their total yield, and the concentrations of methanol and C_2–C_4 alcohols in liquid increase with the pressure. The content of aldehydes in the liquid products decreased monotonically with increasing pressure, whereas the content of organic acids remained almost constant. The concentrations of the other detectable organic products (ethers and ketones), except for methyl ethyl ketone, also increased with the pressure.

The ratio of the total concentration of alcohols to the total concentration of aldehydes in the produced liquid, as in other studies, increased monotonically with the pressure (Fig. 3.9) and almost exactly reproduced the dependence in Fig. 2.9, plotted based on the data from [69].

The total yield of liquid products (Fig. 3.10) and their composition (Fig. 3.11) at pressures from 30 to 230 atm were obtained in [69].

The yield of liquid organic products increases markedly with the pressure up to 100 atm, above which pressure rise has virtually no effect. The methanol concentration in the liquid product shows a similar behaviour and becomes commensurate with the water concentration at 100 atm. The formaldehyde concentration decreases slowly with increasing pressure up to 100 atm and then too remains almost constant.

There are a number of studies of the DMTM process at very high pressures. Several series of experiments in static steel reactors at pressures from 140 to 14 000 atm, oxygen concentrations of 8–9%, and initial temperature of 250–340 °C were carried out in [46]. The authors noted a marked influence of the state of the reactor surface and, tentatively attributing to it a poor reproducibility of the results (Appendix I). Two series of experiments in a static reactor at pressures of 1700 and 3400 atm, temperatures of 270–310 °C, an oxygen concentration of 8%, and a residence time of up to 30 min were performed in [48]. For a typical set of

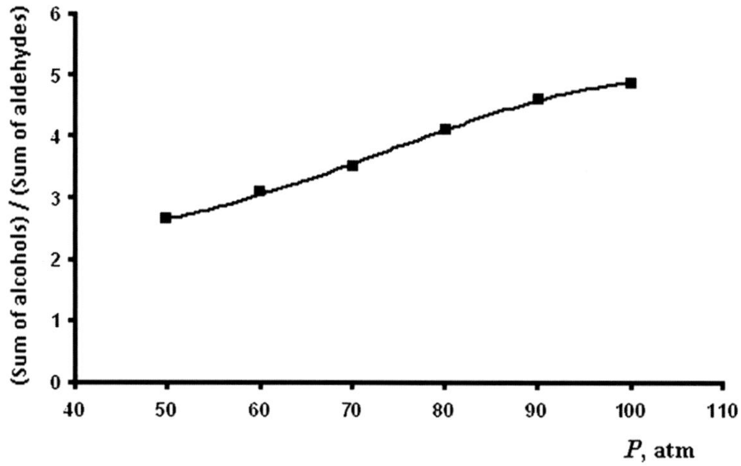

FIGURE 3.9 Pressure dependence of the ratio of the sum of alcohols to the sum of aldehydes. *Based on the data from [76].*

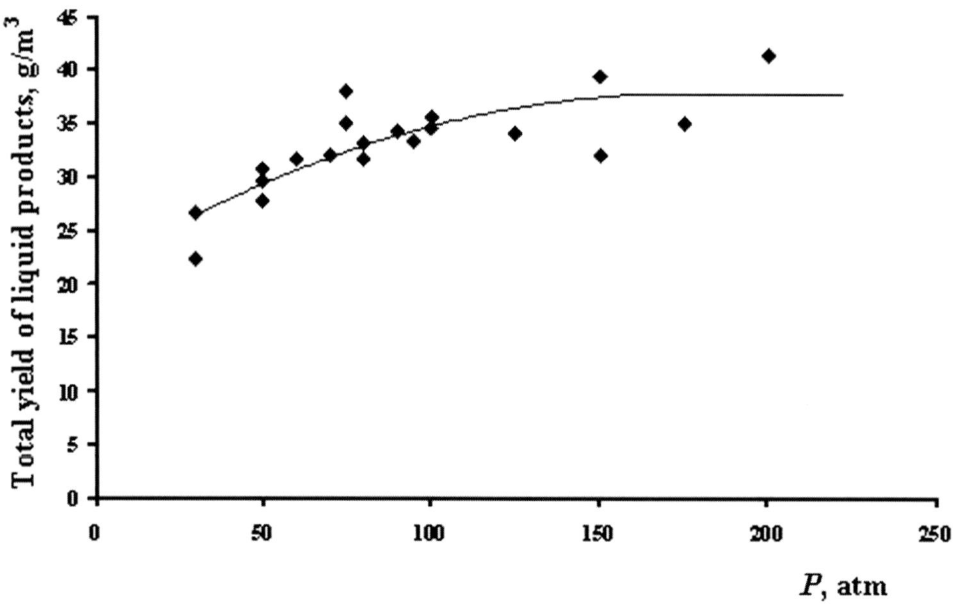

FIGURE 3.10 Pressure dependence of the total yield of liquid oxidation products at $T = 400\,°C$ and $[O_2]_0 = 2.8\%$. [69].

products, as in a number of experiments performed in [46], there was a clear predominance of CO_2 over CO. Of the oxygenates, along with methanol and formaldehyde, trace amounts HCOOH, C_2H_5OH, and CH_3COCH_3 were also detected. It is interesting that, after the reaction essentially stopped, 20% of the oxygen remained unreacted. In general, the data of this

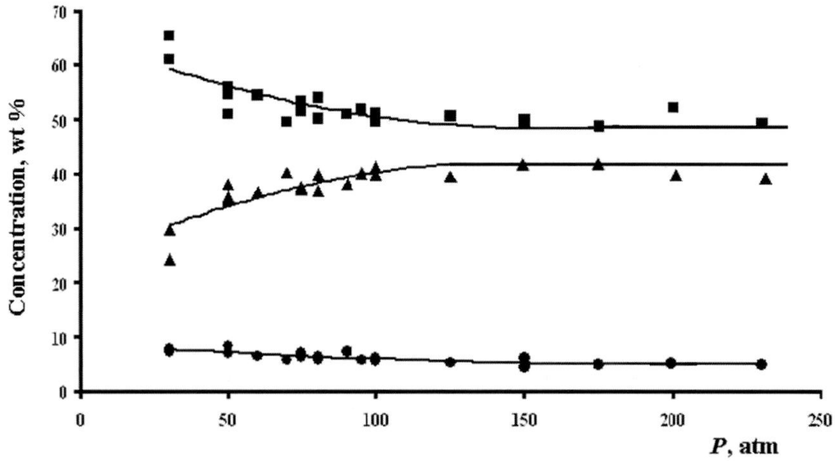

FIGURE 3.11 Pressure dependence of the composition of the liquid products at $T = 400\,°C$ and $[O_2]_0 = 2.8\%$: (■) H_2O, (▲) CH_3OH, and (●) CH_2O [69].

work are difficult to interpret; however, based on the plots presented and the known initial concentration of oxygen, it is possible to estimate the methanol yield at the maximum oxygen conversion: ~1.7% at 1700 atm and ~0.17% at 3400 atm (Appendix I). Thus, the behaviour of the DMTM process at extreme pressure does not show any specific features in the kinetic characteristics or in the composition of the products, contrary to the assumptions on the possibility of the cage effect under these conditions [79]. The same turned out to be true when the reaction was conducted under supercritical conditions (see Chapter 9). Thus, we can claim that the influence of the pressure on the yield of the DMTM products is most pronounced in the range of up to 100 atm, becoming weak with further pressure rise.

Methanol is also formed at atmospheric pressure, for example, in experiments [25] with $CH_4/O_2 = 9{:}1$, 4:1, and 2:1 mixtures at 456 °C in a static quartz reactor washed with nitric acid. The selectivity of methanol formation reached a maximum at the end of the highest-rate period of methane conversion, exceeding 20% for the $CH_4/O_2 = 9{:}1$ and 4:1 mixtures, after which it rapidly decreased, leading to a low integrated methanol yield. Formaldehyde was detected only in trace amounts (0.01–0.1%).

Interesting data on the effect of pressure on the yield of various products at initial reaction temperatures of 550–630 °C, intermediate between those typical of DMTM and oxidative coupling of methane (OCM), were obtained in [81,82] at pressures of 30–100 atm and a reaction time of ~1 s. With increasing temperature, as well as pressure, this transition region featured a marked changeover from oxygen-containing, mainly COx products, to C_2+ products (Fig. 3.12). Because of a high initial concentration of oxygen, $[O_2]_0 = 14\%$, its complete conversion caused a substantial heating of the mixture, up to 500 °C, so that the final temperature could exceed 1000 °C, well deep in the OCM temperature range. Therefore, the effects described below, despite the monotonic decrease of the initial temperature in this series from 630 to 515 °C may be related, at least in part, to the heating of the mixture. However, the results show that the COx/C_2+ ratio changes with the pressure, as does the C_2H_6/C_2H_4 ratio, showing an increase in the relative yield of ethylene (Fig. 3.12).

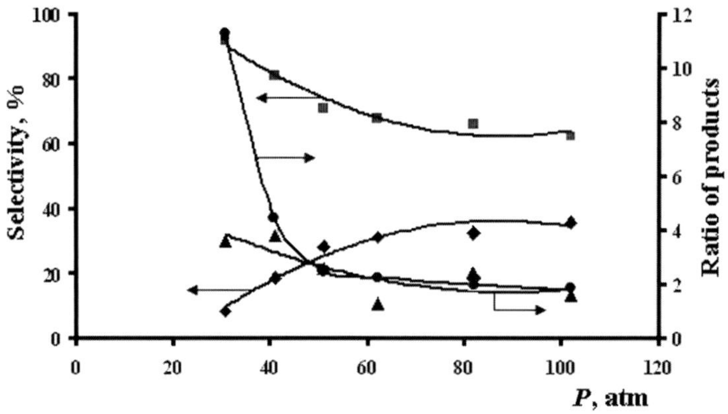

FIGURE 3.12 Pressure dependence of the selectivity S and the ratio of products at $T_0 = 550-630\,°C$, $[O_2]_0 = 14\%$, and $t_r \approx 1$ s: (■) S_{COx}, (◆) S_{C2+}, (●) COx/C_2+, (▲) C_2H_6/C_2H_4. *Based on the data from [81,82].* (For interpretation of the references to colour in this figure legend, the reader is referred to the online version of this book.)

Since the COx/C_2+ ratio at a constant initial oxygen concentration changes with increasing pressure, the methane conversion and $\Delta[CH_4]/[O_2]_0$ ratio increase symbatically in this case (Fig. 3.13).

At higher temperatures, 515–630 °C, when the yields of ethane and ethylene become significant, a pressure rise from 31 to 102 atm increases the yield of C_2 hydrocarbons from 8 to 36%. At the same time, the initial temperature at which the complete conversion of oxygen was achieved decreased by 100 °C.

The pressure dependence of the yield of the main liquid products of the partial oxidation of methane and their content in the liquid products were investigated in [91]. At $P \leq 80$ atm,

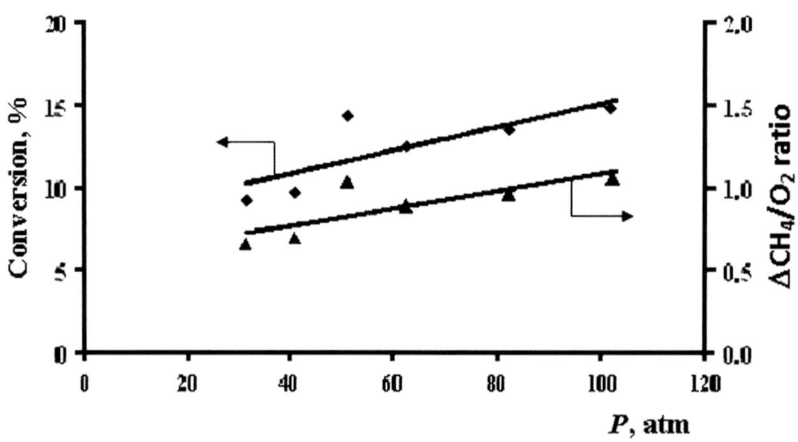

FIGURE 3.13 Pressure dependence of the methane conversion (◆) and $\Delta CH_4/O_2$ ratio (▲) at $T_0 = 550-630\,°C$, $[O_2]_0 = 14\%$, and $t_r \approx 1$ s. *Based on the data from [81,82].*

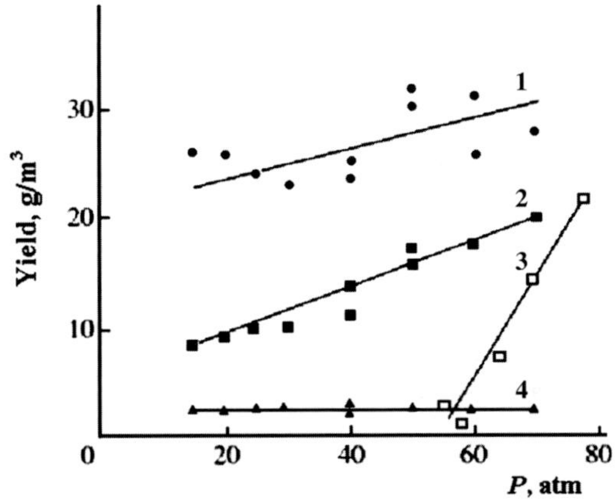

FIGURE 3.14 Pressure dependence of the yields Y (g/m^3) of (1) water, (2, 3) methanol, and (4) formaldehyde for the oxidation of methane in a (1, 2, 4) quartz and (3) stainless steel reactor at $[O_2]_0 = 3.6\%$ and $T = 550\ °C$ [91].

the yield of the sum of liquid oxidation products, that of methanol (Fig. 3.14), as well as the concentration of methanol in the liquid oxidation products (Fig. 3.15), decreased monotonically with the pressure. By contrast, the formaldehyde concentration increased with decreasing pressure, while its yield remained almost constant despite the decline in the overall yield of the liquid products. There was a clear effect of the reactor surface material on the yield of the products. Although the maximum methanol yield at high pressures in quartz and stainless steel reactors is practically the same, a higher initial concentration of oxygen is required to achieve it in a quartz reactor.

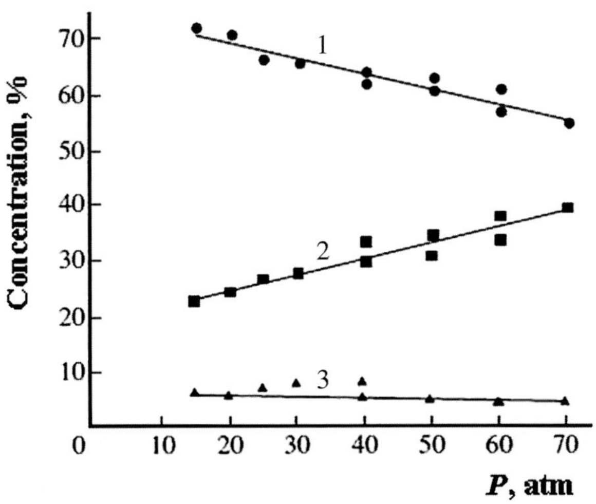

FIGURE 3.15 Pressure dependence of the concentrations of (1) water, (2) methanol, and (3) formaldehyde in liquid products at $[O_2]_0 = 3.6\%$ and $T = 550\ °C$ in a quartz reactor [91].

EFFECT OF TEMPERATURE ON THE YIELD OF THE PRODUCTS

The temperature dependence of the methane conversion at a fixed length of the reactor (or fixed reaction time) exhibits a pronounced S-shaped behaviour (Fig. 3.16) [91]. The transition from a negligibly small conversion to a complete conversion occurs within a very narrow temperature range, reflecting the critical nature of the temperature dependence of this branched-chain process.

With increasing temperature, the reaction-time shortens (the time of complete oxygen conversion, practically identical to the induction period of the reaction). The nearly ideal Arrhenius temperature dependence of the oxygen conversion displayed in Fig. 3.17 gives an effective activation energy of ~46 kcal/mol. The heating of the mixture under these experimental conditions is close to adiabatic with a ~43 °C increase of the temperature per each percent of oxygen in the reaction mixture. These experimental data were well reproduced by kinetic simulations of the process [89].

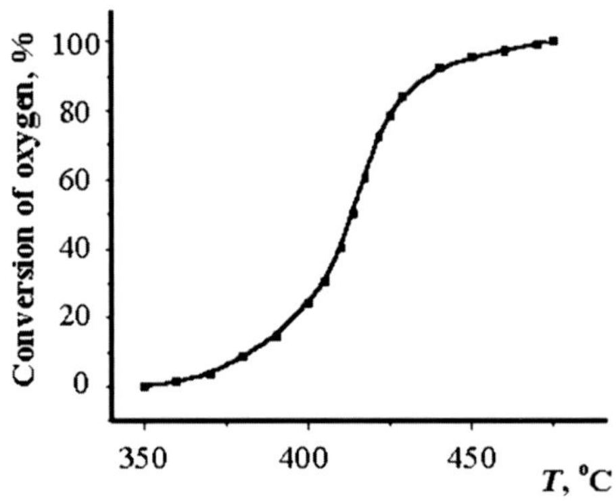

FIGURE 3.16 Dependence of the oxygen conversion in the stainless steel reactor on the initial temperature of the reaction mixture at $P = 80$ atm, $[O_2]_0 = 3.3\%$, and $t_r \approx 1$ s [91].

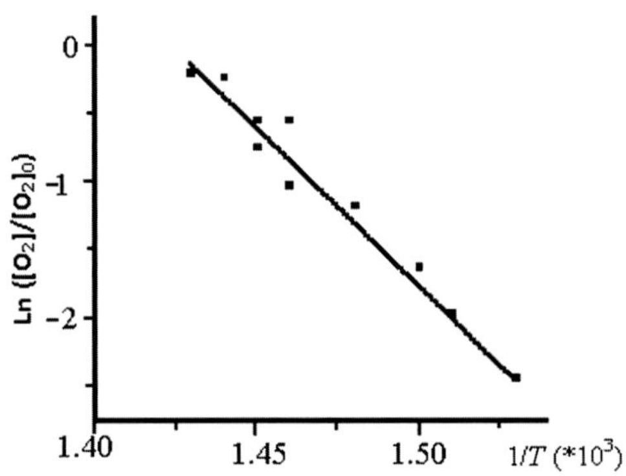

FIGURE 3.17 Temperature dependence of the oxygen conversion in a stainless steel reactor at $P = 80$ atm, $[O_2]_0 = 3.3\%$, and $t_r \sim 1$ s [91].

Another work in which data on the activation energy of the process were obtained is [54], where the effect of the pressure on the ignition delay time for very rich (10–40% air) methane–air mixture ($T \approx 400\,°C$, $P \approx 75$ atm) was studied. It was found, as expected, that the activation energy of the ignition delay time and the activation energy of the reaction rate, quantities with opposite signs, have nearly the same absolute values, 39–45 kcal/mol. A similar activation energy, $E_a = 43.5$ kcal/mol, was obtained in a study of the ignition delay of rich methane–oxygen mixtures with oxygen excess coefficient $\alpha = 0.30$ and 0.35 at an initial pressure of 15 atm and initial temperatures of 500–630 °C [96]. Similar values were reported in a number of works on the slow oxidation of methane at low pressures. A somewhat higher activation energy, $E_a = 56$ kcal/mol, was obtained in experiments [83] in a void Pyrex reactor at $P = 50$ atm, $T = 400$–$450\,°C$, $[O_2]_0 = 4\%$, and a reaction time of $t_r \approx 2$–5 s.

Note that the presence of even a small admixture of heavier hydrocarbons can lead to a significant reduction in the effective activation energy of oxidation. For example, according to [97], at ~6% C_2+ hydrocarbons in the gas mixture, the activation energy E_a decreases to 29–33 kcal/mol.

Moderate increase in the temperature above the temperature of a complete oxygen conversion has little effect on the methanol yield. Only a significant temperature rise causes a reduction in the yield of methanol, due to both a decrease in the selectivity of its formation and the decomposition of part of the methanol on the reactor walls. Nevertheless, the results of a number of studies show that, in general, a lowering of the temperature under the conditions of complete oxygen conversion is beneficial to the yield of oxygenates, although only to a limited extent. Figure 3.18 illustrates this point by showing the temperature dependence of the methanol yield at 200 and 300 atm from [51]. The cause of the decrease of the methanol yield in this study is most likely its decomposition on the heated copper surface of the reactor.

The monotonic decrease of the selectivity of methanol formation with increasing temperatures was observed [45] in a Pyrex flow reactor at temperatures of 375–500 °C, reaction times of 50–300 s, and various pressures (Fig. 3.19) and oxygen concentrations (Fig. 3.20). Note that, with increasing pressure, the temperature dependence becomes flatter. This work is the only one in which, at certain temperatures, a higher selectivity of methanol

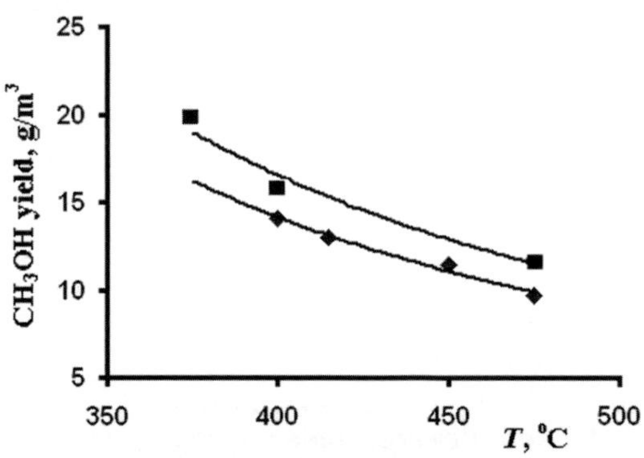

FIGURE 3.18 Temperature dependence of the CH_3OH yield in a copper flow reactor at $[O_2]_0 = 3\%$, $t_r = 1.8$ s, and $P =$ (◆) 200 and (■) 300 atm. *Based on the data from [51].* (For interpretation of the references to colour in this figure legend, the reader is referred to the online version of this book.)

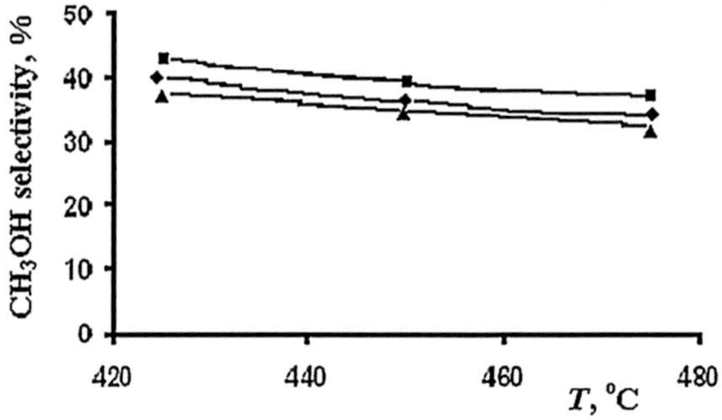

FIGURE 3.19 Temperature dependence of the selectivity of methanol formation at $P = 30$ atm and $[O_2]_0 = $ (◆) 2.5%, (■) 4.9%, and (▲) 6.6%. *Based on the data from [45].*

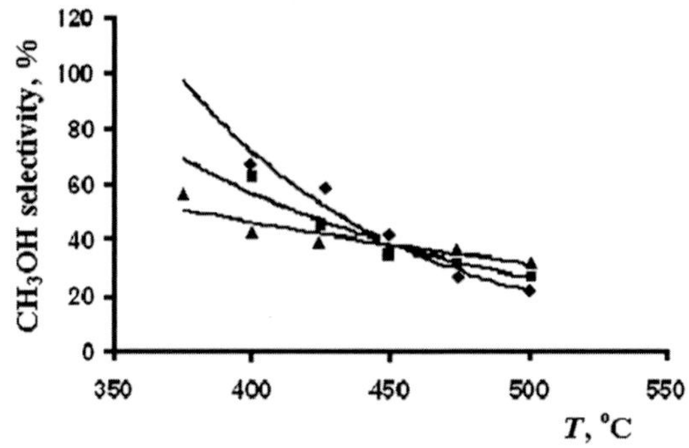

FIGURE 3.20 Temperature dependence of the methanol selectivity at $[O_2]_0 = 2.5\%$, $P = $ (◆) 10, (■) 30, and (▲) 50 atm. *Based on the data from [45].*

formation was obtained at lower pressures (Fig. 3.20). It seems that, at long reaction times and low temperatures and pressures, the contribution from heterogeneous processes on the reactor surface was significant in this work, so that the effect of pressure was less pronounced than in the homogeneous process.

In [98], the methanol yield decreased monotonically, though not very significantly, with increasing temperature (Fig. 3.21, dashed line); however, at temperatures above 600 °C, the decrease becomes steeper (Fig. 3.21, solid line).

According to data obtained in a study [76] of the conversion of natural gas (with C_2+ hydrocarbons content of ~5%) on a pilot plant with a stainless-steel flow reactor (residence time, ~0.2 s; $P = 100$ atm; $[O_2]_0 = 2.8\%$), the selectivity of formation and yield of liquid organic products and the concentrations of methanol and C_2–C_4 alcohols in them pass through a flat maximum at ~410 °C (Fig. 3.22).

In this case, the content of aldehydes and acids increased monotonically with temperature. The concentration of the other detectable organic products (ethers and ketones) also passed through a weakly pronounced maximum at about the same temperature. The yield of aldehydes increased with the temperature: the ratio of the total concentration of alcohols to the

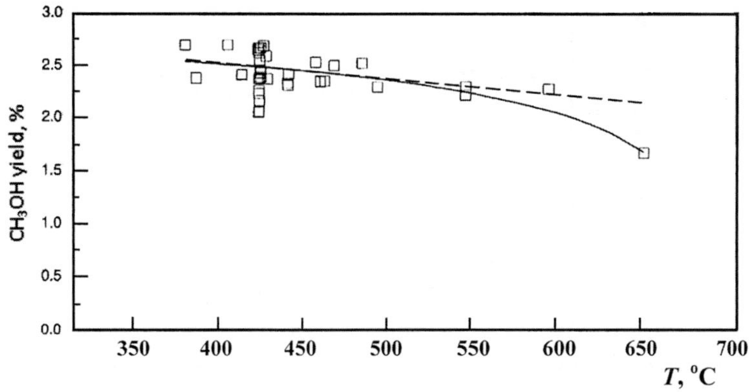

FIGURE 3.21 Effect of temperature on the methanol yield at $P = 91$ atm, $[O_2]_0 = 9\%$, and $t_r = 35$ s [98].

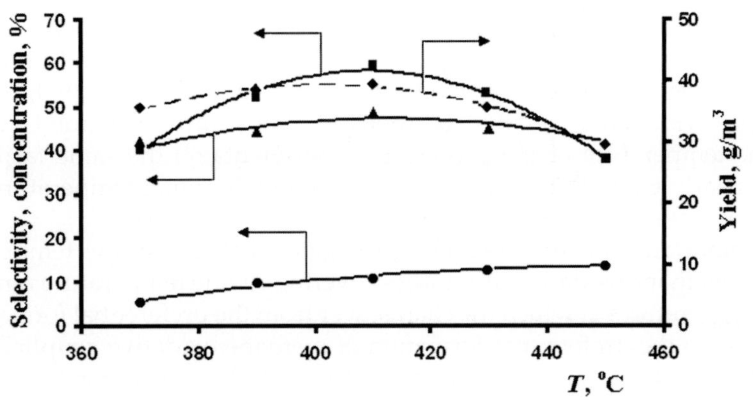

FIGURE 3.22 Temperature dependence of the (■) selectivity of formation and (◆) yield of the sum of liquid organic products and of the concentrations of (▲) methanol and (●) formaldehyde in them at $P = 100$ atm, $[O_2]_0 = 2.8\%$, and $t_r \approx 0.2$ s. *Based on the data from [76].*

total concentration of aldehydes in liquid products decreased monotonically with increasing temperature, presumably, levelling out onto a plateau at $T > 440$ °C (Fig. 3.23).

According to the data from [91] (a flow reactor with a quartz inner surface, $P = 20$–70 atm, and reaction time of ~2 s), after reaching the temperature of complete oxygen conversion, the oxidation process becomes only slightly sensitive to the initial and maximum temperatures of the reaction mixture. The yield of liquid products and the concentration of methanol in the liquid products slightly increased with the maximum temperature of the mixture up to $T \approx 520$ °C, after which the methanol yield decreased gradually. Such a temperature dependence is consistent with that predicted by kinetic simulations [89]. The formaldehyde and ethanol yields remained almost constant at maximum mixture heatings of 480–600 °C. Higher temperatures were not used in this work, since, in this case, the DMTM can be accompanied by noticeable soot formation.

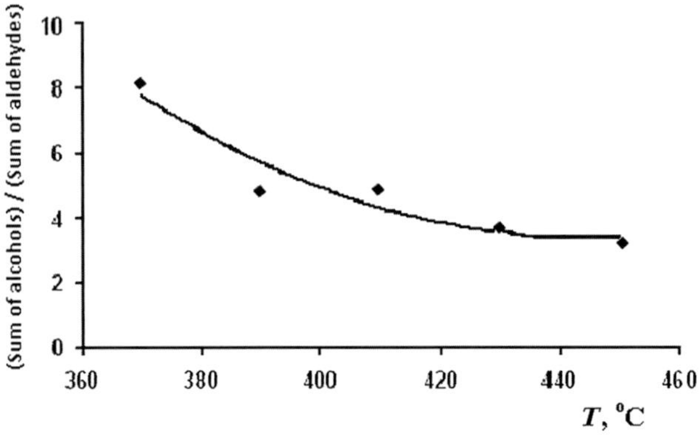

FIGURE 3.23 Temperature dependence of the ratio of the total yields of alcohols and aldehydes at $P = 100$ atm, $[O_2]_0 = 2.8\%$, and $t_r = \sim 0.2$ s. *Based on the data from [76].*

The difference in the optimal temperatures in [91] and [76] is most likely associated with differences in the composition of the gas mixture, which contained a significant concentration of methane homologues in the latter study. As was demonstrated in [97], the changeover from natural gas with ~6% C_2+ hydrocarbons to almost pure methane (~0.2% C_2+) led to a rise in the temperature of the process in experiments on the same facility by almost 90 °C, a value nearly equal to the difference between the optimal temperatures reported in [91] and [76].

Of special importance for understanding the general nature of the temperature dependence of partial methane oxidation are studies covering a wide temperature range, including, in particular, temperatures at which the changeover from the preferential formation of typical DMTM products to the preferential formation of methane oxidative coupling products occurs. One of the most interesting in this regard is a study [80] performed in an alumina-surface flow reactor (Alsint) over a temperature range of 500–750 °C, pressures of 3–40 atm, O_2 concentrations of <9%, and reaction times of 0.1–1.0 s. Under these conditions, variation of pressure within 10–25 atm caused only minor changes in the selectivity of products formation, whereas a temperature rise resulted in fundamental changes. While at temperatures below 500 °C, only oxygenated products were formed, the rise above this temperature was accompanied by the continuously growing selectivity of formation of oxidative coupling products (Fig. 3.24), with the selectivity change being actually consistent with the temperature dependence of the equilibrium constant of the reaction

$$CH_3^{\cdot} + O_2 \Leftrightarrow CH_3OO^{\cdot} \qquad (3.2)$$

At low temperatures, the equilibrium is shifted to the right, so that the subsequent reactions of methylperoxy radicals yield mainly oxygenates. At higher temperatures, the enhancing of concentration of methyl radicals $CH_3^{\cdot}$ in the gas phase leads to the formation of C_2H_6, C_2H_4, and H_2.

However, with the increasing methane conversion, the selectivity of formation of C_2 products decreased rapidly, mainly due to their conversion to carbon oxides COx (Fig. 3.25). That the product composition and methane conversion change monotonically with increasing

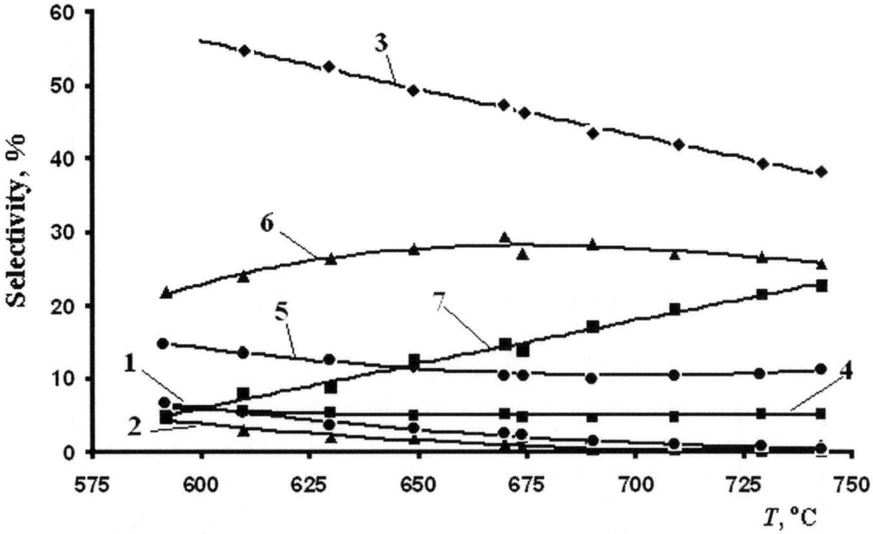

FIGURE 3.24 Temperature dependence of the selectivity of formation of partial methane oxidation products: 1 – CH_3OH (●), 2 – CH_2O (▲), 3 – CO (◆), 4 – CO_2 (■), 5 – H_2 (●), 6 – C_2H_6 (▲), 7 – C_2H_4 (■). *Based on the data from [80].*

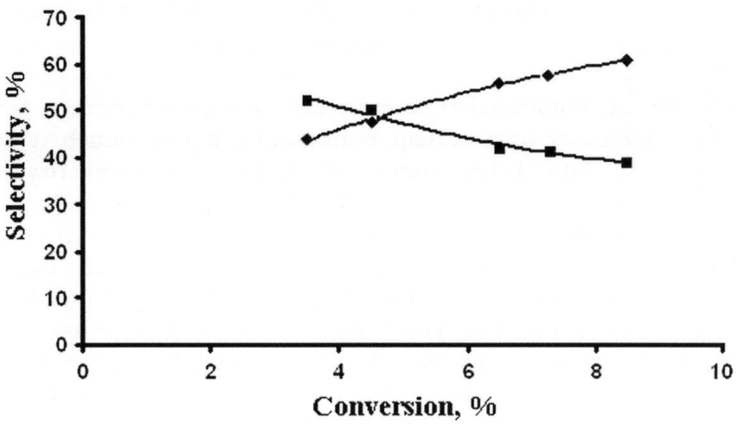

FIGURE 3.25 Dependence of the selectivity of formation of the sum of COx (◆) and the sum of C_2 products (■) on the methane conversion. *Based on the data from [80].*

temperature is one of the major arguments in favour of a general mechanism of partial methane oxidation at moderate temperatures, $T < 1000\,°C$.

Experiments at intermediate temperatures, 550–630 °C, carried out in two studies [81,82] at pressures of 30–100 atm and a reaction time of ~1 s, showed a marked changeover from oxygenated products to C_2+ products not only with increasing pressure but also with increasing temperature. However, as discussed above, due to the high initial concentration of oxygen, $[O_2]_0 = 14\%$, the heating of the mixture upon complete oxygen conversion in these

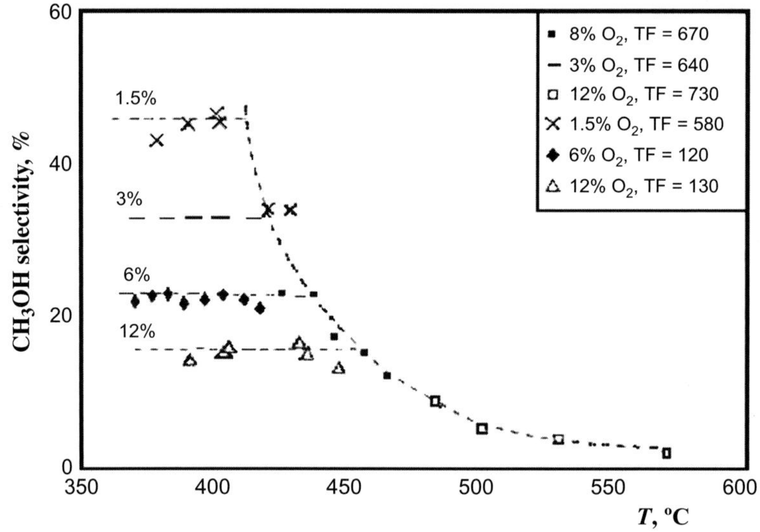

FIGURE 3.26 Methanol selectivity as a function of the reactor temperature at different oxygen concentrations; TF is the total flow of the reactants, cm^3/min [84].

experiments could be as high as 500 °C, so the final temperature could exceed 1000 °C. Therefore, many of the above effects could be related, at least partially, to the heating of the mixture.

The authors of [84] demonstrated that, with decreasing temperature, the selectivity of methanol formation increases to a certain temperature below which it becomes nearly temperature-independent (at reaction times sufficient for complete oxygen conversion) (Fig. 3.26).

Thus, the available experimental data, supported by the kinetic modeling [89] suggest that, at short reaction times, ~1 s, the optimum temperature range for the DMTM process is 480–550 °C. At temperature above 550 °C, a significant change of the yields of the reaction products in favour of ethane and ethylene is observed. At temperatures above 600 °C, not only the yield of methanol decreases, but also soot formation begins [91], contaminating the liquid products. Given the high heat effect of the reaction, this imposes limitations on the amount of oxygen supplied into the reactor, making the problem of heat removal urgent at O_2 concentrations above 3%.

On the other hand, a combination of high temperatures and relatively low pressures favours formaldehyde formation. As already noted, the homogeneous oxidation of methane at $T = 625$ °C, $P = 5$ atm, methane/air = 1:5, and a residence time of 2.3 s provided a formaldehyde yield of 3.5%, selectivity of its formation of $S_{CH2O} \approx 50\%$, and a methane conversion of 7% [57].

The authors of [99] rightly pointed out the need to distinguish between the reactor surface temperature, usually controlled in experiments, and the temperature of the reaction mixture itself, which often cannot be determined unequivocally because of a large heat effect of the reaction. Figure 3.27 graphically illustrates as, with increasing temperature of the walls of

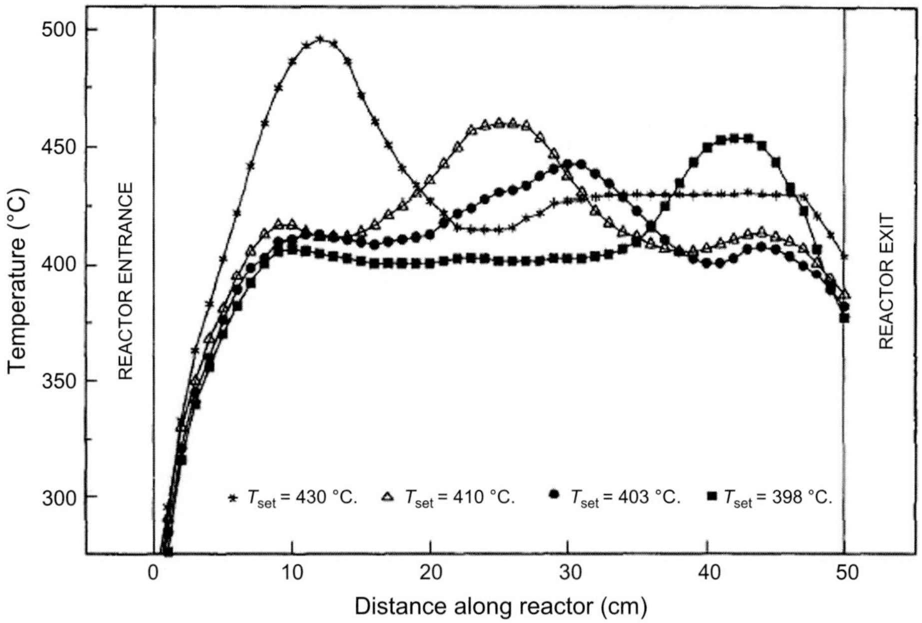

FIGURE 3.27 Temperature profiles along tubular reactor for various preset reactor wall temperatures at a flow rate of 1000 ml/min, $P = 3.0$ MPa, and $[O_2]_0 = 5\%$ [99].

the reactor, the temperature profile in it becomes increasingly nonuniform, with the maximum heat-up zone, corresponding to the maximum heat release in the reaction, naturally shifting to the reactor entrance, in accordance with the shortening of the induction period. Thus, strictly speaking, it is not always clear what constitutes the reaction temperature and how it corresponds to the values reported in experimental works and then used in kinetic simulations.

EFFECT OF THE OXYGEN CONCENTRATION (CH_4/O_2 RATIO) ON THE SELECTIVITY AND YIELD OF THE PRODUCTS

A strong dependence of the selectivity of methanol formation on the oxygen concentration (CH_4/O_2 ratio for the initial mixture) at high pressure was apparently observed for the first time in [15]. Experiments at $P = 100$ atm, $T = 500$ °C, and reaction time of $t_r = 10$ s in a flow reactor with an inner surface coated with enamel demonstrated a sharp decrease in the selectivity of CH_3OH formation with increasing oxygen concentration (Fig. 3.28).

As described in the patent [40], experiments with injection of preheated oxidant (oxygen, air, or air diluted with nitrogen) through a narrow nozzle (at velocity 76–268 m/s) into a flow (1.3–2 m/s) of preheated methane showed a less sharp decrease in the methanol selectivity with increasing oxygen concentration (Fig. 3.29). After reacting for 0.1–1.0 s in the heated part of the reactor, the mixture entered a water cooler located in the second part of the reactor

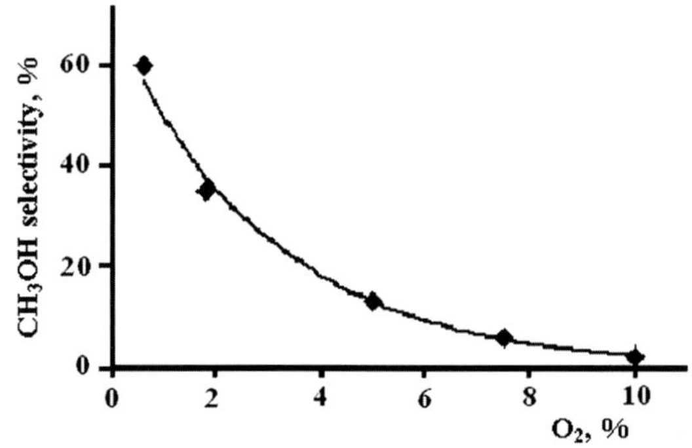

FIGURE 3.28 Dependence of the selectivity of methanol formation on the oxygen concentration at $P = 100$ atm, $T = 500$ °C, and $t_r = 10$ s. *Based on the data from [15].*

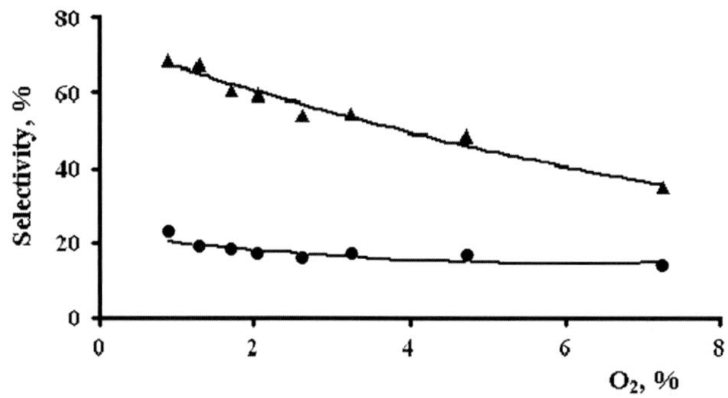

FIGURE 3.29 Dependence of the selectivity of formation of (▲) methanol and (●) formaldehyde on the oxygen concentration at $P = 40$ atm, $T = 440$ and 460 °C, and $t_r = 0.6$–1.0 s. *Based on the data from [40].*

at a distance of 120 cm from the nozzle, where it was cooled through the outer wall of the reactor. Despite some distinctions from most other works in the organization of the process, the results (Appendix I) correlate well with the main set of studies carried out at short reaction times. The selectivity of CH_3OH and CH_2O formation decreases monotonically with increasing oxygen concentration, the selectivity of methanol formation being close to 50% at $[O_2]_0 = 4$–5%, as in most of the other studies.

The conversion of methane (Fig. 3.30) was proportional to the oxygen concentration, while the $\Delta CH_4/O_2$ ratio in most experiments remained constant and practically equal to 1, indicative of the invariability of the mechanism and the formation of the main products according to gross equation (2.1). In this case, the methanol yield was ~2% at a total yield of oxygenates of ~3% (Appendix I).

In experiments [76] on natural gas oxidation in stainless-steel flow reactor (at a residence time of ~0.2 s, $P = 100$ atm, and $T = 410$–450 °C), an increase in the oxygen concentration from 2 to 4% raised the yield of all liquid products (Fig. 3.31).

The selectivity of formation of the sum of liquid organic products and the methanol concentration in the liquid oxidation product passed through a maximum at an initial oxygen

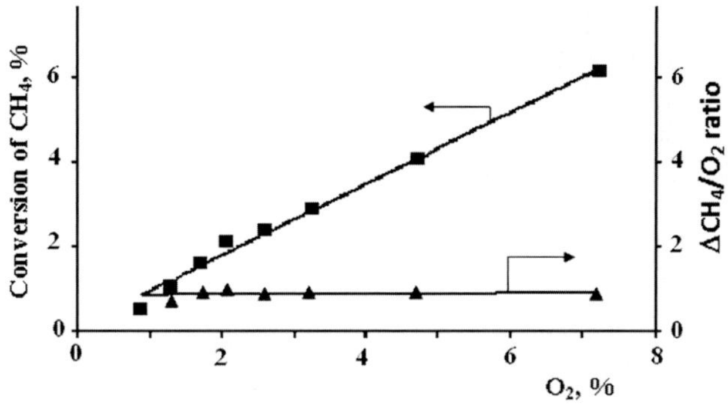

FIGURE 3.30 Dependence of the (■) methane conversion and (▲) $\Delta CH_4/O_2$ ratio on the oxygen concentration at $P = 40$ atm, $T = 440$ and 460 °C, and $t_r = 0.6–1.0$ s. *Based on the data from [40].*

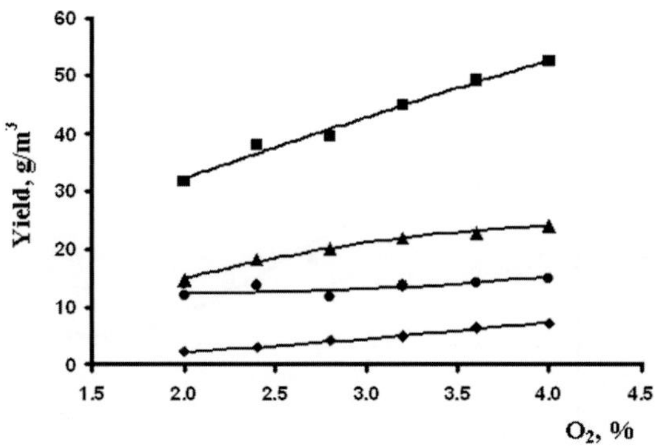

FIGURE 3.31 Dependence of the total yield of (■) liquid products, (▲) methanol, (●) water, and (◆) formaldehyde on the concentration of oxygen at $P = 100$ atm, $T = 410–450$ °C, and $t_r = \sim 0.2$ s. *Based on the data from [76].*

concentration of 2.8–3.0%. The concentration of $C_2–C_4$ alcohols featured a maximum at a lower oxygen concentration, ~2.5%. At the same time, the concentration of aldehydes increased monotonically with the oxygen concentration (Fig. 3.32). The content of acids in this oxygen concentration range remained almost constant. The concentrations of the other detected organic products (ethers and ketones), except for that of methyl ethyl ketone, also increased with the oxygen concentration.

The ratio of the total concentration of alcohols to the total concentration of aldehydes in the liquid monotonically decreased with increasing oxygen concentration in this range, from ~7 at low O_2 concentrations to ~3 at higher O_2 concentrations (Fig. 3.33).

The results of [84] also demonstrate a rapid decrease in the methanol selectivity with increasing initial concentration of oxygen. The dependence plotted based on these data (Fig. 3.26) is shown in Fig. 3.34.

The only work in which nonmonotonic decrease in the selectivity of methanol formation with increasing oxygen concentration was observed is [45] (Fig. 3.35). However, as discussed above, at long reaction times (50–300 s) and relatively low pressures and temperatures ($P = 30–50$ atm, $T = 375–500$ °C) used in this study, the reaction seems to be substantially

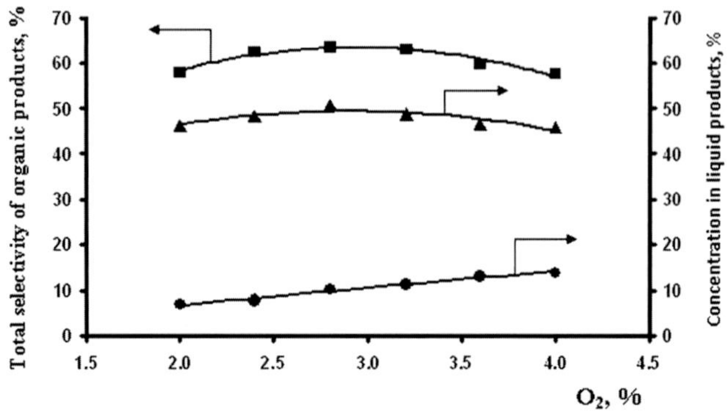

FIGURE 3.32 Dependence of (■) the selectivity of formation of organic products and the concentrations of (▲) methanol and (●) formaldehyde in the liquid products on the oxygen concentration at $P = 100$ atm, $T = 410–450$ °C, and $t_r \approx 0.2$ s. *Based on the data from [76].*

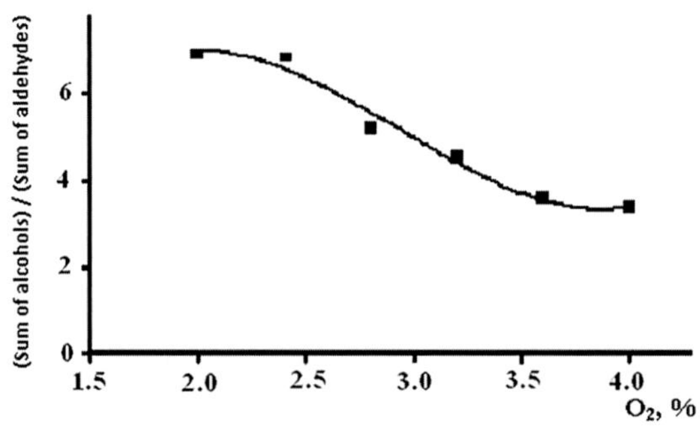

FIGURE 3.33 Dependence of the ratio of the yields of alcohols and aldehydes on the oxygen concentration at $P = 100$ atm, $T = 410–450$ °C, and $t_r \approx 0.2$ s. *Based on the data from [76].*

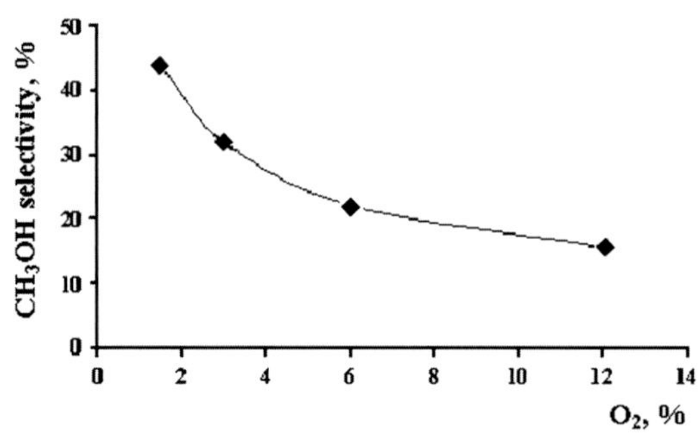

FIGURE 3.34 Dependence of the selectivity of methanol formation on the oxygen concentration at $P = 50$ atm, $T = 400–450$ °C, and $t_r \approx 15$ s. *Based on the data from [84].*

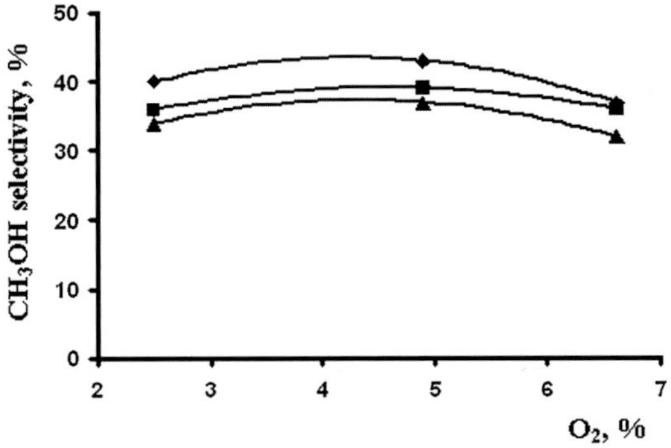

FIGURE 3.35 Dependence of the selectivity of methanol formation at $T = $ (◆) 425, (■) 450, and (▲) 475 °C on the concentration of oxygen at $P = 30–50$ atm and $t_r = 50–300$ s. *Based on the data from [45].*

influenced by processes on the Pyrex surface of the reactor, as evidenced by the methanol selectivity decreasing with increasing temperature.

Clear and convincing dependences of the main parameters of the DMTM process in a quartz flow reactor at $P = 91$ atm, $T = 427$ °C, and a residence time of $t_r = 35$ s were presented in [98]. The experiments were performed with natural gas containing up to 4% ethane and 1% C_3+ hydrocarbons. Figure 3.36 shows dependences plotted based on the averaged data from this work. As can be seen, over a wide range of oxygen concentrations, from 1.5 to 12.5%, the methane conversion increased linearly with the oxygen concentration (Fig. 3.36).

In this case, with increasing CH_4 conversion, the $\Delta CH_4/O_2$ ratio decreased from 1.4 to 1.0, which most likely reflects the observed reduction of the methanol selectivity and growth of the CO_2 selectivity. Interestingly, in these experiments, the CO selectivity remained constant, while the selectivities of formation of methanol and CO_2 changed almost strictly antibatically, so that the fraction of carbon they contain also remained almost constant (Fig. 3.37). No appreciable amounts of formaldehyde and other products were detected.

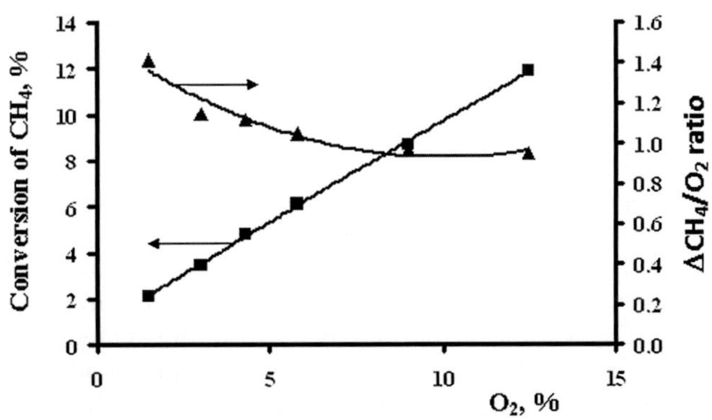

FIGURE 3.36 Dependence of the (■) methane conversion and (▲) $\Delta CH_4/O_2$ ratio on the oxygen concentration at $P = 91$ atm, $T = 427$ °C, and $t_r = 35$ s. *Based on the data from [98].*

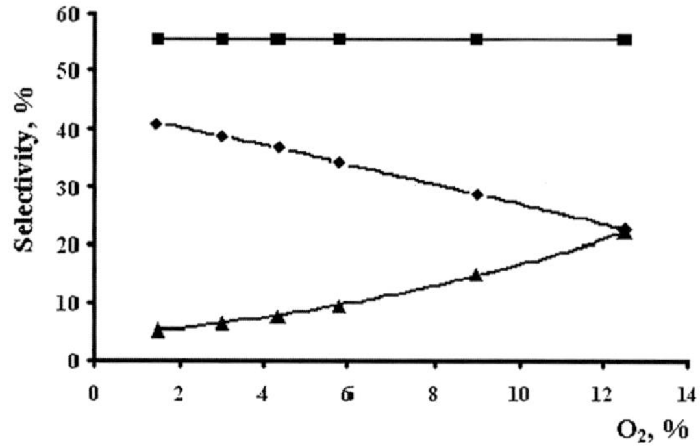

FIGURE 3.37 The dependence of the selectivity of (◆) CH_3OH, (■) CO and (▲) CO_2 on the concentration of oxygen. $P = 91$ atm, $T = 427\,°C$, $t_r = 35$ s. *Based on the data from [98].*

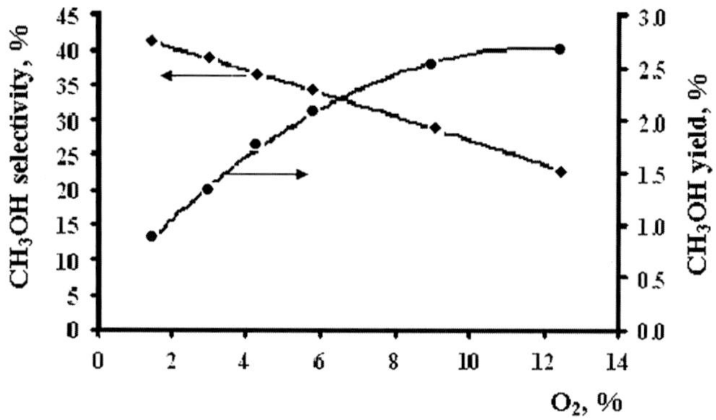

FIGURE 3.38 Dependence of the (◆) selectivity and (●) methanol yield on the oxygen concentration at $P = 91$ atm, $T = 427\,°C$, and $t_r = 35$ s. *Based on the data from [98].*

While the selectivity of methanol decreases almost linearly with increasing oxygen concentration, its yield levels out onto a 2.7% plateau at $[O_2] > 8\%$ and a methanol selectivity of ~30% (Fig. 3.38). The maximum methanol selectivity, 42%, was achieved at $[O_2]_0 = 1.5\%$, the lowest oxygen concentration used in [98].

That the selectivity of formation of liquid organic products decreases with increasing oxygen concentration naturally results in an extreme behaviour of their yield. The yield of liquid DMTM products passing through a maximum with increasing oxygen concentration (decreasing CH_4/O_2 ratio) was observed in [100] (Fig. 3.39).

How the total yield of the main liquid oxidation products and their individual concentrations depend on the oxygen concentration was examined in detail in [91]. The maximum concentration of methanol in the liquid products formed in a stainless steel, 42%, was reached at an initial oxygen concentration of ~3.5% (Fig. 3.40). The maximum yield of methanol, ~20 g/m^3 of passed gas, was observed at a somewhat higher initial oxygen concentration (~4.5%), due to the liquid product's yield increasing with the conversion. In a quartz reactor, the maximum yield of methanol was about the same, but it was reached at a higher initial oxygen

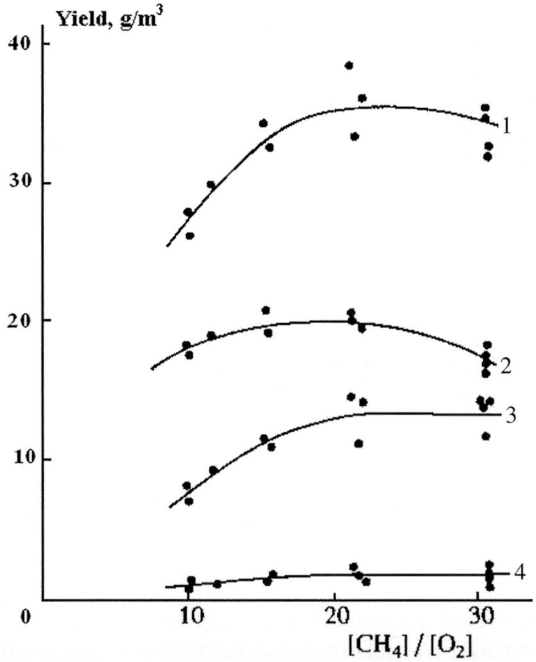

FIGURE 3.39 Dependence of the yield of liquid products on the CH_4/O_2 ratio at $P = 100$ atm and $T = 400\,°C$: (1) total liquid products yield, (2) H_2O, (3) CH_3OH, and (4) CH_2O [100].

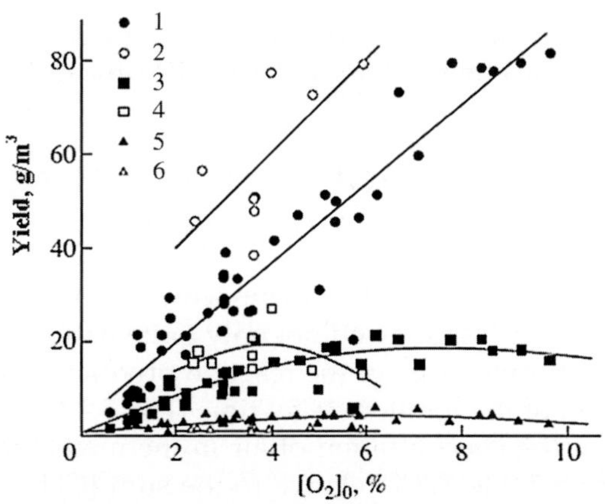

FIGURE 3.40 Dependence of the yield Y (g/m^3) of (1, 2) the sum of liquid products, (3, 4) methanol, and (5, 6) formaldehyde on the initial oxygen concentration for the oxidation in a quartz (1, 3, 5) and stainless steel reactor (2, 4, 6) at $P = 80$ atm and $T = 550\,°C$ [91].

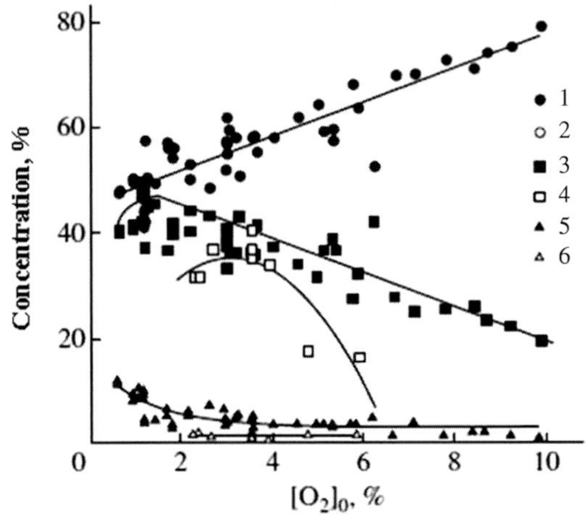

FIGURE 3.41 Dependence of the concentrations of water (1, 2), methanol (3, 4), and formaldehyde (5, 6) in the liquid on the initial oxygen concentration for the oxidation in a quartz (1, 3, 5) and stainless steel (2, 4, 6) reactors at $P = 80$ atm and $T = 550\,°C$ [91].

concentration, 6%, after which it decreased gradually. Up to an initial oxygen concentration of 10%, the overall yield of liquid products increased almost linearly, mainly due to the formation of water.

The concentration of methanol in the liquid oxidation products formed in the quartz reactor decreases almost linearly with increasing oxygen concentration from its maximum value of ~42% at $[O_2]_0 = 2-3\%$ (Fig. 3.41). The nature of a small decrease of the methanol concentration at initial oxygen concentrations below 2% is not clear. The formaldehyde concentration decreases monotonically with increasing initial oxygen concentration from ~10% at $[O_2]_0 \approx 1\%$ to about 5% at $[O_2]_0 \approx 6\%$ and continues to decrease with further growth of the oxygen concentration. A smoother dependence on the oxygen concentration for the quartz-surface reactor is more consistent with the results of the kinetic simulations [89], confirming the assumption of a substantial effect of the surface on the process in small-diameter reactors.

INFLUENCE OF THE OXYGEN CONCENTRATION ON THE REACTION TEMPERATURE AND REACTION RATE

A detailed study of the slow reaction of methane with air during the ignition delay period in a static stainless steel reactor [54] showed not only quite natural increases in the methane conversion and in the final temperature of the reaction products with the oxygen concentration, but also an elongation of the ignition delay time (Fig. 3.42). This means that, in the induction period, increasing the concentration of air in the methane–air mixture from 10 to 40% (increase of $[O_2]_0$ from 1.9 to 8.8% relative to the sum $[CH_4]_0 + [O_2]_0$) clearly reduces the rate of the $CH_4 + O_2$ reaction.

Both the ignition delay time and the mixture-heating temperature grew linearly with the oxygen concentration (Fig. 3.43). Processing these results ($T_0 = 406\,°C$, $P = 72-83$ atm) gives

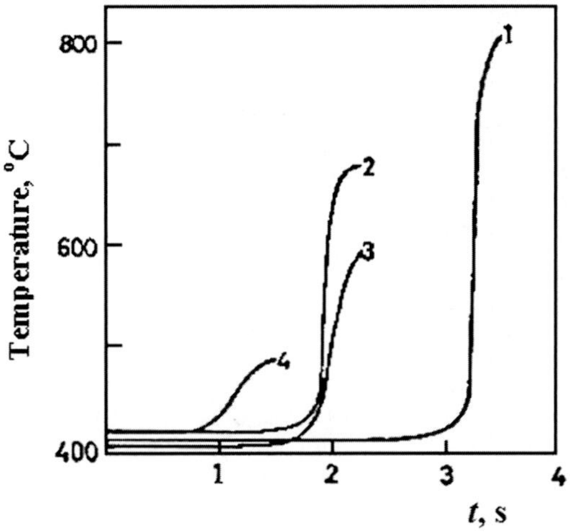

FIGURE 3.42 Time history of the temperature for the ignition of methane-air mixtures at $P = 103$ atm and various air concentrations: (1) 40%, (2) 29%, and (3) 21%, and (4) 10% [54].

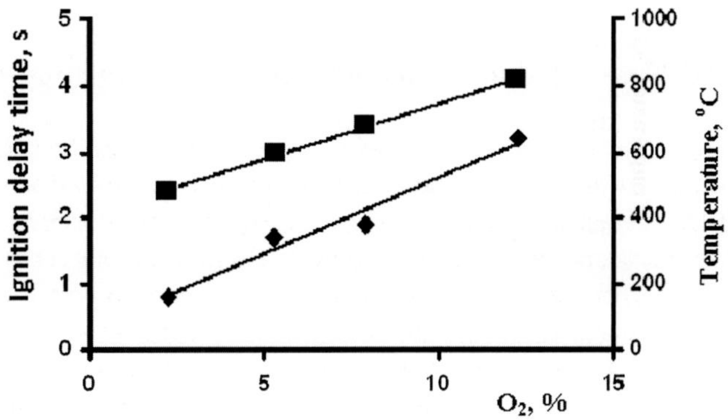

FIGURE 3.43 Dependence of the ignition delay time (◆) and heating temperature (■) for the ignition of methane–air mixtures on the oxygen concentration (relative to $[CH_4]_0 + [O_2]_0$) at $P \approx 72–83$ atm and $T_0 \approx 406$ °C. *Based on the data from [54].*

$\tau_{ig} \sim [O_2]^{0.6}$ (Fig. 3.44). The oxygen concentration exponent in this expression only weakly depended on the pressure, varying within 15%. The reason why the rate of the slow reaction in rich methane–oxygen mixtures features a negative dependence on the oxygen concentration will be discussed below.

The authors of [101] found that the reaction ignition temperature increased from 401 °C at $[O_2]_0 = 2.5\%$ to 411 °C at $[O_2]_0 = 9.5\%$. The paper [79] also reports that the minimum temperature required for complete oxygen conversion rapidly increases with the initial oxygen concentration (Fig. 3.45).

Experiments performed in [102] showed that an increase in the initial oxygen concentration within $[O_2]_0 = 2.7–5.0\%$ leads to an increase in the temperature of its conversion

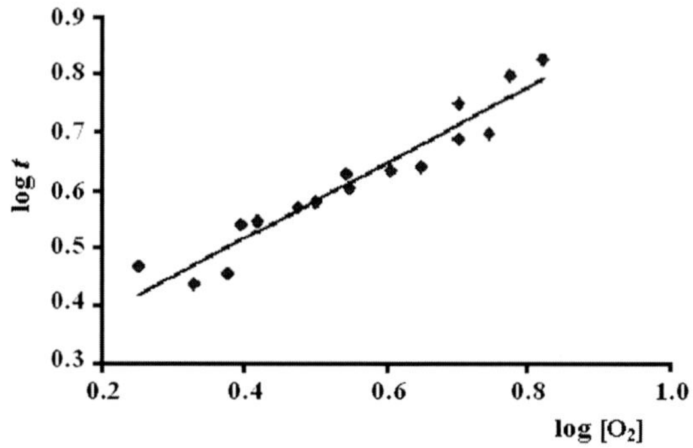

FIGURE 3.44 Dependence of the ignition delay for the ignition of methane–air mixtures on the oxygen concentration (relative to $[CH_4]_0 + [O_2]_0$) at $P = 72–83$ atm and $T_0 \approx 406\,°C$. *Based on the data from [54].* (For colour version of this figure, the reader is referred to the online version of this book.)

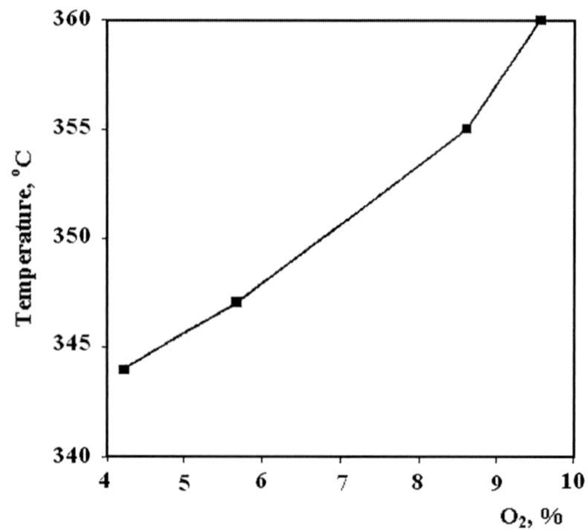

FIGURE 3.45 Dependence of the minimal temperature required for complete oxygen conversion on the initial oxygen concentration at $P \approx 81$ atm [79].

($P = 10$ atm, $t_r = 11–15$ s). Also given are the results of kinetic simulations of the DMTM process (Fig. 3.46), confirming that the time of complete oxygen conversion increases with the oxygen concentration. The nonmonotonic behaviour of the 95% oxygen conversion time at high initial concentrations of oxygen can be explained by an increase in self-heating of reaction mixture. This effect leads first to the thermal acceleration of the process (minimum in the curve) and then, at higher concentrations of O_2, to its slowdown, as the system enters the area of negative temperature coefficient of the reaction rate (second maximum).

Another set of results confirming a weak or negative dependence of the DMTM process rate on the oxygen concentration was presented in [97]: for the oxidation of natural gas in a pilot plant at a constant initial temperature, the reaction time was practically independent of oxygen concentration variation within 1–3%.

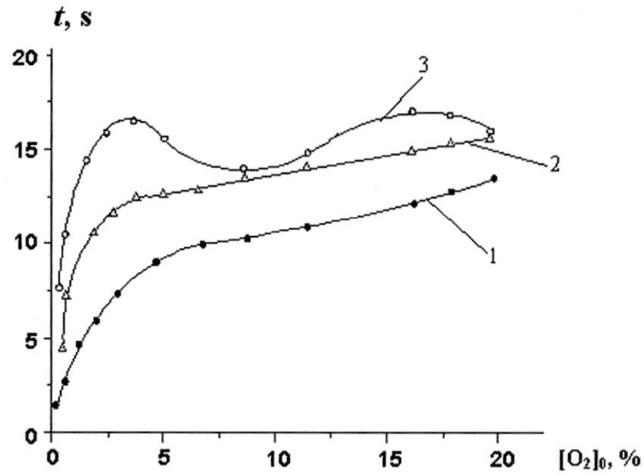

FIGURE 3.46 Dependences of the times of (1) 5%, (2) 50%, and (3) 95% oxygen conversion in the reaction mixture at $T = 693$ K and $P = 10$ atm [102].

REACTION TIME

As noted in the above overview of the research on the DMTM process, the duration of the process is of fundamental importance from the point of view of the relative roles and the mutual influence of the homogeneous and heterogeneous steps of the process. At high rates of flow of the reactants and a time of their residence in the reactor is not much longer than the time of their diffusion to the surface, within a few seconds, the process is expected to be essentially homogenous, adequately described by gas-phase models. At low rates of flow of the reagents and, consequently, a long residence time (a few minutes), i.e., much greater than the time of their diffusion to the surface, the role of heterogeneous steps becomes significant, which makes the kinetics of the process more complex and difficult to analyze. In any case, however, it is desirable to limit the residence time of the mixture in the reactor to that required for complete oxygen conversion. Within each of the two specified groups of conditions in reactors with a relatively inert surface (silica, Pyrex, stainless steel), a small change in the residence time after complete oxygen conversion only slightly influences the yield of methanol and even of formaldehyde, a compound less stable under these conditions. According to the published data, even a quick stop of the reaction by cooling (quenching) the mixture at the reactor outlet has not succeeded in a substantial increase of the methanol yield.

An interesting example of the partial oxidation of methane on the millisecond scale time, yielding a set of primary oxidation products, is presented in [49], where experiments were performed in an adiabatic-compression reactor, more specifically, in a cylinder of an engine driven by an electric motor. In this case, the initiation of the high-temperature reaction by compression and its rapid quenching during the subsequent expansion of the mixture occurs near the top dead centre of the stroke. The temperature of the mixture reached ~1500 F (~800 °C) at a degree of compression of 45. The reaction time was estimated to be ~0.02 s. The optimum oxygen concentration was reported to be 8% at the methane concentration of 75%, so that the concentration of oxygen with respect to $[CH_4] + [O_2]$ was 9.7%. A deviation from the optimum oxygen concentration resulted in a sharp decrease in

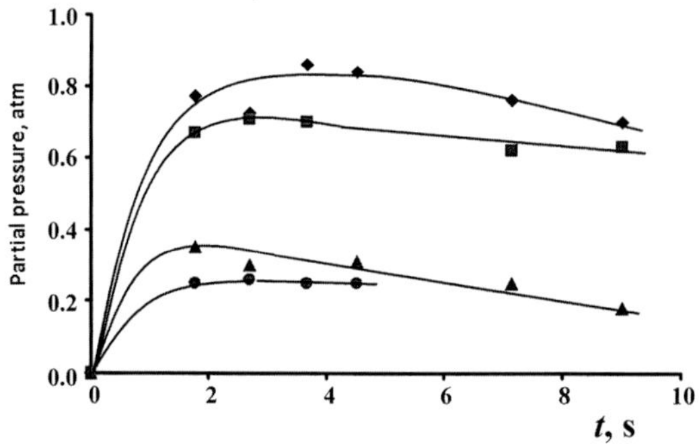

FIGURE 3.47 Time evolution of the yield of various products in copper reactor at $P = 300$ atm and $T = 375\,°C$: (◆) CH_3OH, (■) CO, (▲) CH_2O, and (●) CHOOH. *Based on the data from [51].*

the yield of C_1 products, which were practically undetectable at oxygen concentrations below 4% and above 15%. At the optimal concentration of oxygen, its conversion was ~44%, with the $\Delta CH_4/\Delta O_2$ being 1.25. A typical set of products was obtained, with carbon selectivities of $S(CH_3OH) \approx 22.5\%$, $S(CH_2O) \approx 15.2\%$, $S(HCOOH) \approx 3.4\%$, and $S(CO) \approx 33.3\%$. Hydrogen was also observed, but no CO_2. The concentration of C_2+ hydrocarbons remained virtually unchanged. That a substantial amount of formaldehyde was found in the products is probably due to a high temperature, relatively low pressure, and short duration of the reaction, with the latter two factors being favourable for its preservation. When 0.5% NO_2 was added as a promoter, the yield of C_1 products almost doubled.

Of the works in which the kinetics of the process at low reaction times was examined, it is worthwhile to single out [51]. According to the reported results (Fig. 3.6), even at pressures of 100–300 atm, an increase in the residence time of the reaction mixture in a reactor with an active (copper) surface from 1.8 to 9.0 s significantly reduced the yields of methanol and formaldehyde. The kinetics of accumulation of the reaction products (Fig. 3.47) shows that, after almost simultaneously achieving a maximum, the concentrations of the main products decrease monotonically. Kinetic modeling [5,63–66] showed that the maximum concentration of oxygenates is reached at the time of complete oxygen conversion corresponding to the cessation of the branched-chain oxidation reaction. The subsequent decline in the concentration of oxygenates is apparently due to both heterogeneous decomposition and ongoing radical processes in the bulk, with the latter led by hydroxyl radicals produced in the decay of hydrogen peroxide accumulated in the course of the branched-chain process. The second factor seems predominant, since, after a rapid initial decline, the subsequent decrease of the concentrations proceeds slowly.

Interestingly, at atmospheric pressure (static quartz reactor washed with nitric acid, $T = 456\,°C$, mixtures with $CH_4/O_2 = 9{:}1$, 4:1, and 2:1), the methanol selectivity also reached a maximum at the end of the maximum rate of methane conversion [25]. Note that, at this moment, the methanol selectivity exceeded 20%, but its integral yield was low. At the same time, only trace amounts (0.01–0.1%) of formaldehyde were detected. A similar observation was made in a study of the reaction during the induction period [54].

Only a few papers presented measured temperature profiles for the initial stage of the reaction, which could be compared to the predictions of kinetic simulations. In addition to the

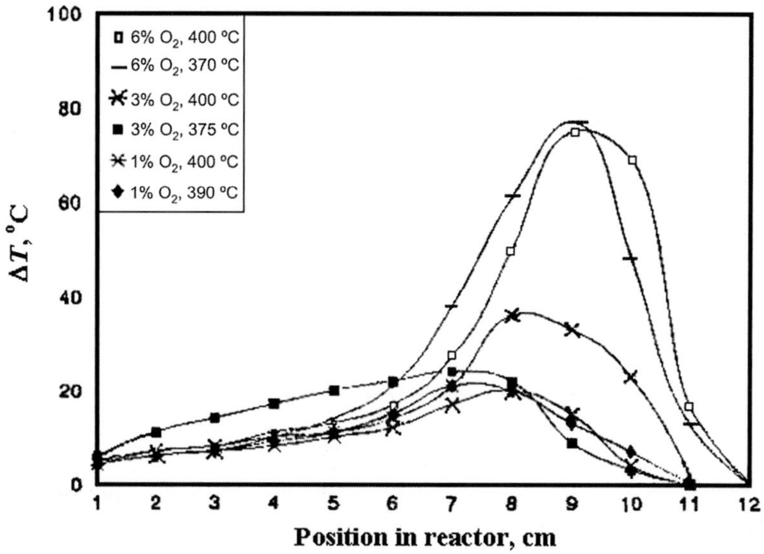

FIGURE 3.48 Temperature profiles in the reactor at $P = 50$ atm and various initial temperatures and oxygen concentrations [84].

data from [99] (Fig. 3.27), temperature profiles of the reaction are given in [84]. Although the residence time of the reagents in the reactor was not specified, the available data on the flow of the reagents and the reactor volume make it possible to estimate it as ~15 s. As can be seen from Fig. 3.48, with increasing oxygen concentration, a characteristic temperature profile of the reaction arises, with a pronounced induction period followed by a rapid heating of the mixture.

Using a flow reactor lined inside with alumina (Alsint) and equipped with five sequentially mounted thermocouples, the authors of [67] measured the temperature profiles (Fig. 3.49), which turned out to be qualitatively consistent with the results of kinetic simulations. In the absence of an oxygen conversion, the preheated reactants cooled down monotonically, whereas when the reaction occurred, an induction period followed by rapid heating was observed. A specific feature of these experiments was the instability and incessant fluctuations of the temperature profile. For example, the maximum heat-up temperature fluctuated up and down by ~5 °C. It was also pointed out that the results depended on the spatial position of the maximum heat-up zone, with the best results being obtained when the reaction zone was as close to the water-cooled quenching unit as possible.

In [97], the temperature profile of the reaction was measured in a pilot plant operating on natural gas fed at a flow rate of 760 m^3/h (stainless steel reactor with an inside diameter of 25 mm and a length of 10 m). The oxygen concentration was varied from 1 to 3%. The gas contained only ~0.2% of C_2+ methane homologues. The kinetic curves (Fig. 3.50) demonstrate a long induction period, constituting ~80% of the total reaction time. The predictions of the kinetic model [63–66] turned out to be in quite satisfactory agreement with these experimental temperature profiles. However, the changeover to a crude gas containing ~6% heavy hydrocarbons C_2+, drastically reduced the time (or temperature) of complete oxygen

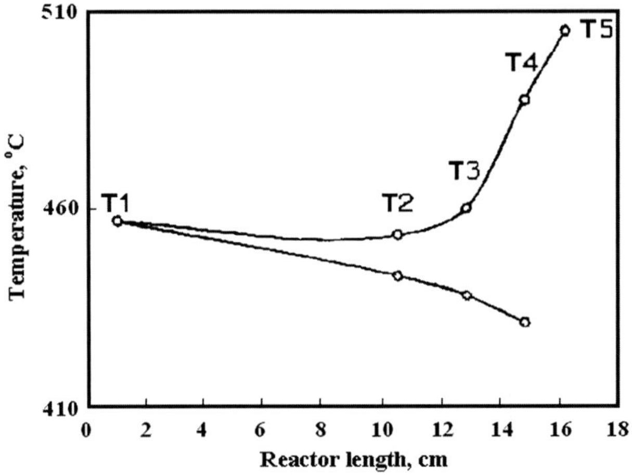

FIGURE 3.49 Experimental temperature profile in the reactor without (1) and with (2) reaction at $P = 40$ atm, $[O_2]_0 = 2.8\%$, $t_r = 1$ s. T1–T5 are the positions of the thermocouples [67].

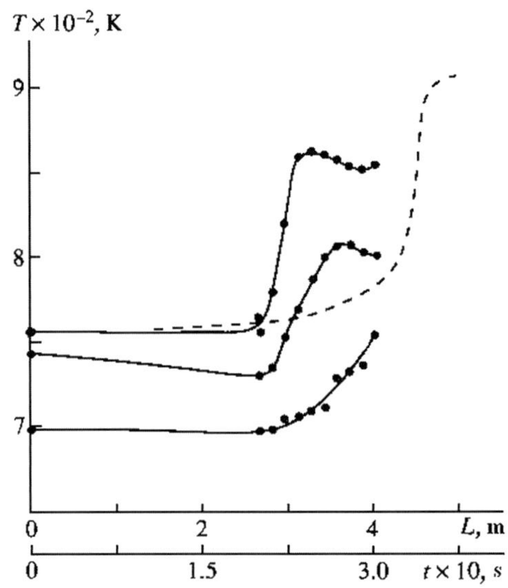

FIGURE 3.50 Temperature profiles of the reaction at $P = 100$ atm, $[O_2]_0 = 3\%$, and various initial temperature of the gas mixture [97]. The dashed line represents the simulation results at $T = 755$ K obtained using the model from [63–66].

conversion, since the presence of even small impurities of heavier hydrocarbons in methane substantially affected the rate of the process.

At a long (50–300 s) residence time of the reactants in a Pyrex reactor ($T = 450$ °C, $P = 50$ atm), its effect on the yield and selectivity of methanol in the DMTM process was investigated in [45]. Both the dependences obtained (Fig. 3.51) are nearly identical with the only difference that the selectivity of methanol formation after reaching maximum remains

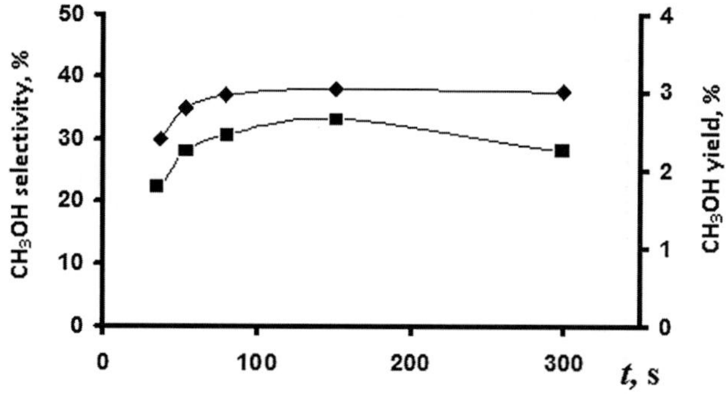

FIGURE 3.51 Dependence of the selectivity of formation (◆) and yield of methanol (■) on the residence time in the reactor at $T = 450\,°C$, $P = 50$ atm, and $[O_2]_0 = 2.5\%$. *Based on the data from [45].*

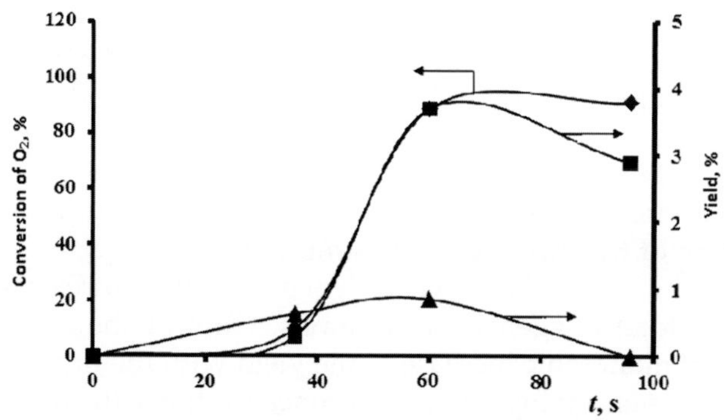

FIGURE 3.52 Dependence of (◆) the oxygen conversion and the yields of (■) CH_3OH and (▲) CH_2O on the reaction time at $P = 34$ atm, $T = 400\,°C$, and $[CH_4]/[O_2] = 16$. *Based on the data from [85].*

unchanged, while the methanol yield decreases with increasing reaction time. The decrease of the methanol yield at long residence times is apparently due to the observed decomposition of methanol on the surface of the reactor.

The dependence of the parameters of the DMTM process on the reaction time was also investigated in [85]. Clearly seen are the induction period and subsequent sharp rise in the oxygen conversion, as well as a significant reduction in the CH_3OH yield after the maximum (achieved simultaneously with complete oxygen conversion) and a sharp decline, almost to zero, of the CH_2O yield after complete oxygen removal (Fig. 3.52). At the same time, the $\Delta[CH_4]/\Delta[O_2]$ varies slightly over time, confirming the absence of significant changes in the reaction mechanism.

SPECIFICS OF THE ORGANIZATION OF THE PROCESS

An analysis of the above works on the DMTM process suggests that, despite frequent references to the design features as a cause of discrepancies between the results of different studies, it is true only for experiments with a long residence time of the reagents in the reactor, seriously complicated by heterogeneous processes. Nearly all the differences in the results of experiments at a short residence times are, as a rule, well explained by differences in experimental conditions, with the results themselves being satisfactorily described within gas-phase kinetic models of the process. Along with most numerous flow-reactor experiments, this conclusion is confirmed by experiments in a rapid compression machine [49,50] and experiments with supply of the reagents in the form of individual jets [40].

As noted, the quenching of the reaction mixture by cooling at the reactor outlet produces only a slight increase in the methanol yield. For example, the authors of [67] used a finger-shaped quenching unit positioned coaxially at the end of the reactor and cooled by water or air. Although the best results were obtained at the maximum approach of the reaction zone to the quenching unit, the methanol selectivity was generally not much different from that found in similar studies but without quenching.

A more effective measure to increase the methanol yield is to supply oxidant at several points along the length of the reactor. The effect of stepwise oxidation was studied in [78] on a pilot plant with a stainless-steel flow reactor into which the same amount of oxidant was supplied at a gas flow rate of $\sim 1000\ m^3/h$ but through a different number of inlets (from 1 to 4) positioned along the reactor. Contrary to kinetic model predictions, according to which the methanol yield should increase monotonically, although slightly, with the number of air-supply inlets (Table 3.1), the best effect was achieved for two inlets; for four inlets, however, it was even poorer than for one. This discrepancy is likely associated with the disregard of the real heating of the mixture in the simulation.

Figure 3.53 shows how the methanol yield depends on the number of air-supply inlets in the reactor at various total oxygen concentrations. Only at the minimum concentration, $[O_2]_0 = 2\%$, a small, but monotonic increase in the yield with the number of air-supply inlets was observed. As can be seen, at higher oxygen concentrations, the methanol yield decreases with increasing number of inlets in excess of two. The maximum methanol yield, $28.2\ g/m^3$ ($P = 100$ atm) was obtained at $[O_2] = 4\%$ and two inlets. That there is no effect of spread oxidant supply at high oxygen concentrations is most likely due to an overheating of the reaction mixture, which has no time to cool down after the conversion of the next portion of oxygen within a short (~ 0.2 s) reaction time under these conditions. At an efficient heat removal between the oxidant-supply inlets [103], one can expect a greater effect of spread oxidant supply.

TABLE 3.1 Methanol Yield as a Function of the Number of Oxidant-Supply Inlets ($[O_2]_0 = 4\%$, $P = 100$ atm) [78]

Number of Oxidant-Supply Inlets	Methanol Yield, g/m^3 of Gas	
	Calculation	Experiment
1	22.64	23.68
2	24.90	28.20
3	27.20	24.60
4	29.36	19.54

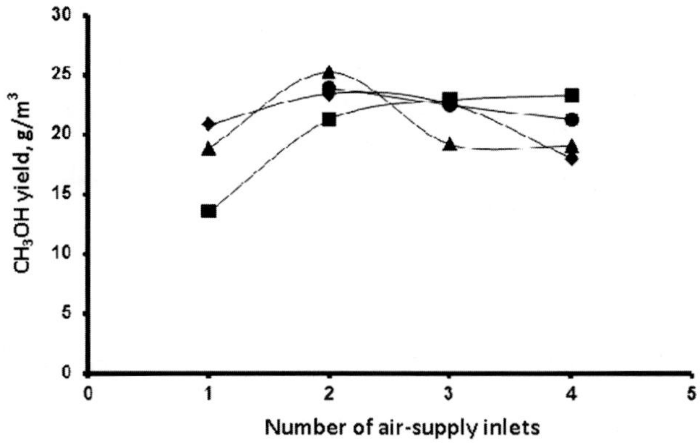

FIGURE 3.53 Dependence of the methanol yield on the number of oxidant-supply inlets at $P = 65$ atm and various oxygen concentrations ($[O_2]_0$ in %): (■) 2, (▲) 3, (◆) 4, and (●) 5. *Based on the data from [78].*

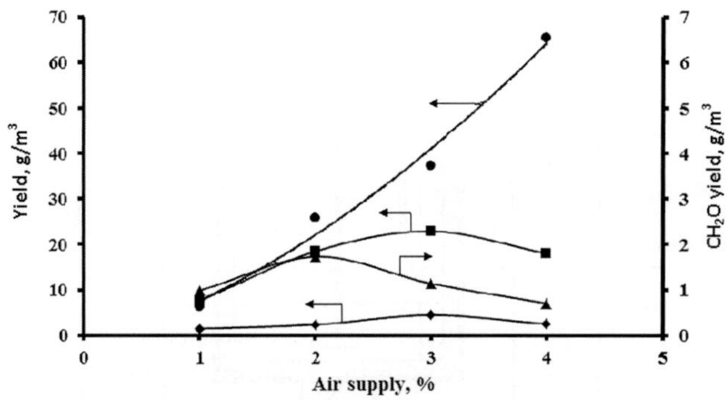

FIGURE 3.54 Dependence of the yields of (●) water, (■) methanol, (◆) ethanol, and (▲) formaldehyde on the O_2, introduced in 1% portions to a total of $[O_2]_0 = 4\%$, at $P = 65$ atm. *Based on the data from [78].*

Figure 3.54, plotted based on the data of the same paper, demonstrates how the yields of the main liquid products changes after introduction of the next portion of oxidant into the reactor (4% O_2, through four inlets). The yields of methanol and ethanol reach maximum values at $[O_2]_0 = 3\%$ and then decrease. Propanol, butanol, and acetone, detected in small concentrations, behave similarly. The aldehyde yield passes through a maximum earlier. The water output grows monotonically.

The effect of the mode of oxygen supply into the DMTM reactor was investigated in detail in [84]. Of several designs, that with distributed oxygen supply through eight holes of diameter 0.6 mm positioned in pairs over a length of 30 mm (Fig. 3.55) turned out to be most successful. All other things being equal, this mode of oxygen supply provided the maximum methanol yield, 1.8% as compared to 1.2% in the usual mode of oxidant supply, through one inlet.

To eliminate the effects of spatial inhomogeneities in the concentrations of reagents and products, inherent to flow reactors, jet-stirred continuously stirred tank reactors (CSTRs) are often used. In addition, such organization of the process makes its modeling simpler.

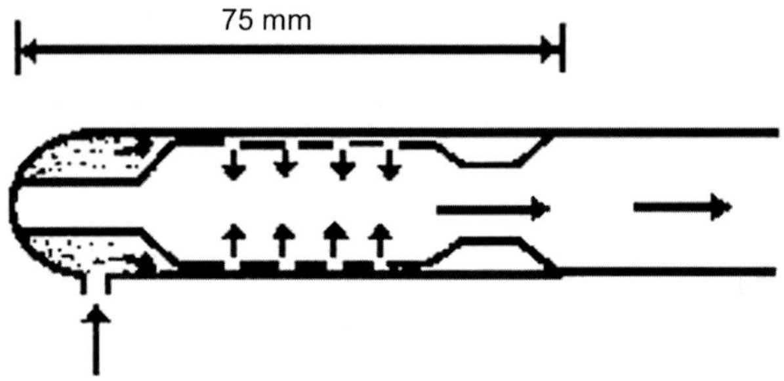

FIGURE 3.55 Schematic diagram of a reactor with distributed oxygen supply [84].

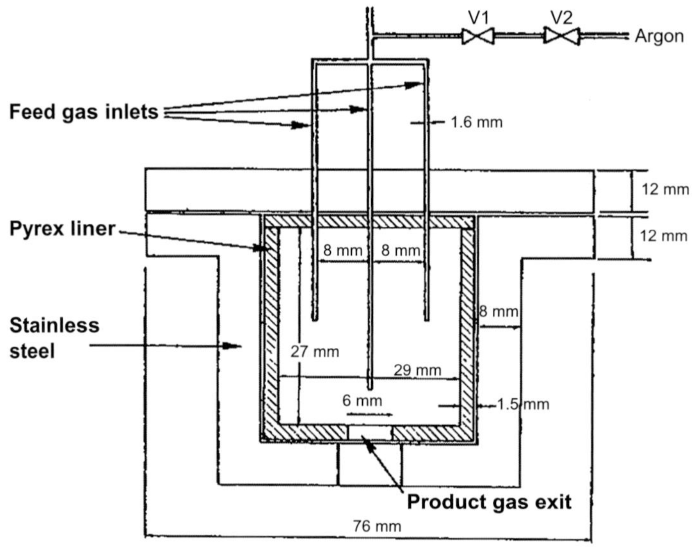

FIGURE 3.56 Schematic diagram of a Continuously Stirred Tank Reactor (CSTR) [101].

In [101], a cylindrical CSTR with an inner Pyrex liner was used (Fig. 3.56). Experiments in this reactor showed that the reaction has a thermal hysteresis, the parameters of which are in qualitative agreement with estimates obtained in [104] within the framework of a nonisothermal kinetic model of the process. It was demonstrated that the residence time of the mixture in the CSTR after complete oxygen conversion does not affect the selectivity and yield of methanol. As in flow reactors, at a constant rate of reagents consumption, the selectivity and yield of methanol increase with the pressure, in agreement with the simulation results. Generally, the results obtained in CSTRs are consistent with the results obtained in flow reactors.

One of the important technical problems is a significant carry-over of liquid products by the exhaust gas flow. According to phase-equilibrium calculations, when the outgoing gas is cooled with ambient-temperature water, ~5–10% of the methanol formed is carried away. In fact, this percentage may reach 15% or more, as follows from the results of [51]. A more complete extraction of condensable products is one of the most feasible ways to increase the methanol yield. Note, however, that, in order to decrease the fraction of carried-away methanol to less than 1%, it is necessary to cool the exhaust gas to a temperature close to 0 °C.

To increase the yield and concentration of methanol, the reaction products were passed through a trap filled with silica treated with boric acid [105]. At a temperature of 120 °C, such a trap proved effective in extracting methanol from the reactants flow. After a certain period, the trap was disconnected from the reactant's flow and connected to a steam generator to extract adsorbed methanol. The extraction efficiency was as high as 80%, so it was possible to obtain a concentrated methanol solution.

Along with the well-known attempts to increase the conversion of natural gas while avoiding a deep oxidation of the target products by using two-stage schemes with intermediate conversion of natural gas into more easily convertible products, such as syngas, methyl halides, or bisulfate, various methods of isolation of the products have been applied. These include continuous extraction in some way or the binding of the products, as well as the separation of CH_4 and O_2 prior to their direct interaction, for example, on a catalytic permeable membrane, etc. Unfortunately, no feasible methods for the selective adsorption or membrane separation of methanol at temperatures close to the temperature of its formation in the DMTM process have been proposed.

An interesting attempt to implement one of these ideas was described in [106], where a simulated countercurrent moving-bed chromatographic reactor was used. Such a device enables to combine the processes of formation and separation of the products. In this work, methanol was extracted using three successively switchable absorption columns packed with Supelcoport/10% Carbowax, followed by recirculation of methane. This approach made it possible to achieve a more than eightfold increase in the overall conversion of methane, up to 50% in the most favourable conditions, as compared to 6% per pass through the reactor. The selectivity of methanol formation in a $CH_4/O_2 = 16.3$ mixture diluted by ~60% with helium at a total pressure of $P = 100$ atm, $T = 477$ °C, and $t_r = 10$ s in a small ($d = 1$ mm) fused silica capillary was 40–50%, i.e., the yield was over 20%. The implementation of such a process on a larger scale is obviously technically very complex, but may prove promising.

A number of studies have pointed to the importance of good mixing of the gas entering the reactor. As described in the patent [44], methane and oxygen were separately fed into a special premixing chamber filled with Teflon chips. A theoretical analysis justifying the need for the rapid mixing of the reagents to achieve a high conversion, selectivity, and stable operation of the reactor was carried out in [107]. Kinetic simulation of the DMTM process with consideration given to molecular and turbulent diffusion of the separately fed reactants showed that the rapid mixing of the reactants within a time interval much shorter than the reaction time ($t_{mix} << t_r$), is essential for obtaining a high selectivity of methanol formation. In addition, slow mixing produces large amounts of a mixture capable of self-ignition, making the conversion process unstable.

CHAPTER

4

Effect of the Gas Composition

HYDROCARBONS

The major components that influence the nature of the partial oxidation of methane and the yield of the products are heavier homologues of methane, nearly always present as impurities. Since the oxidation of methane homologues and mixtures with a high content thereof will be considered separately, here we will confine ourselves to an analysis of the impact of their small amounts, at the level of several percent. In general, the effect of these impurities is reduced to an increase in the rate of oxidation or a decrease in the temperature of the process, and can be quite substantial. The influence of small amounts of heavier hydrocarbons on the yield of the products is substantially weaker.

As reported in [26], the temperature of onset of the reaction in carefully purified methane (0.36% C_2H_6) was not less than 520 °C, even at 135 atm, whereas for unpurified methane (2.1% C_2H_6, 6.3% C_3H_8), it was 370 °C. A similar changeover from natural gas to methane showed an increase in the temperature of onset of the process from 350 to 425–475 °C [87]. Addition of 5% ethane to methane caused a 50 °C lowering of the temperature at which the products became detectable but had no effect on the selectivity of methanol formation [45]. The experiments described in [97] demonstrated that the changeover from natural gas with ~6% C_2+ hydrocarbons to almost pure methane (~0.2% C_2+) led to an increase in the temperature of the process by ~90 °C, to a value practically identical to that obtained by kinetic simulations of the DMTM.

The observed decrease in the process temperature is caused by a marked decline in the effective activation energy of oxidation in the presence of heavy homologues of methane. Based on the experimental dependence of the time of complete oxygen conversion on the initial temperature for a gas containing ~6% C_2+ hydrocarbons (4.1% C_2H_6, 1.15% C_3H_8, and 0.45% C_4H_{10}) in [97], the activation energy can be estimated as $E_a = 29–33$ kcal/mol, a value significantly lower than that typical of methane oxidation. It is worthwhile to note that an increase in the oxygen concentration from 1 to 3% caused a decrease in the effective activation energy from 33.1 to 28.8 kcal/mol. It is the difference in the content of C_2+ hydrocarbons in the test gas that is most likely responsible for the discrepancies between the optimum temperatures of the process reported in different works, see, e.g., [76,92].

http://dx.doi.org/10.1016/B978-0-444-63253-1.00004-0

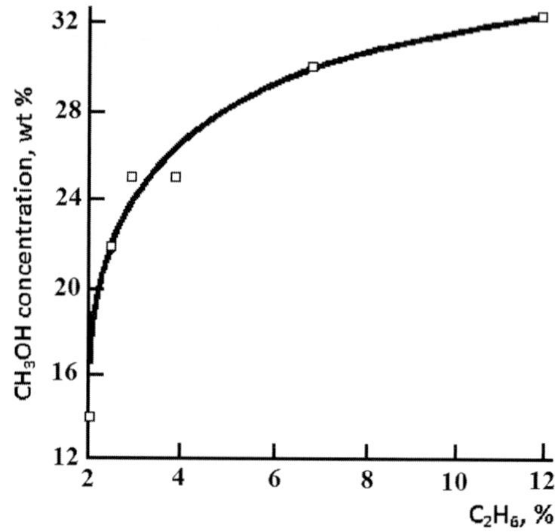

FIGURE 4.1 Effect of ethane addition to natural gas on the methanol concentration in the liquid oxidation products at $P = 140.6$ atm and $[O_2] = 6\%$ [26].

There is evidence [26] that the methanol concentration in the liquid products increases substantially in the presence of significant ethane additives (Fig. 4.1). However, given the low conversion of hydrocarbons in the DMTM process, it is difficult to exclude the oxidation of ethane itself as the main source of the observed changes in the methanol yield. For example, according to [83], upon addition of 5.34% C_2H_6 to methane at $P = 50$ atm, $T = 433$ °C, and $[O_2] = 4\%$, the conversion of ethane at the time of complete oxygen consumption was 21%, corresponding to half of the entire conversion in terms of carbon. In this case, the selectivity of ethylene formation reached 3.5–6.4%, whereas the selectivity of methanol formation reduced. Unfortunately, no systematic studies of the influence of other hydrocarbons on the DMTM process have been performed.

Experiments [98] with natural gas ($P = 91$ atm, $T = 427$ °C, $t_r = 35$ s) containing, along with ethane, ~1% C_3+ hydrocarbons showed that increasing the ethane concentration in the mixture from 1 to 5% led to an increase in the overall degree of conversion of hydrocarbons, but, at the same time, the methane conversion was substantially lower, which is quite natural, given the higher reactivity of ethane. However, variation of the ethane concentration within the indicated range caused no noticeable changes in the selectivity of formation of the products.

The most comprehensive study of the influence of small (3%) additives of heavier alkanes on the methane conversion and methanol yield was performed in [53] at $T = 300$–500 °C, $P = 41$ atm, $[O_2] = 3\%$, and $t_r = 1$ s. As might be expected, the main result was a reduction in the temperature of conversion: ~60, ~50, and ~30 °C for isobutane, propane, and ethane, respectively (Fig. 4.2).

The issue of how hydrocarbon additives influence the selectivity of methane conversion to methanol is more complicated. Although the selectivity of methanol formation increases noticeably (Fig. 4.3), a significant portion of this gain can be associated with the conversion of the added alkane.

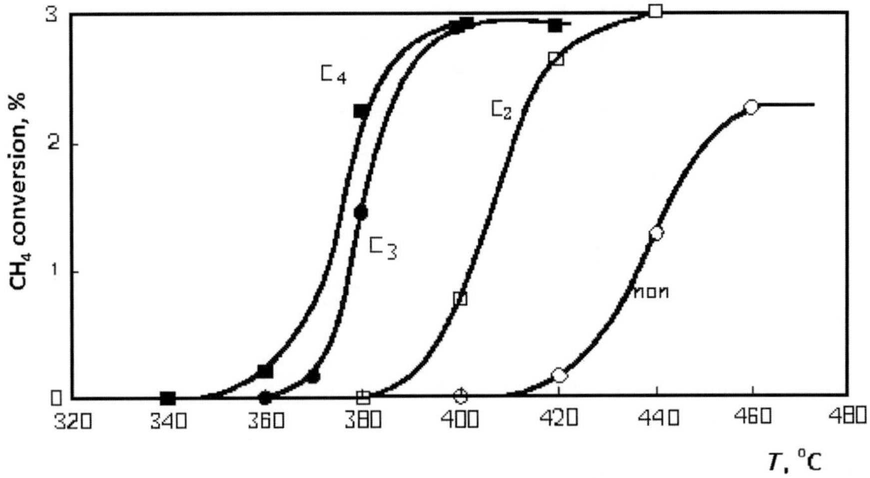

FIGURE 4.2 Effect of addition of 3% C_2–C_4 hydrocarbons on the methane conversion at $P = 41$ atm, $t_r = 1$ s, and CH_4:O_2:N_2 = 30:1:4 [53].

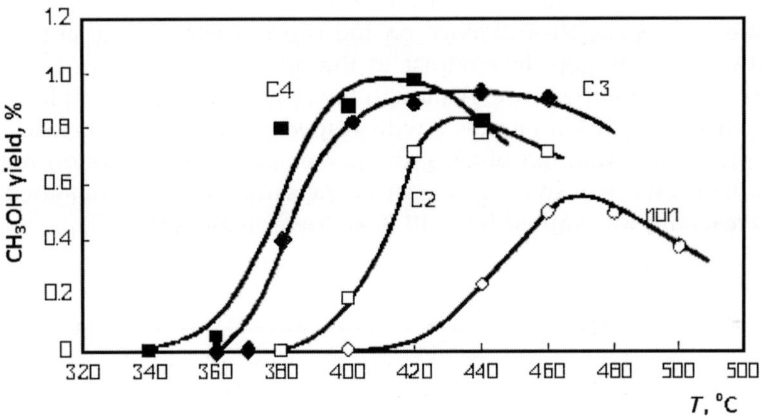

FIGURE 4.3 Effect of addition of 3% C_2–C_4 hydrocarbons on the methanol yield at $P = 41$ atm, $t_r = 1$ s, and CH_4:O_2:N_2 = 30:1:4 [53].

The effect of heavy methane homologues on the temperature and rate of the reaction is naturally explained by their high reactivity with respect to oxygen in the gas-phase reactions (their kinetic parameters given in Table 4.1) of initiation of the radical process:

$$C_nH_{2n+2} + O_2 \rightarrow C_nH_{2n+1}^{\bullet} + HO_2^{\bullet} \quad (4.1)$$

In the first approximation, this effect can be considered as the promotion of methane oxidation; however, because of the complex interaction of the oxidation products, the real picture is more complicated [108]. Nevertheless, at temperatures typical of the DMTM, the

TABLE 4.1 Kinetic Parameters for the Reactions of C_1–C_4 Alkanes with Oxygen $k(T) = A\ \exp(-E_a/RT)$

Reaction	*A*, cm³/molecule s	E_a, kJ/mol	Temperature range, K	*k* (700 K), cm³/molecule s
$CH_4 + O_2$	6.59×10^{-11}	238	500–2000	7.58×10^{-29}
$C_2H_6 + O_2$	1.00×10^{-10}	217	500–2000	6.45×10^{-27}
$C_3H_8 + O_2$	6.59×10^{-11}	200	300–2500	7.88×10^{-26}
$C_4H_{10} + O_2$	6.59×10^{-11}	184	300–2500	1.23×10^{-24}

http:/kinetics.nist.gov/kinetics/index.jsp.

initiation rates for ethane and propane are, respectively, two and three orders of magnitude higher than that for methane.

HYDROGEN

Hydrogen is one of the major gaseous products of the DMTM process; in addition, it is similar to methane regarding the typical conditions and mechanism of oxidation, which makes the issue of its possible influence on the process very important. As reported in [53], the addition of hydrogen, in contrast to the addition of alkanes, did not lead to a lowering of the reaction temperature. Although the authors argue that, at low temperatures, 420–460 °C, hydrogen addition (4 mol % with respect to methane) increases the methanol yield and suppress the formation of CO_2, the presented graphic data do not support this conclusion. In the presence of hydrogen, exactly the same yield was obtained, but the temperature of its reaching was shifted by ~10 °C to lower temperatures.

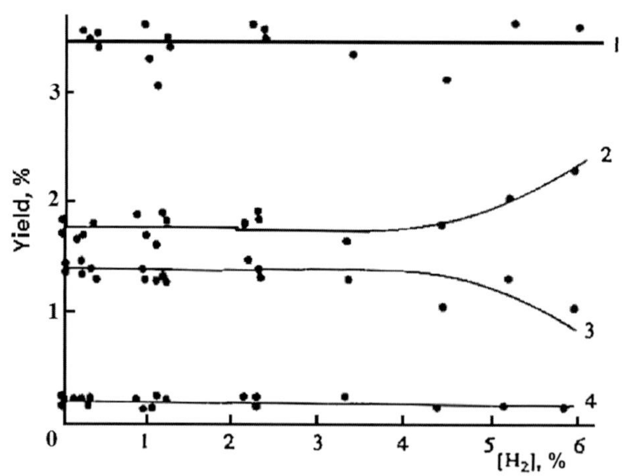

FIGURE 4.4 Dependence of the yield of liquid oxidation products on the hydrogen concentration at $P = 100$ atm and $T = 400$ °C: (1) total yield of liquid products, (2) H_2O, (3) CH_3OH, and (4) CH_2O [100].

Modeling the impact of addition of 10% H_2 at 34 atm, 430 °C, $[CH_4]/[O_2] = 16$, and almost complete oxygen conversion [85] showed a slight decrease in the selectivity of methanol formation without a noticeable change in the methane conversion.

The effect of hydrogen on the DMTM process was also investigated in [100] (Fig. 4.4). This study demonstrated that the presence of up to 5% of hydrogen in the mixture has practically no effect on the yield of liquid oxidation products. At higher hydrogen concentrations, the total amount of liquid products remained unchanged, but there was a noticeable reduction in the methanol yield with the respective increase in the yield of water. The concentration of the hydrogen remained almost the same as at the reactor inlet or increased slightly, not exceeding, however, the hydrogen concentration produced in experiments without hydrogen addition.

CARBON MONOXIDE

Carbon monoxide is the main gaseous product of the process, rather reactive under these conditions, so its influence can be very significant, especially at high methane conversion. How the yields of methanol and carbon oxides depend on the initial concentration of CO in the mixture was studied in [51] (Fig. 4.5) at $T = 375$ °C, $P = 300$ atm, $t_r = 9$ s, and $[O_2] =$ 3.0–3.8%. As can be seen, the final concentration of carbon monoxide in the products, $[CO]_f$, increases linearly with the initial concentration $[CO]_0$. Note that the concentration of 'oxidized carbon oxide' $[CO]_{ox}$, equal to the sum of $[CO]_{rn}$, CO formed in the reaction without its addition, and $[CO]_0$ (added CO) minus the final concentration $[CO]_f$ ($[CO]_{ox} = [CO]_{rn} + [CO]_0 - [CO]_f$) almost exactly corresponds to the yield of CO_2. Based on these data, one can conclude that CO_2 is mainly formed by the oxidation of CO. The yield of methanol decreases monotonically with increasing concentration of added CO, $[CO]_0$, since, at a given initial concentration of oxygen, its ever increasing fraction is spent for the

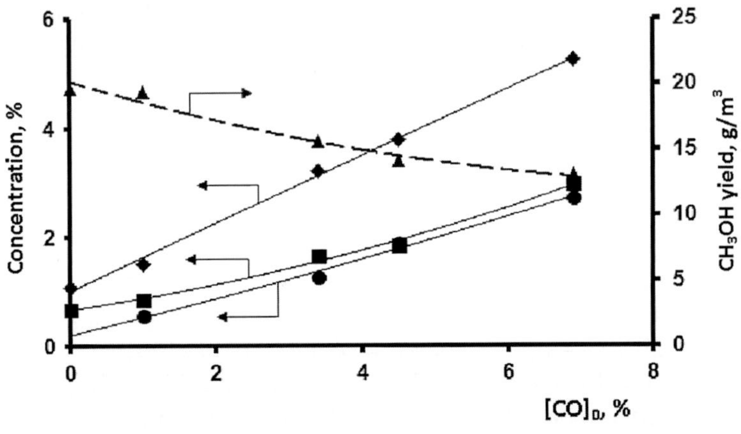

FIGURE 4.5 Dependence of (◆) the final concentration $[CO]_f$, the concentrations of (●) "oxidized" CO $[CO]_{ox}$ and (■) formed CO_2 $[CO_2]_f$, and (▲) the methanol yield on the initial concentration $[CO]_0$ [51].

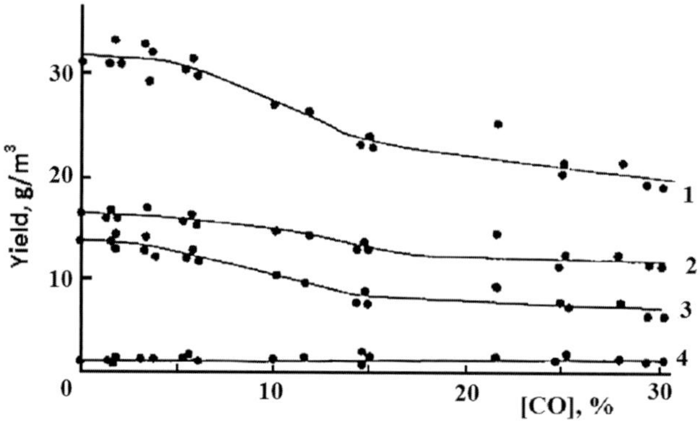

FIGURE 4.6 Dependence of the yield of liquid oxidation products on the CO concentration at $P = 100$ atm and $T = 400$ °C: (1) total yield of liquid products, (2) H_2O, (3) CH_3OH, and (4) CH_2O [100].

oxidation of CO to CO_2 and, accordingly, the methane conversion decreases. It can be assumed that the correspondence between the CO conversion and the CO_2 yield in a reactor with an active copper surface (used in this work) is mainly associated with the heterogeneous oxidation of CO in the presence of hydrocarbons.

According to [100], added in concentrations of up to 5%, carbon monoxide, like hydrogen, has little effect on the yield of partial methane oxidation products (Fig. 4.6). A noticeable decline in the yield of liquid products occurs as the CO concentration increases from 5 to 15%. Note, however, that, at high CO concentrations, this effect is associated, at least in part, with a decrease in the methane concentration in the mixture and, consequently, in the CH_4:O_2 ratio, since these experiments were performed at a constant oxygen concentration. In the same concentration range, the difference between the initial and final CO concentrations decreased monotonically. Apparently, at such concentrations, a significant amount of CO reacts, thereby reducing the fraction of oxygen spent on methane oxidation. At even higher concentrations of CO, a significant fraction of it was consumed in the reaction. According to [106], an admissible concentration of CO in the reaction mixture is 12%.

INERTS

Since the ultimate goal of most works was to achieve the highest degree of conversion of natural gas, relatively few experiments have been carried out with mixtures substantially diluted with inert components. Given that, at complete oxygen conversion, temperature only slightly influences the process, dilution with inerts (while maintaining the total pressure and the $[CH_4]/[O_2]$ ratio) apparently has little effect on the reaction.

Comparison of the results described in the patent [40], where oxygen, air, and air diluted with nitrogen were used as oxidants, does not enable to identify any systematic differences in the conversion of methane (Fig. 4.7) or the methanol yields (Fig. 4.8) within the experimental (rather large) error.

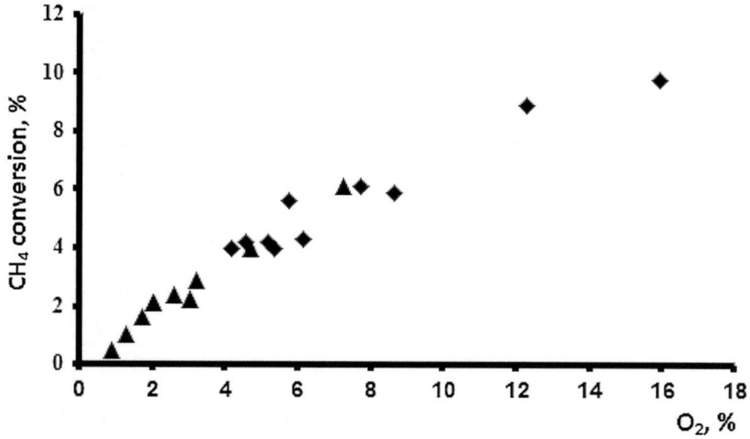

FIGURE 4.7 Dependence of the methane conversion on the oxygen concentration for the oxidation with (▲) oxygen and (◆) air at $P = 40$ atm, $T = 440–460$ °C, and $t_r = 0.6–1.0$ s [40].

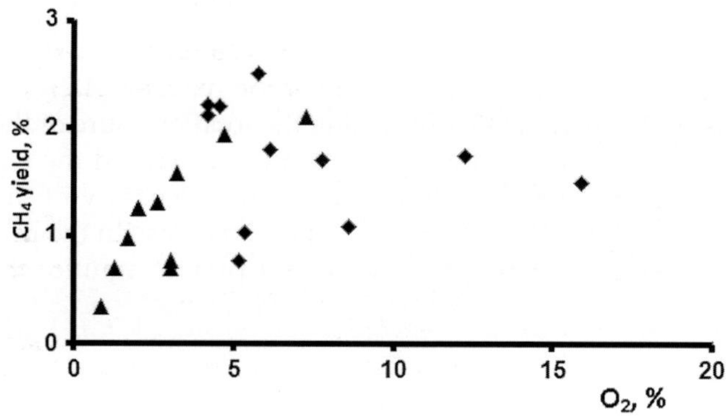

FIGURE 4.8 Dependence of the methane conversion on the oxygen concentration for the oxidation with (▲) oxygen and (◆) air at $P = 40$ atm, $T = 440–460$ °C, and $t_r = 0.6–1.0$ s [40].

The authors of [45] found no difference in the case of mixtures with and without nitrogen, whereas [83] showed that, at a fixed total pressure of $P = 50$ atm, dilution with nitrogen (up to 40%), CO, or CO_2 did not influence the selectivity of methanol formation.

It was demonstrated [84] that, at $P = 50$ atm and $[O_2] = 6\%$, dilution of the reaction mixture with helium (the degree of dilution is not given) led only to a barely noticeable decrease in the selectivity of conversion to methanol (Fig. 4.9).

According to [53], the addition of 10 mol % of CO_2 to methane increases the methanol yield. However, the presented graphic data do not support, but rather refute this conclusion. These data suggest that the addition of CO_2, like the addition of hydrogen, virtually does not influence the yield – only the temperature at which it is reached lowers by ~20 °C. The authors' attempt to explain the allegedly observed promoting effect of CO_2 by the occurrence of the reaction

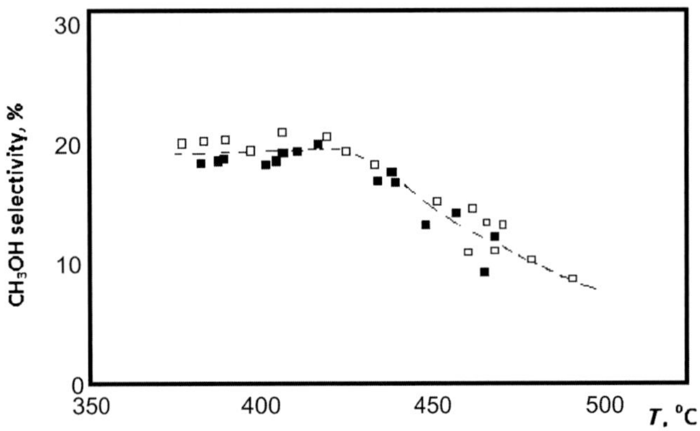

FIGURE 4.9 Temperature dependence of the selectivity of methanol formation in a $CH_4 + 6\%\ O_2$ mixture (■) diluted with helium and (□) undiluted at $P = 50$ atm [84].

$$CO_2 + H \rightarrow CO + OH \quad (4.2)$$

is inconsistent, at least, because the rate of reaction (4.2) under these conditions is low, which, together with the fact that the concentration of hydrogen atoms in the DMTM process is low, does not give grounds to expect a significant role of the reaction (4.2).

In the experiments performed in [43], increasing the total pressure of the methane mixture with 6.5% O_2 from 10 to 50 atm by diluting it with CO_2 decreased the temperature of complete oxygen conversion and significantly increased the selectivity of methanol formation; however, these effects may be associated with a general increase in the pressure. In addition, in these experiments, heterogeneous processes could play a significant role, as discussed above.

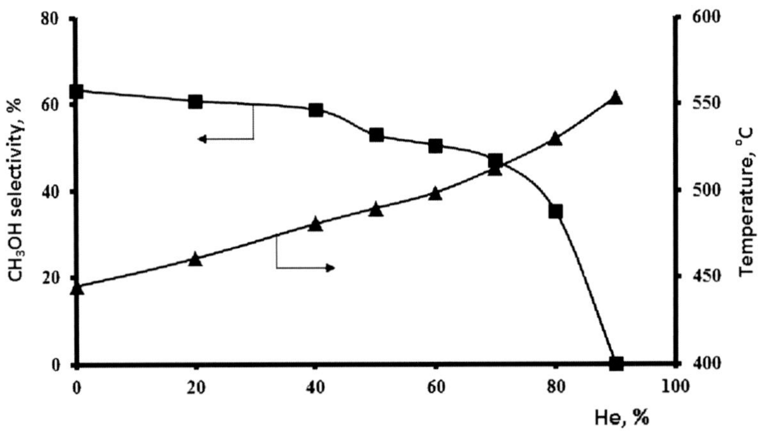

FIGURE 4.10 Effect of dilution of a $[CH_4]/[O_2] = 16.3$ mixture with helium on (■) the selectivity of methanol formation and (▲) temperature of onset of the process in a small ($d = 1$ mm) fused silica capillary at a total pressure of $P = 100$ atm and $t_r = 10$ s [106].

The paper [106] reports data on the effect of dilution with helium on the selectivity of methanol formation and complete oxygen conversion temperature for the DMTM reaction in a small ($d = 1$ mm) fused silica capillary at $P = 100$ atm and $t_r = 10$ s. At a constant total pressure and reaction time, the increase of the helium fraction (decrease in the concentrations of the reactants) in the mixture was accompanied by the monotonic growth of the complete oxygen conversion temperature and the reduction of the selectivity of methanol formation to zero at [He] = 90% (Fig. 4.10).

Given the specificity of these experiments, including a significant change in the partial pressure of the reactants, the results are difficult to compare with other data on the effect of inert diluents; however, in general, they show that this effect is insignificant.

CHAPTER

5

Key Features of the Mechanism

MECHANISM OF THE GAS-PHASE OXIDATION OF METHANE IN THE MEDIUM-TEMPERATURE RANGE

In early works (in the 1930s—1950s), the emphasis in studying the DMTM process was on experimental methods, with the results being interpreted and commented based on the then prevailing theories. In the later period, in the 1970s, when the radical-chain nature of the process has no longer been questioned, even fragmentary and unsystematic experimental data often became an occasion for extensive speculations about the details of the DMTM mechanism. Comments were mostly focused on individual reactions or groups of reactions in an attempt to explain the formation of various products without regard to the mechanism of the oxidation process as a whole. The situation changed dramatically with the advent of the possibility of modeling complex kinetic mechanisms including tens of species and hundreds of elementary steps. Kinetic simulations of this complex nonlinear process showed a close interplay of multiple parallel reactions of interconversion of intermediates, a feature that makes the use of demonstrable speculative interpretations and simple graphic schemes unpromising. Therefore, at present, speculative and interpretative aspects, which occupied a large space of the works of that period, are only of historical interest and, therefore, are not analyzed in the present review.

Studies in which real efforts have been made to model the DMTM are relatively scarce and were undertaken only by a few teams [63—67,84,89,102,109—112], as well as the recent works [113,114].

The first and most successful attempt at a comprehensive analysis of the kinetic mechanism of the DMTM was apparently made in a series of papers by Vedeneev et al. [63—66,109]. It is the model proposed in these works, relatively compact, but comprehensive, that enabled to understand the specific features of the process and to describe its kinetics. The survey of the main kinetic features of the DMTM presented below relies heavily on the analysis conducted and conclusions drawn in this series of works, as well as on the subsequent studies within the framework of this model and on the experimental verification of its findings [89,102,110].

As now firmly established, the oxidation of methane at moderate temperatures ($T < 600\,°C$) occurs mainly through the formation of the methylperoxy radical $CH_3O_2^{\bullet}$:

$$CH_4 \rightarrow CH_3^{\bullet} \rightarrow CH_3O_2^{\bullet} \qquad (5.1)$$

http://dx.doi.org/10.1016/B978-0-444-63253-1.00005-2

However, the subsequent transformations of these radicals are less evident. Kinetic simulations of the DMTM process in the framework of the mechanism developed by Vedeneev with co-workers [63–66], have demonstrated that an important role is played not only by the interaction of this radical with molecular products to form peroxides, the decomposition of which provides an effective branching in this reaction – no less important are the radical–radical reactions involving $CH_3O_2^{\bullet}$.

Along with temperature, an important parameter in the oxidation of hydrocarbons is pressure. Technological processes are normally conducted at above-atmospheric pressures. The pressure influences gas-phase reactions through several mechanisms. As the pressure increases, so does the frequency of intermolecular collisions and, hence, the rate of the process, a factor that makes it possible to lower the temperature of the reaction. This, in turn, increases the probability of 'survival' of some intermediates that decompose or undergo further transformations at high temperature. Pressure also shifts the equilibrium in reversible reactions, the most important of which in the oxidation of hydrocarbons at moderate temperatures are reactions of the type

$$R^{\bullet} + O_2 \leftrightarrow RO_2^{\bullet} \quad (5.2)$$

A pressure variation can lead to a change in the relative importance of the different channels in multichannel reactions, for example, in reactions of biradicals with unsaturated hydrocarbons [115], to an increase in the yield of radical recombination products, and to a deactivation of excited molecules. Lastly, the pressure is a critical factor for branched-chain reactions. Some of the authors [79] also discussed the possibility of the appearance of fundamentally new reaction channels associated with the manifestation of the cage effect when the resulting short-lived molecular complex has time to interact with other agents before decaying.

The characteristic features of methane oxidation at high pressure are determined by the initial stage of the process (Table 5.1), the mechanism of which was analyzed in detail in a series of papers [63–66,109].

The set of reactions shown in Table 5.1 accounts for the formation of the main oxidation products: methanol, formaldehyde, and water, but does not provide for their further transformation, since only the initial stage of the process is considered. Nevertheless, its analysis can explain the main qualitative features of the DMTM process, although its quantitative modeling requires much more complex, open-type models that would take into account the totality of homogeneous and heterogeneous elementary steps important in this range of conditions. This means that the model includes all the relevant elementary steps and, if required, it can be readily extended and that all the kinetic parameters are taken from independent databases. Thus, these parameters can and should be modified based only on the subsequent recommendations of these databases.

The predictive possibility of the model should not be limited to describing a particular group of experiments, but the totality of relevant experiments. First, the poor accuracy of the relevant kinetic parameters or their absence fundamentally limits the predictive power of any model. Secondly, actual chemical processes are greatly influenced by a variety of difficult to control but very important macrokinetic parameters of the system, such as the gradients of temperature, pressure, concentration, mass and heat transfer, including velocity gradients and the difference in the nature of the species flows, the boundary conditions

TABLE 5.1 The Mechanism of the Initial Stage of the DMTM Process

Initiation:

$$CH_4 + O_2 \xrightarrow{(I)} CH_3^{\bullet} + HO_2^{\bullet} \longrightarrow CH_3O_2^{\bullet}$$

Branching:

$$CH_3O_2^{\bullet} + \left\{\begin{array}{l} CH_4 \xrightarrow{(II)} \\ CH_2O \xrightarrow{(III)} \\ CH_3OH \xrightarrow{(IV)} \\ H_2O_2 \xrightarrow{(V)} \end{array}\right\} CH_3OOH \xrightarrow{(VI)} CH_3O^{\bullet} + OH^{\bullet}$$

$$HO_2^{\bullet} + \left\{\begin{array}{l} CH_4 \xrightarrow{(VII)} \\ CH_2O \xrightarrow{(VIII)} \\ CH_3OH \xrightarrow{(IX)} \end{array}\right\} H_2O_2 \xrightarrow{(X)} OH^{\bullet} + OH^{\bullet}$$

Propagation:

$$CH_3O_2^{\bullet} + HO_2^{\bullet} \xrightarrow{(XI)} CH_3OOH + O_2$$

Termination:

$$CH_3O_2^{\bullet} + CH_3O_2^{\bullet} \xrightarrow{(XII)} 2CH_3O^{\bullet} + O_2$$

$$CH_3O_2^{\bullet} + CH_3O_2^{\bullet} \xrightarrow{(XIII)} CH_2O + CH_3OH + O_2$$

$$HO_2^{\bullet} + HO_2^{\bullet} \xrightarrow{(XIV)} H_2O_2 + O_2$$

$$CH_3O_2^{\bullet} \xrightarrow{(XV)} \text{heterogeneous termination}$$

on the reactor surface, and processes at interphase interfaces. Therefore, the 'ideal' conditions of modeling do not always meet the specific and difficult to consider experimental conditions.

Note that the main objective of modeling is not a quantitative agreement with the experimental results, but a qualitative analysis and prediction of how the main characteristics of the process depend on its parameters. In modeling, the focus should be on understanding, not on particular numerical results [7]. However, with the development of methods and practical skills of kinetic modeling, a decline in the desire for understanding and analyzing the chemical nature of the processes is unfortunately observed in favour of achieving a 'quantitative' description of concrete experimental results.

An analysis of the concise model presented in Table 5.1 shows that the sequence of reactions (II) and (VI) leads to chain branching. In principle, the dissociation of hydrogen peroxide to $OH^{\bullet}$ radicals is a chain-branching step also; however, because of a long characteristic time of this process, more than 10^3 s under these conditions, chain branching via reaction (X) can be ignored. At high pressures, due to the slow diffusion of radicals to the reactor walls and a high concentration of methane, the rate of branching is usually higher than the rate of the heterogeneous decay of radicals:

$$\varphi = k_{II}[CH_4] - k_{XV} > 0 \,. \tag{5.3}$$

Consequently, at the initial stage, the process develops according to the branched-chain mechanism, with an exponential growth of the concentration of radicals and, hence, the reaction rate.

Since the relationship

$$K_{VI}\,[CH_3OOH] \geq k_{II}[CH_4]\,[CH_3O_2^{\cdot}]\,, \tag{5.4}$$

usually holds; i.e., the rate of decomposition of methyl hydroperoxide is higher than the rate of its formation, with the branching rate being controlled by the interaction of the methylperoxy radical with methane. Thus, at the initial stage, the kinetics of the process is mainly determined by reactions (I), (II), (VI), and (XV) (given by in the parentheses are the subsequent fast reactions of radicals, which do not limit processes of their conversion):

$$\begin{gathered}
CH_4 + O_2 \rightarrow CH_3^{\cdot} + HO_2^{\cdot}\ (CH_3^{\cdot} + O_2 \rightarrow CH_3O_2^{\cdot}) \\
CH_3O_2^{\cdot} + CH_4 \rightarrow CH_3OOH + CH_3^{\cdot} \\
CH_3OOH \rightarrow CH_3O^{\cdot} + OH^{\cdot}\ (OH^{\cdot} + CH_4 \rightarrow H_2O + CH_3^{\cdot}) \\
CH_3O_2^{\cdot} \rightarrow \text{termination}
\end{gathered} \tag{5.5}$$

However, the initial exponential autoacceleration of the reaction does not lead to an explosion, as in most other branched-chain reactions, due to the rapid growth of the rate of quadratic recombination of $CH_3O_2^{\cdot}$ radicals (negative interaction of chains, nonlinear termination of chains) in reaction (XIII). Therefore, after the initial stage, the process becomes quasistationary, remaining so for some time (Fig. 5.1).

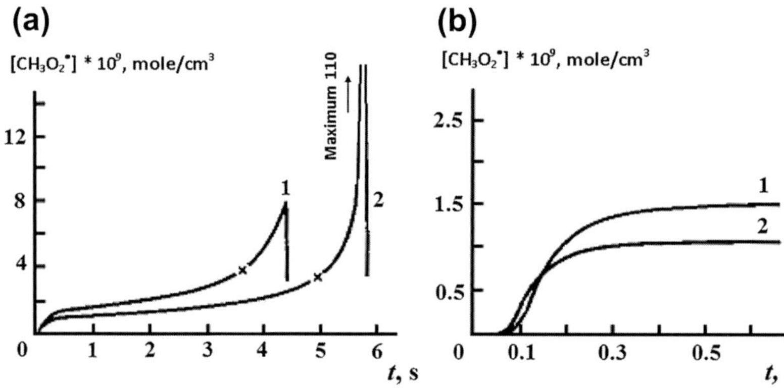

FIGURE 5.1 Calculated time histories of the $CH_3O_2^{\cdot}$ radical concentration for (a) the process as a whole and (b) the first stage and beginning of the second at $P_{CH4} = 82$ atm, $T = 406\,°C$ and (1) $P_{O2} = 1.8$ atm and (2) $P_{O2} = 8.5$ atm [66].

The scheme in Table 5.1 takes into account the most important reactions of quadratic interaction of radicals, since the concentrations of methylperoxy and hydroperoxy radicals are comparable and orders of magnitude greater than the concentrations of all other radicals. The onset of the steady-state mode is markedly influenced by both pathways of the quadratic interaction of methylperoxy radicals, (XII) and (XIII), especially (XIII), since reaction (XII) alone does not provide the establishment of the steady-state mode, as do not individual reactions (XI) and (XIV), although they affect the kinetics of transformations of peroxide radicals.

The initial autoacceleration of the reaction is completed in a very short time (Fig. 5.1), at a negligible degree of conversion of the reactants, and is not observed experimentally. The induction period in the oxidation of methane at high pressures is, in fact, a manifestation of a branched-chain process occurring in the steady-state mode, which is characterized by the approximate equality of the rates of chain-branching steps (II)–(IX) and quadratic chain-termination steps (XIII)–(XIV).

Even an early stage of the reaction produces a large number of various intermediates, including almost all the major stable molecular products (Fig. 5.2). As the molecular products accumulate, the rate and importance of the processes with their participation grow, leading to a gradual increase in the branching rate. At some point (marked with 'x' in Fig. 5.1), the steady-state mode gives way to a rapid growth of the rate of the process until a complete oxygen conversion. In experiments, just this moment of time is normally considered as the end of the induction period, although in terms of the reaction mechanism it marks only a change in the set of main reactions providing chain branching (from reaction (II) to (III)–(V)).

It is also worth noting the work [67], where a relatively compact model composed of 61 elementary reactions is presented. The model is limited to processes involving C_1 species

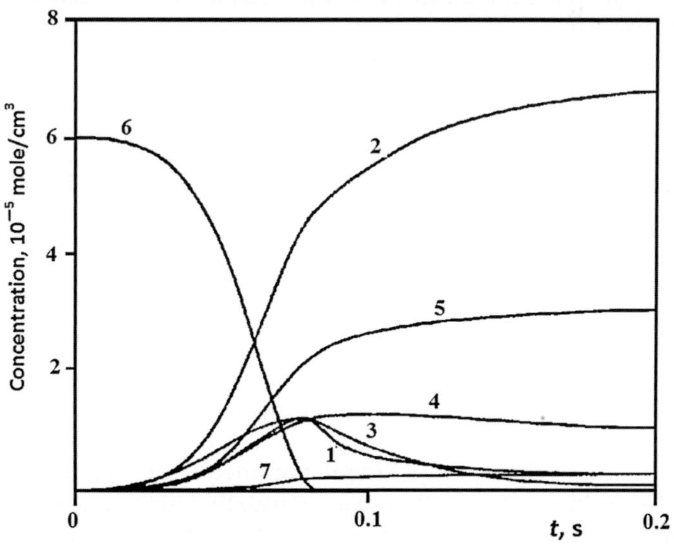

FIGURE 5.2 Kinetics of the formation of partial methane oxidation products at $T=783$ K, $P=85$ atm, CH_4:$O_2=21$:1: (1) CH_2O, (2) H_2O, (3) H_2O_2, (4) CH_3OH, (5) CO, (6) O_2, (7) CO_2 [70].

and does not consider heterogeneous processes on the reactor walls, which is hardly realistic. Nevertheless, it enables to qualitatively describe a number of features of the process, including the emergence of a Negative Temperature Coefficient (NTC) region with increasing temperature. The latter can be explained by an increased role of $CH_3^{\cdot}$ radicals and their subsequent recombination to C_2H_6, factors that decrease the rate of formation of the species providing chain branching. The CH_2O/CH_3OH ratio was demonstrated to increase rapidly with the oxygen concentration, which can be accounted for by the competition between the reactions of methylperoxy radical $CH_3O_2^{\cdot}$ with oxygen and methane. It was also shown that the selectivity of formation of oxygenates at very low O_2 concentrations decreases due to an increase of the contribution of the competing reaction of recombination of $CH_3^{\cdot}$ radicals to ethane. The resulting dependence of the selectivity of CH_3OH formation on the pressure reflects the rapid growth of this quantity with increasing pressure in the range of ~5–~50 atm.

The results of analysis of these and other previously published models diverge substantially from the results of analysis of the model presented in [113] and [114]. Without disputing and commenting on the basic conclusions of Vedeneev and co-workers [63–66] on the stepwise branched-chain nature of the process, the leading role of $CH_3O_2^{\cdot}$ radicals, and their participation in branching, recombination, and quadratic chain termination, the authors assign the key role in the formation of products to the $CH_3O^{\cdot}$ radical. They suppose that this radical is largely formed by the radical–radical reaction

$$CH_3O_2^{\cdot} + CH_3^{\cdot} \rightarrow CH_3O^{\cdot} + CH_3O^{\cdot}, \tag{5.6}$$

which is surprising enough, given the relatively low concentration of $CH_3^{\cdot}$ radicals as compared to that of $CH_3O_2^{\cdot}$ radicals, let alone to molecular products, such as CH_4, CH_3OH, CH_2O, etc., which quickly react with $CH_3O_2^{\cdot}$.

According to these works, the balance between methanol and formaldehyde is determined by the reactions

$$CH_3O^{\cdot} + CH_4 \rightarrow CH_3OH + CH_3^{\cdot} \tag{5.7}$$

$$CH_3O^{\cdot}\ (+M) \rightarrow CH_2O + H^{\cdot}\ (+M), \tag{5.8}$$

while, in the mechanism proposed in [63–69], this balance is determined by reactions (5.7) and (5.9)

$$CH_3O^{\cdot} + O_2 \rightarrow CH_2O + HO_2^{\cdot} \tag{5.9}$$

Given the difference in activation energy, a low temperature of the process, and a relatively high O_2 content at 100 atm, reaction (5.9) seems much more preferable even by rough estimates. Generally, in view of a low reaction rate and, consequently, a low concentration of most radicals, it is unclear why radical–radical processes play an unrealistically prominent role in the above-mentioned studies.

At the current quality of data on the kinetic parameters of most elementary process, attempts to reduce the problem of modeling to a quantitative description of the yield of a small number of products of a process in which dozens of species and hundreds of reactions are involved without any analysis of the totality of the phenomena accompanying the process and the results obtained, are rather pointless. All the more so that any experiment has its own features, mostly ignored in modeling. As we have repeatedly emphasized, the main objective of modeling is not a numerical coincidence with a set of experimental results, but the understanding of the mechanism of the process. The model developed in [63–69] enabled not only to qualitatively describe almost all the main features of the DMTM process, but also to make a number of predictions, experimentally confirmed later, which constitutes its principal value. This model obviously requires a thorough update of the underlying rate constants and a significant extension with the objective of gradually transforming it into an 'open model' [7]. However, any model that claims to be its replacement should at least provide an adequate interpretation of the phenomena that have found a reasonable explanation within the framework of the previous model.

MAIN KINETIC FEATURES OF THE DMTM PROCESS

Based on numerous experimental studies and kinetic analysis of the DMTM process, it is now firmly established that the partial oxidation of methane at high pressure is a degenerate branched-chain process with a very short chain length and a significant induction period. Rather intense chain branching in this reaction is provided by several parallel pathways that lead to the formation and subsequent decomposition of the CH_3OOH and HOOH peroxides. A compact detailed mechanism includes 70 elementary reactions [63–66]. The reactions that play the most important role at the initial stage of the process and account for its main features are listed in Table 5.1.

The process consists of three successive stages (Fig. 5.1). The oxidation begins as a typical branched-chain reaction, with a well-pronounced exponential growth of the concentrations of the radicals with time. However, the accumulation of methylperoxy radicals $CH3O_2^{\bullet}$ completes quickly due to an intense quadratic chain termination via the interaction of these radicals with each other, mainly via reaction (5.10) (reaction (XIII) in Table 5.1):

$$CH_3O_2^{\bullet} + CH_3O_2^{\bullet} \rightarrow CH_3OH + CH_2O + O_2 \tag{5.10}$$

In the second stage, a steady-state branched-chain reaction the branching in which is counterbalanced by quadratic chain termination, intermediate products accumulate, especially methanol and formaldehyde, and the reaction mixture heats up slowly. When the critical concentration of these species, much more reactive than methane, is reached, the third stage begins, the final explosive self-acceleration due to the branching associated with the interaction of methylperoxy radicals with the products and due to the thermal heating, which ends in a complete oxygen conversion. It is the quadratic interaction of methylperoxy radicals, ensuring the occurrence of a branched-chain process, that can be viewed as its most characteristic feature. The first two stages taken together can be considered as the induction period of the ignition that takes place in the third stage due to the growth of the branching rate during the interaction of methylperoxy radicals with the products formed in the second stage.

The nonlinear processes of interaction of the chains determine the existence of two quasi-stationary modes of oxidation: simple chain and steady-state branched-chain modes, with a difference in the rates of $\sim 10^4$ (Fig. 5.3) and a critical transition between them upon variation of the reaction parameters, especially, the pressure (Fig. 3.2).

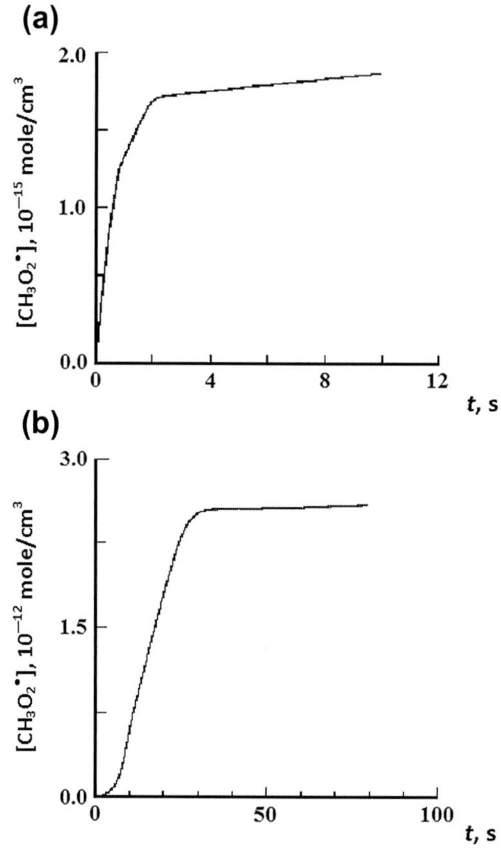

FIGURE 5.3 Kinetics of the formation of $CH_3O_2^{\bullet}$ radicals at $T = 650$ K and $P = $ (a) 3 and (b) 4 atm [94].

Increasing the pressure transfers the reaction from the slow chain mode (Fig. 5.3(a)) to a much faster, but nevertheless steady-state branched-chain mode (Fig. 5.3(b)), and provides a further increase in the oxidation rate. Note the different shape of the curves, in particular, the S-shaped behaviour of the curve in Fig. 5.3(b), characteristic of branched-chain processes, as well as substantially different scales of both axes in Fig. 5.3(a) and (b).

It is the occurrence of a fast steady-state branched-chain reaction that enables to realize a noncatalytic gas-phase process and create a stationary technological process on its basis. The quasi-steady-state branched-chain mode provides a high rate of a noncatalytic reaction at relatively low temperatures, whereas the absence of a solid phase (catalyst) minimizes the influence of heterogeneous processes, which lead to the formation of deep oxidation products. In addition to the critical transition between the oxidation modes, other manifestations of the nonlinear nature of the process, such as cool flames, NTC region, reaction rate temperature hysteresis, and oscillatory regimes, have been observed.

An important feature of the process, discovered experimentally and confirmed by kinetics simulations, is that its rate decreases with increasing oxygen concentration (Figs 3.42–3.46);

i.e., formally, oxygen acts as an inhibitor in this oxidation process. This is another factor that dictates the need for a low initial oxygen concentration to ensure a high rate of the process.

Simulations of the process also show that, in order to provide a high yield of oxygenates, it is necessary to minimize the heating up of the mixture (quasi-isothermal mode) and to maximize the uniformity of the initial mixture, i.e., provide a rapid and efficient mixing of reagents supplied to the reactor [5]. Simulations of the DMTM process also enabled to explain the poor performance of heterogeneous catalysts in the given branched-chain process by competition between heterogeneous and branched-chain initiation of chains – an issue considered in detail in the next chapter.

CHAPTER

6

The Role of Heterogeneous Processes in the Partial Oxidation of Methane to Oxygenates

Catalytic processes of partial oxidation of methane to oxygenates is not the subject of this monograph; however, a close interplay of homogeneous and heterogeneous reactions in the oxidation of light alkanes requires an analysis of the effect of heterogeneous processes on their gas-phase oxidation. Already at an early stage of studying the DMTM process, a wide variety of catalysts was patented; however, most of these patents were not based on systematic research. Now, after several decades of intense studies of the DMTM, it became clear that its main features are associated with high-temperature homogeneous processes, but in parallel, the feasibility of heterogeneous catalytic processes has always been examined.

We have already mentioned a detailed analysis of the work on the catalytic DMTM process in the reviews [4,34–37,116,117], as well as in the monograph [5]. Although it has long been clear that, at high pressure, it is difficult to expect a significant influence of the catalyst on the rate of methane oxidation to methanol [116], sensational results appear from time to time. However, the reproducibly achievable selectivity of CH_3OH and CH_2O formation in catalytic processes has not yet reached values of practical interest [117]. The yield of target products on stably operating DMTM catalysts does not exceed 3% [36], which corresponds to the most reliable values obtained in the absence of a catalyst. For the catalytic oxidation of methane to formaldehyde, the yield of the target product was reported to be higher, up to 4%.

ON THE INTERPLAY BETWEEN THE HOMOGENEOUS AND HETEROGENEOUS-CATALYTIC PROCESSES OF METHANE OXIDATION

The existing and developing processes of oxidative conversion of natural gas are largely based on catalytic processes proceeding at relatively high temperatures, $T > 400\,°C$. The specificity of these processes is that they are real heterophase processes occurring under conditions of intense competition between heterogeneous and concurrent gas-phase reactions. This has become especially clear from studies of the oxidative coupling of methane, in which

Direct Methane to Methanol
http://dx.doi.org/10.1016/B978-0-444-63253-1.00006-4

the heterogeneous reaction of initiation of methyl radicals and their subsequent homogeneous transformation comprise an integrated mechanism of the process.

In the presence of oxide catalysts, the known reactions of methane oxidative conversion have a number of common features, which suggests that the first and rate-limiting step in them is the interaction of CH_4 with the catalyst surface to form $CH_3^{\bullet}$ radicals, capable of escaping into the bulk of the reactor. This assumption is confirmed by direct experimental observations of the release of methyl radicals from the catalyst surface, estimates of the rate of their catalytic production, and data on the role of the gas phase in these processes. Thus, in reality, these are multistage heterogeneous–homogeneous processes partly occurring on the catalyst surface, partly in the gas phase. Moreover, as was demonstrated by a comparison of experimental data and the results of kinetic modeling of the oxidative coupling of methane [118,119], the role of the catalyst in these reactions is apparently limited mainly to the generation of methyl radicals.

The catalytic reactions of methane reforming to syngas, oxidative coupling of methane to ethane and ethylene, direct oxidation of methane to methanol and formaldehyde occur at relatively high temperatures, 400–1000 °C, i.e., can be qualified as high-temperature catalytic oxidation processes. Numerous studies of these reactions, different in many ways, showed, however, that they have a number of common features, the most important of which are:

1. High and similar activation energies (~200 kJ/mol);
2. The proportionality of the reaction rate to the methane partial pressure to the first power,

$$w = kP_{CH4} \tag{6.1}$$

3. The weak dependence of the reaction rate, if any, on the partial pressure of oxygen;
4. The manifestation of the kinetic isotope effect of the reaction rate (in most cases, $k_{CH4}/k_{CD4} > 1$);
5. A limited output of products (with the exception of complete oxidation and conversion to syngas).

For example, the catalytic oxidation of methane to methanol is characterized by a very low yield. At atmospheric pressure, a typical methanol yield is 0.5–1.5%. Even at a pressure of up to 5 MPa, close to the pressure of the gas-phase oxidation of methane to methanol, the maximum yield on the best catalysts (Fe–Na–sodalite and ferrisilicate) is 4% at 410–430 °C. For the catalytic oxidation of methane to formaldehyde, the maximum yield is higher, 2–4%. The best catalyst for this process is evidently V_2O_5/SiO_2, which enables to obtain such yields at 550–600 °C. The activation energy of this process on the MoO_3/SiO_2 catalyst is 189 kJ/mol. Reports of much higher yields in these two processes have not been confirmed in later publications. In the majority of the studies, the kinetics of these reactions obeys Eqn (6.1) [119].

The data on the kinetics of the oxidative condensation of methane are more numerous than on the other reactions of its oxidation. They show that the yield of C_2 hydrocarbons in this reaction does not exceed 25%. As for the aforementioned processes, reports of higher yields, up to 30% or even higher, have not been confirmed. It has been shown that the reaction proceeds via a redox mechanism involving the interaction of methane with O^- or O_2^{2-} oxygen centres on the catalyst surface and the formation of $CH_3^{\bullet}$ radicals and surface OH_s

groups, followed by the reoxidation of the surface by oxygen. For this reaction, Eqn (6.1) holds and the kinetic isotope effect with $k_{CH4}/k_{CD4} > 1$ takes place. The experimental values of the activation energy for the oxidative condensation of methane range within 170–290 kJ/mol.

These data, taken together, suggest that, in high-temperature catalytic oxidation of methane on oxide catalysts, the first and rate-limiting step is the reaction of the CH_4 molecule with the surface to form a $CH_3^{\bullet}$ radical:

$$CH_4 + O_s \rightarrow CH_3^{\bullet} + OH_s \tag{6.2}$$

Then, the surface is reoxidized in a catalytic cycle via the interaction of surface OH groups with O_2:

$$4OH_s + O_2 \rightarrow 2H_2O + 4O_s, \tag{6.3}$$

In most cases, the reoxidation rate does not limit the overall rate of the process. The high and similar activation energies of these reactions, proportionality of their rates to the methane partial pressure, and manifestations of the kinetic isotope effect with $k_{CH4}/k_{CD4} > 1$ confirm the validity of the assumption that the rate-limiting step in these oxidation processes is due to the cleavage of the C–H bond on the catalyst surface.

In contrast to the other considered reactions of partial oxidation of methane, the steam, carbon dioxide, and oxygen reformings of methane to synthesis gas proceed up to almost 100% conversion. It is generally accepted that they occur completely heterogeneously on metal surfaces that contain no surface oxygen O_s capable of producing methyl radicals. However, a number of features of these reactions make them similar to the other reactions of high-temperature methane oxidation. For the three reforming processes, the kinetics of the reaction on completely different catalysts obeys Eqn (6.1). The activation energy for the steam reforming of methane on the NiO/MgO catalyst reaches 240 kJ/mol. The kinetic isotope effect, k_{CH4}/k_{CD4}, on Ni/Al_2O_3, Ni/La_2O_3, and other catalysts can reach a value of 2.0. Typical temperatures of the oxygen, steam, and carbon dioxide reformings of methane are, respectively, 700–800, 800–900, and 900–1000 °C; however, it was shown [120] that, on the Rh/MgO catalyst, all three reformings proceed to completion at the same temperature, 500 °C.

The current mechanisms of methane conversion to syngas postulate that the methane molecule on a metal surface quickly loses hydrogen in the successive transformations $CH_4 \rightarrow CH_3 \rightarrow CH_2 \rightarrow CH \rightarrow C$ without release of methyl radicals into the gas phase. However, some data suggest that these reactions occur via a heterogeneous–homogeneous mechanism, with possible release of $CH_3^{\bullet}$ radicals into the gas phase. Accordingly, the deep oxidation of methane to CO_2 and H_2O – processes lowering the selectivity and yield of the main products of its catalytic conversion – presumably involves three processes: gas-phase reactions of free radicals, mainly $CH_3O_2^{\bullet}$, radical chain termination on the surface, and deep heterogeneous oxidation on the catalyst surface with the participation of lattice oxygen O_2^{2-}. The latter two factors are especially important for catalysts with highly developed surface [119].

To justify the assumption that the cleavage of the C–H bond on the catalyst surface is the primary and rate-limiting step of the considered oxidation processes, followed by $CH_3^{\bullet}$ radical release in the reactor volume, the authors of [118,119] simulated the effect of

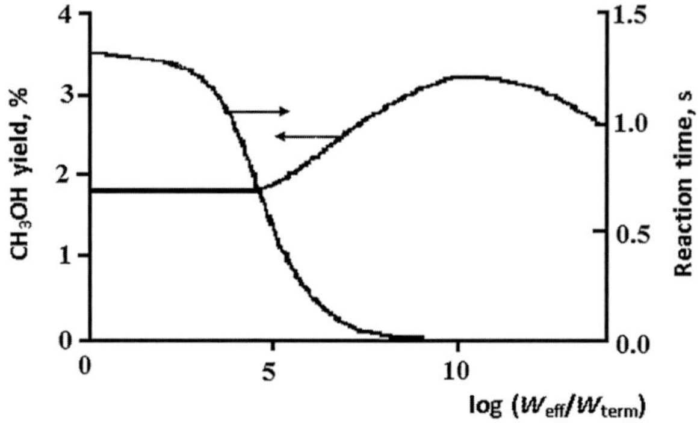

FIGURE 6.1 Dependence of the calculated methanol yield and reaction time t_r on the ratio of the effective homogeneous generation rate of CH_3 radicals to the rate of their thermal formation in the gas phase W_{eff}/W_{therm} at $T = 410\,°C$, $P = 100$ atm, and $[CH_4]/[O_2] = 19$ [119].

additional production of methyl radicals on the gas-phase oxidative coupling of methane and its oxidation to methanol and formaldehyde. Simulations of methane oxidative coupling showed that, under conditions typical of this process, an increase in the rate of initiation of methyl radicals increases the yield of C_2 hydrocarbons, but only up to a certain maximum value, ~22%, in the particular model, which is close to the best results obtained in experiments on the catalytic coupling of methane. This maximum yield is achieved at a rate of formation of $CH_3^{\cdot}$ radicals exceeding the rate of their thermal generation in the gas phase by approximately six orders of magnitude. Thus, for the oxidative coupling of methane, the role the catalyst is confined largely to the generation of methyl radicals.

The oxidation of methane to methanol at high pressures, above 5 MPa, occurs mainly in the homogeneous mode, so that the introduction of a catalyst causes no significant changes in the kinetic characteristics of the process. At high pressure, the conversion is a quasi-steady-state branched-chain reaction, the rate of generation of radicals in which is determined not by the slow thermal formation in the bulk or on the surface, but by much more rapid chain-branching steps occurring at a rate of ~10^4 times higher than that of the initiation reactions [2,121]. Clearly, under these conditions, the catalyst can significantly influence the reaction kinetics only if the rate of formation of radicals on the surface is comparable to the rate of their generation in the branched-chain reaction. The results of kinetic calculations displayed in Fig. 6.1 support this conclusion.

While the rate of the additional generation of methyl radicals W_{eff} exceeds the rate of their thermal generation W_{therm} by less than 10^4 times, the additional initiation of radicals has little effect on the process. However, at $\log(W_{eff}/W_{therm}) \approx 5$, the reaction time begins to decrease sharply, accompanied by the increase of the methanol yield. According to the calculations, the maximum increase in the methanol yield due to the additional generation of methyl radicals can reach ~50%; however, this requires that the rate of generation of radicals would exceed the rate of their thermal generation by about 10 orders of magnitude.

Simulations of the direct oxidation of methane to formaldehyde at atmospheric pressure demonstrated similar regularities (Fig. 6.2). The additional generation of methyl radicals

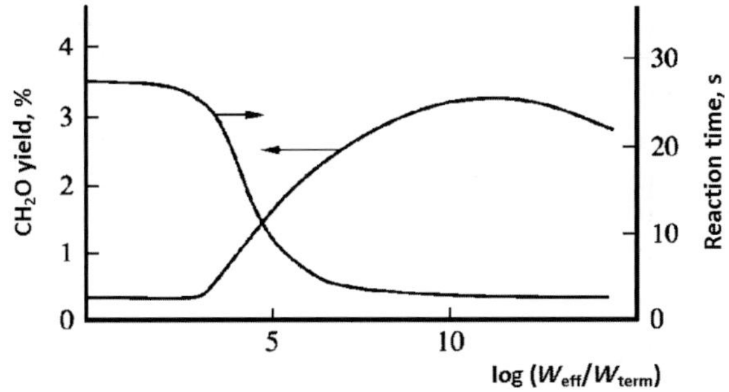

FIGURE 6.2 Dependence of the CH_2O yield and reaction time t_r on the $\log(W_{eff}/W_{therm})$ at $T = 753$ K, $P = 1$ atm, and $CH_4{:}O_2 = 9{:}1$ [119].

also produced an appreciable effect only at $\log(W_{eff}/W_{therm}) \geq 5$. The sharp decrease in the reaction time is accompanied by the increase of the formaldehyde yield from ~0.3%, typical of the noncatalytic gas-phase process, to a maximum value of more than 3%, which is practically identical to the best results achieved in catalytic experiments.

These results support the assumption on a close interaction between the processes in the gas phase and on the catalyst surface during the oxidation of methane. In all cases, the additional generation of methyl radicals makes it possible to obtain in calculations the best values of the yield of the products achieved in catalytic experiments. On the other hand, an analysis of the diffusion of atoms and free radicals in porous catalysts showed that, in the catalytic oxidation of methane, methyl radicals are the only active species that pass into the gas phase and initiate all the subsequent reactions involved in the conversion process. Thus, the results suggest that the role of heterogeneous processes in these reactions is mainly to provide the necessary rate of generation of methyl radicals. The hypothesis that the catalyst has no other specific role in the high-temperature oxidation of methane simplifies kinetic analysis, enabling to consider separately the heterogeneous oxidative generation of methyl radicals, diffusion of radicals near the surface of the catalyst, and homogeneous radical processes of oxidative conversion of methane, for which there are fairly well-developed kinetic models.

The assumption that, in most high-temperature catalytic processes of oxidative conversion of methane, the role of the catalyst, at least in the first approximation, is limited to its ability to generate (and remove) radicals suggests that an effective catalyst must provide a sufficiently high rate of generation of methyl radicals – more specifically, several orders of magnitude higher than the rate of their thermal formation in the gas phase. The maximum yield of the target products is determined by the kinetics of the gas-phase process and cannot be raised by further increasing the rate of generation of radicals. This restriction evidently does not apply to processes of catalytic oxidation of methane in which the role of the catalyst is not confined to its ability to generate and remove radicals. If the catalyst plays an active role not only in the formation and removal of radicals, but also in their transformations, it is possible, in principle, to achieve higher product yields. However, this is evidently realizable only in a nearly completely heterogeneous process. This suggests that, in cases where the maximum yield of the products is determined by the gas-phase reaction, the choice of the catalyst for a particular industrial process should be motivated largely by its stability, mechanical properties, availability, and cost rather than by its specific ability to generate methyl radicals.

An interesting example of a close interplay between heterogeneous and homogeneous reactions and an ambiguous role of the catalyst in the high-temperature heterogeneous–homogeneous oxidation of methane is described in [122]. While, under certain conditions, a first portion of catalyst promotes the homogeneous reaction, much more rapid but having a higher activation energy, the introduction of a second portion of the same catalyst downstream of the first, by contrast, causes its inhibition by transforming the branched-chain process into a slower heterogeneous catalytic reaction.

In spite of a short reaction time and relatively low rate of diffusion of the reactants at high pressure, even the reactor surface can produce a marked catalytic effect on a fast branched-chain process. That the steam reforming of methane proceeds by the homogeneous–heterogeneous mechanism and that the catalyst plays an dual role is confirmed by the data reported in [123]. In this work, the reaction rate was observed to decrease with increasing amount of Ru/Al_2O_3 or Ni/Al_2O_3 catalyst, since an increase in the catalyst volume and the corresponding reduction in the free volume of the reactor transforms the homogeneous–heterogeneous process into a slower heterogeneous process. Another interesting example is described in [124], where the homogeneous gas-phase combustion of methane at a pressure of 10 atm was inhibited by the introduction of a catalytic surface, which manifested itself, in particular, through an increase in the fraction of unburnt products.

The oxidation of methane to formaldehyde by the heterogeneous-homogeneous mechanism at temperatures above 600 °C in the presence of a surface containing SiO_2 (surfaces of the reactor wall, silica packing material, and/or SiO_2-containing catalyst) was observed in [125]. The process is initiated by the formation of $CH_3^{\bullet}$ radicals on the surface of SiO_2, which then escape into the bulk and ensure its chain-mode continuation in the gas phase due to rapid free-radical reactions.

A complex nature of interaction of heterogeneous and homogeneous processes in high-temperature reactions was demonstrated in [126]. Although the heat produced by the catalytic reaction initiates the gas-phase oxidation, the depletion of the mixture by the intense heterogeneous oxidation, in turn, leads to the suppression of the gas-phase reaction. Therefore, the intensity of the process decreases with increasing catalyst activity in the series Pt > Pt–10% Rh > Ni. A similar phenomenon was observed for the partial oxidation of methane to methanol: carrying out the process in a reactor with a catalytically active surface (stainless steel) did not increase the yield of incomplete oxidation products. The reason is the competition for highly deficient (<10%) oxygen between the homogeneous reaction of methanol formation and the catalytic formation of deep-oxidation products on the surface [127].

That the heterogeneous and homogeneous processes in the oxidation of methane are closely related was shown in [128–130]. In experiments with a sponge Ni catalyst and a Pt/ZrO_2 catalyst at a low contact time (<0.01 s) and temperatures above 600 °C, a hot spot arose and moved quickly towards the gas into the frontal layer and even beyond it. When a quartz wool wad was placed in front of the catalyst layer, its temperature (~900 °C) exceeded that of the catalyst. Clearly, the catalyst generates reactive species (most likely free radicals) that diffuse upstream and initiate reactions in the gas phase in front of the catalyst, where the mixture is still rich in O_2, but its content is too low to maintain purely gas-phase combustion. In this case, the process proceeds selectively with predominant formation of CO and H_2.

An analysis of the kinetics of high-temperature homogeneous–heterogeneous processes should take into account the possibility of a complex nonlinear interaction of the processes on the catalyst surface and in the gas phase. The current practice of using catalysts in industrial processes is based on the idea of the invariability of their properties during catalysis. However, virtually any catalyst behaves in the same way as biological enzymes, transforming its structure not only during the catalytic event, but also under the influence of changes in the gas phase with which it interacts. Vivid examples of such mutual transformations are changes in the structure and catalytic activity of ZnO- or SiO_2-supported Cu clusters with a diameter of 3–6 nm under the influence of variations in the characteristics of the gas phase [131] and the dynamics of the growth of carbon deposits (whiskers) on a nickel catalyst with a particle of nickel at the end of the growing carbon formation [132], as indicated by electron microscopy. The only driving force of these processes, as all natural processes, is the tendency to minimize the energy of the system, which causes it to change its shape according to changes in the chemical environment. Thus, the gas phase and the processes occurring in it can transform the surface no less significantly than the surface can transform the gas phase. Therefore, at present, experimental research and modeling of the kinetics of reactions involving a catalyst surface shift to studying and simulating the kinetics of mutual transformations of the catalyst and the gas-phase substrate in a single chemical process.

In connection with the above, it is appropriate to single out high-temperature reactions of partial oxidation of methane and its closest homologues with the formation of CO, H_2, CH_3OH, CH_2O, C_2H_4, etc. that can occur in the presence of a catalyst and in homogeneous conditions into a special group of homogeneous–heterogeneous processes. There is a good reason to believe that an adequate analysis and simulation of such processes in the presence and absence of a catalyst require a full accounting of the elementary processes in both phases. Simulations can be performed based on the unified mechanism of the medium-temperature ($T < 1000\,°C$) oxidation of methane [7].

INFLUENCE OF THE REACTOR SURFACE MATERIAL

The influence of the reactor surface material and its pretreatment on the kinetics of DMTM and the yield of the products has been observed in many studies, so that this issue has received much attention. While earlier studies were performed mainly in reactors with metal surfaces (stainless steel, copper), present-day experiments are carried out in reactors in which the metal surface is more or less isolated from the reaction volume by inserts made of more or less inert material, such as quartz, Pyrex, and various forms of alumina. However, the results of many studies indicate that quartz, alumina, and other similar materials also cannot be considered as inert with respect to the reactants under conditions typical of the DMTM process. For example, according to [133], at ~600 °C in the presence of silica and oxygen, up to 4.5% of the methane was converted into formaldehyde. The authors of [134] point out that Vicor glass and quartz reactor walls exhibit a significant activity in the formation of CH_2O. At higher temperatures, above 620 °C, these materials were observed to catalyze the

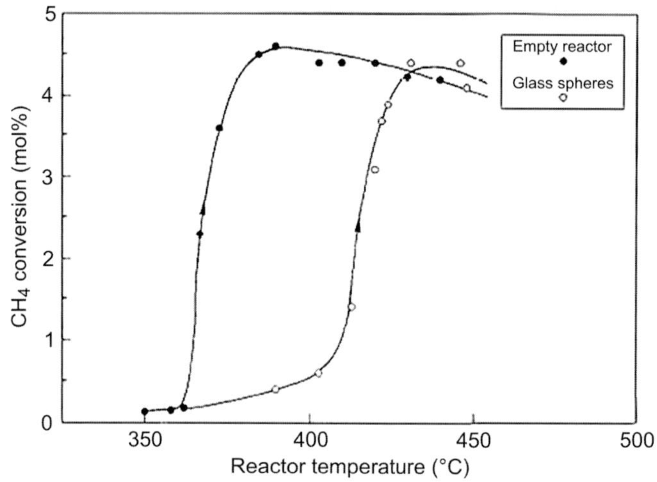

FIGURE 6.3 CH_4 conversion versus reactor wall temperature in the (○) presence and (●) absence of quartz filling spheres at a gas flow rate of 200 ml/min, $P = 3.0$ MPa, and $[O_2]_0 = 5\%$ [99].

formation of methane coupling products, mainly C_2H_6, a result indicative of the generation of radicals $CH_3^{\bullet}$ on these materials. The MgO and γ-Al_2O_3 oxides are even more active than quartz, but they produced only CO and CO_2.

A study [135] of the heterogeneous–homogeneous oxidation of methane on silica and alumina at 600–700 °C demonstrated that, for both materials, the reaction occurs partly on the surface and partly in the bulk, with the heterogeneous reaction on alumina being much more intense than on silica, leading, therefore, to much higher heating of the surface. In this case, methane is almost completely oxidized on alumina, whereas its oxidation on silica gel yields CO and CH_2O. An increase in the initial concentration of oxygen in the mixture led to an increase in the contribution from the heterogeneous reaction. The main result of this work is the direct EPR detection of $HO_2^{\bullet}$ and $RO_2^{\bullet}$ free radicals in the gas phase, which are generated by methane oxidation on the surface of these materials and released into the reactor volume, where they can participate in gas-phase chain reactions.

According to [99], an increase in the internal surface area of a quartz reactor by introducing quartz or Pyrex packing spheres raises the reaction initiation temperature (Fig. 6.3).

Formally, this effect can be attributed to a reduction in the residence time of the mixture in the reactor; however, the authors believe that quartz cannot be considered in this case as an inert material, and the effect should be attributed to an increase in the rate of loss of radicals on the surface. The effect becomes more pronounced with decreasing size of the balls. However, if the reaction begins, then it proceeds as in the empty reactor, but with the difference that the lower selectivity of methanol formation in the packed reactor corresponds to a higher reaction temperature (Fig. 6.4).

A high activity of pure SiO_2 in the conversion of methane to formaldehyde was noticed in [133] and [136], with the selectivity of formaldehyde formation being lower when the SiO_2 surface is coated with a metal oxide catalyst [137]. It is assumed that the catalytic oxidation of methane proceeds through the formation of the CH_3OM methoxide on catalyst's active

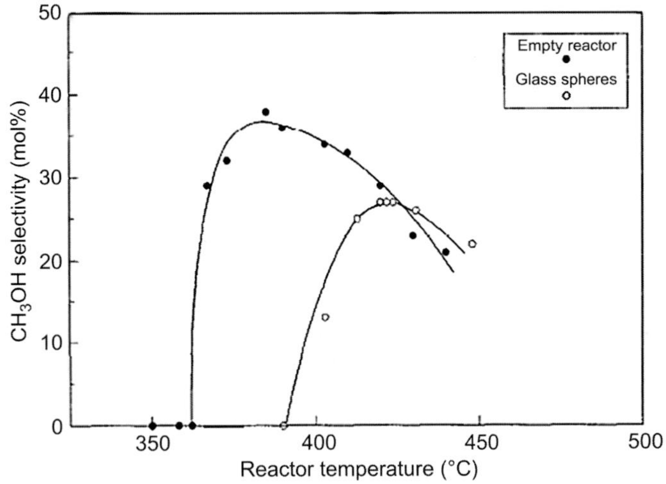

FIGURE 6.4 CH_3OH selectivity versus reactor wall temperature in the (○) presence and (●) absence of quartz filling spheres at a gas flow rate of 200 ml/min, $P=3.0$ MPa, and $[O_2]_0=5\%$ [99].

sites (M) with its subsequent oxidation. At 600 °C, in the presence of MoO_3/SiO_2, the methoxide is oxidized to $MOCH_2OM$ and further to CO_2 or to CH_2OM (adsorbed formaldehyde) and then to CO [138].

The partial oxidation of methane on a SiO_2 surface was analyzed by semiempirical quantum chemical methods [139]. Calculations showed that the activation of methane by the lattice oxygen involved in strained siloxane bonds is characterized by a high activation energy and, therefore, is thermodynamically unfavourable. For this reason, the formation of methyl radicals on a quartz surface proceeds with great difficulty. On the contrary, in the presence of gaseous oxygen, methane readily interacts with 'activated' oxygen adsorbed on the quartz surface to form adsorbed methanol, an exothermic reaction with a moderate activation energy (41–57 kcal/mol). Then, with an activation energy of only 13 kcal/mol and a large exothermic effect (96 kcal/mol), formaldehyde is desorbed into the gas phase, a process accompanied by the destruction of the siloxane bond, the oxygen of which is replaced by the two remaining hydrogen atoms.

According to [54], covering the inner surface of a stainless steel reactor with sodium silicate has virtually no effect on the ignition delay of methane–air mixtures at pressures of 58–110 atm and initial reaction temperatures of 350–410 °C. A 12-fold change of the surface-to-volume ratio of the reactor by filling it with steel balls suppressed the ignition of richer mixtures, but did not affect the activation energy, ignition delay time, and rate of ignition. This suggests that the filling of the reactor with balls (increasing the surface-to-volume ratio) seems to increase the decay of radicals, but has little effect on the initiation rate and the overall contribution to the surface reaction.

The authors of [73] systematically examined the differences in the characteristics of flow reactors with different surfaces at $P=1.4$ atm, $T=300$–560 °C, $[CH_4]/[O_2]$ from 0.5 to 10, and a reaction time of 0.6–3.6 min. It was found that the surface of the reactor, especially steel, significantly catalyzed the oxidation. In their catalytic ability, the studied materials can be arranged in the following order: copper > steel > aluminium > Pyrex. In addition, the effect of surface pretreatment on the reaction clearly manifested itself: the reactor surface after seasoning was more active than the fresh one.

A comparative study of the DMTM process at a reaction time of 50–300 s, $P = 1–50$ atm, and $T = 375–500$ °C revealed a strong dependence of the selectivity of methanol formation on the material of the reactor's inner surface [45]. The methanol formation selectivity obtained on the studied materials decreased in the order: Pyrex > quartz > stainless steel. Note, however, that, with increasing pressure, the effect of the surface material declines, and the values of the selectivity converge (Fig. 6.5).

The filling of the reactor with copper chips led to a sharp decrease in the selectivity of methanol formation, which, according to the authors' opinion, can be explained by a deep oxidation of methane on the copper surface and by the oxidation of methanol. In this work, the influence of other metallic and nonmetallic catalysts was also studied; however, in all cases, the selectivity of methanol formation was lower than in the empty reactor.

On the other hand, studying the influence of the surface material on the DMTM process in one [140] of a series of works, where a very high selectivity of methanol formation ($P = 30$ atm, T ≈ 350 °C, $t_r > 100$ s) was observed, revealed no significant differences in the selectivity and yield of methanol in reactors with different surfaces, such as Pyrex, Teflon, stainless steel, silver, and copper. In these experiments, the selectivity of methanol formation on all surfaces reached values close to 90% or more, whereas the methanol yield was as high as 10.7%. However, the authors do not exclude the possibility of influence of the surface material on the temperature and reaction time. In particular, it was found that the maximum selectivity in reactors with metal surfaces was achieved at temperatures higher by nearly 50 °C. This is probably due to the fact that, at lower temperatures in the presence of a metal surface, the oxidation occurs mostly in the heterogeneous mode with the formation of mainly deep-oxidation products.

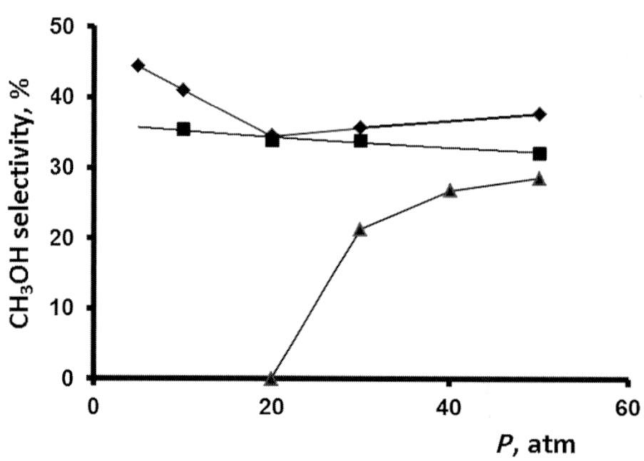

FIGURE 6.5 Pressure dependence of the selectivity of methanol formation for reactors with (◆) Pyrex, (■) quartz, and (▲) stainless steel surfaces; $[O_2] = 2.5\%$, $t_r \approx 200$ s [45]. (For colour version of this figure, the reader is referred to the online version of this book.)

Another work by the same authors [77] dealt with the effect of filling of the reactor with balls made of different materials. In all cases, this reduced the selectivity of methanol formation compared to that characteristic of the empty reactor; in the order of decreasing selectivity of methanol formation, these materials can be arranged as glass > stainless steel > silver > nickel = zinc = tin > carbon > copper. Note, however, that the decrease in the selectivity for all materials other than carbon and copper was negligible. It was reported that, of the investigated metals, the highest corrosion resistance to the reaction products was exhibited by nickel and stainless steel. In general, the results of [77] and [140] contradict the results of other studies, in particular [45], which revealed, under the same conditions, a sharp decrease in the methanol yield in the presence of metallic, especially copper, surfaces.

A strong influence of the reactor surface material on a slow ($t_r \sim 50–100$ s) DMTM process was demonstrated in [41,86]. The authors used a stainless steel reactor with an inner quartz insert firmly planted on O-rings, which eliminated the contact of the heated reactants and products to the metal surface. The products were quickly carried away from the hot zone to stop the radical reactions and prevent further oxidation. The data show that, in this case, the selectivity of methanol formation at $P = 50$ atm, $T = 450$ °C, and a mixture composition of $[CH_4]:[O_2]:[N_2] = 100:10:10$ reached a very high value, 62.3%, with an yield of 8.2%. When the O-rings were removed, a modification that presumably allowed the reagents to penetrate into the gap between the quartz insert and the stainless steel surface, the selectivity and yield dropped to values typical of these conditions, 31.6 and 2.9%, respectively.

The influence of the surface area on the homogeneous DMTM process was investigated in [141]. The experiments were carried out in quartz reactors with diameters of 8 and 27 mm, both empty and filled with quartz pieces of various sizes ($T = 400$ °C, $P = 20$ atm, $[O_2] = 10\%$, $t_r = 90–100$ s). Figure 6.6 shows that, for the 27-mm-diameter quartz reactor, the methanol selectivity decreases monotonically with increasing S/V ratio.

That the methanol yield depends strongly on the reactor surface material even at high pressures (100–300 atm) and short (<2 s) residence times of the reactants in the reactor is illustrated by a comparison of its values obtained under very similar conditions in a copper [51] and a stainless steel reactor [76] (Appendix II). For the copper reactor, the methanol yield obtained in the stainless steel reactor at 100 atm was reached only at 300 atm, a pressure at which the surface has no any noticeable effect on the process.

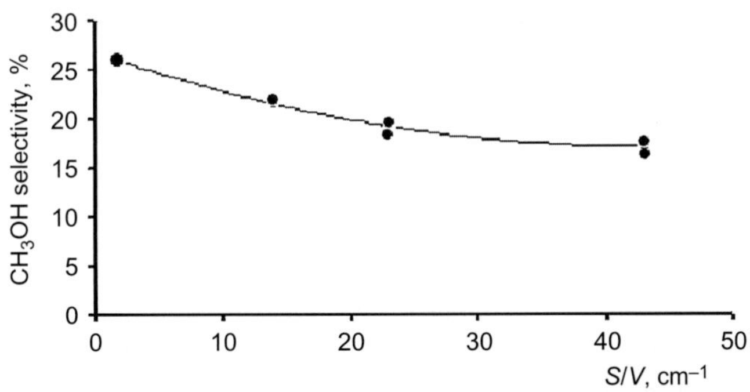

FIGURE 6.6 Dependence of the methanol selectivity on the S/V ratio at $T = 400$ °C, $P = 20$ atm, $[O_2] = 10\%$, and $t_r = 90–100$ s. *Based on data of [141].*

A well-pronounced dependence of the yield of DMTM products on the reactor surface material (quartz vs. stainless steel) at a short residence time (~2 s), especially strong at lower pressures, was observed in [91] (Fig. 3.14). Only at high pressures, close to 80 atm, the difference becomes insignificant. At this pressure, both materials provided nearly the same maximum yield of methanol, which was achieved, however, at different oxygen concentrations: ~3.5% O_2 for the stainless steel reactor and 6–8% O_2 for the quartz reactor (Fig. 3.40). Such a sharp distinction can be explained not only by different rates of the heterogeneous activation of methane and decomposition of the products on these surfaces, but also by different rates of the heterogeneous oxidation of methane to deep-oxidation products, CO_2 and H_2O.

Another significant manifestation of the influence of the reactor surface is that the CO_2 yield measured in most of the experiments is substantially higher than that predicted by kinetic simulations for the purely gas-phase process [89]. The formation of 'extra' CO_2 is likely associated with the heterogeneous processes of formaldehyde decomposition, CO oxidation, and deep oxidation of methane on the reactor surface.

DECOMPOSITION OF THE PRODUCTS ON THE REACTOR SURFACE

The nature of the reactor surface affects not only the kinetics of the formation of the products during the reaction, but also their subsequent transformations, especially in experiments with a long residence time of the reactants in the reactor. Note, however, that, even at residence times not exceeding a few seconds, the selectivity and yield of methanol and, especially, formaldehyde decrease with increasing residence time after oxygen-conversion completion. However, the available data on the rate of decomposition of the products are rather contradictory, strongly dependent on the nature of the reactor surface material.

In [142], the oxidation of methanol on the surface of a quartz reactor in the presence of oxygen, mainly according to the scheme $CH_3OH \rightarrow CH_2OH \rightarrow CH_2O \rightarrow HCO \rightarrow CO \rightarrow CO_2$, was observed only at temperatures above 700 °C, i.e., outside the DMTM temperature range. On the surface of metals and metal alloys, methanol can decompose even in the absence of oxygen, in particular with the formation of carbon deposits. However, on the surface of most of the nickel alloys investigated in [143], no methanol decomposition was observed at temperatures up to 490 °C.

An analysis of the kinetics of methanol decomposition after completion of DMTM [45] showed that the decomposition of methanol is particularly effectively promoted by a copper surface, the presence of which quickly leads to an almost complete disappearance of methanol from the products. Methanol is more stable with respect to the decomposition on a stainless steel surface; however, at ~450 °C and a residence time of t_r ~ 100 s, more than 50% of the methanol formed is lost in the presence of this surface. Methanol was practically stable in a Pyrex reactor, without any noticeable decomposition even at 500 °C (Fig. 6.7).

A similar study of methanol decomposition on the surface of a quartz reactor was carried out in [144]. A methane flow containing ~13% methanol was introduced into a heated reactor of special design intended to prevent the interaction of the gaseous reactants with

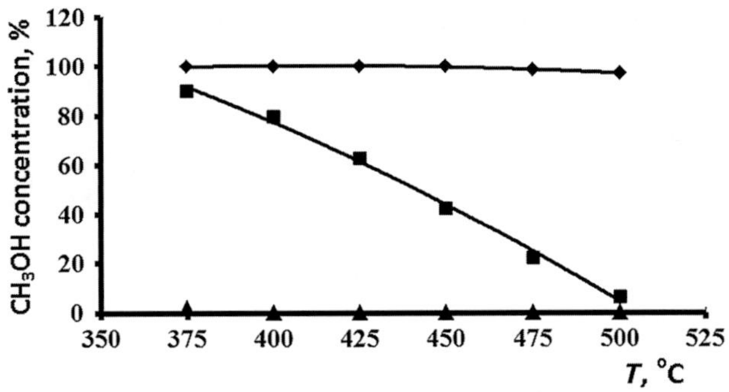

FIGURE 6.7 Temperature dependence of the degree of decomposition of CH_3OH on the surface of (▲) copper, (■) stainless steel, and (◆) Pyrex at $P = 30–50$ atm, $[O_2] = 2.5–4.9\%$, and $t_r = 200$ s. *Based on data of [45].*

the heated metal surface. It was found that, in the absence of oxygen, the degree of methanol conversion on the quartz surface of the reactor at 200–500 °C and a residence time of ~5 min did not exceed 5%. The main product was carbon monoxide.

Marked differences in the rates of decomposition of the oxidation products on various surfaces in the absence of oxygen were reported in [73]. At a reaction time of ~40 s, the decomposition of methanol on an oxidized aluminum surface begins to occur at 350 °C and proceeds to a full completion at 450 °C, with the main products being H_2 and CO in a proportion of 2:1:

$$CH_3OH \rightarrow 2H_2 + CO \quad (6.4)$$

By contrast, in a copper reactor, a complete conversion of methanol within 40 s occurred over a temperature range of 300–500 °C. The reaction over a well-oxidized copper surface predominantly produced CO_2 and H_2O in a ratio of 2:1, whereas the main products formed over the reduced surface were CO and H_2. Among the other products, the reaction over the reduced copper surface yielded dimethyl ether, which is usually not detected in the DMTM products. In a Pyrex reactor heated to 300 °C, no decomposition of methanol was observed, only its adsorption. At 350 °C, however, some decomposition occurred, as judged by the appearance of trace amounts of CO_2 and H_2O. The decomposition of acetaldehyde took place on surfaces of all the above types.

The oxidation of methanol on a quartz surface at $T = 270–330$ °C proceeded slowly, with the main product being carbon dioxide [145]. At higher temperatures (400–600 °C) [146], the quartz surface showed a higher activity, which increased sharply with the temperature. At 400 °C, the main products were HCHO and CO_2; however, with increasing temperature, the selectivity of their formation declined, whereas that of CO rose. At temperatures above 500 °C, CO became the main product (Fig. 6.8). The fact that the selectivity of HCHO formation at zero methanol conversion is close to 100% suggests that chemisorbed methoxy radicals are dehydrogenated to form formaldehyde, the primary oxidation product, which is further oxidized to CO and CO_2.

A special study of CO conversion under the same conditions showed that, although the conversion increases with temperature, it remains much lower than the observed CO_2 yield. The activation energy of CO oxidation on a quartz surface was found to be 20.4 kJ/mol, while

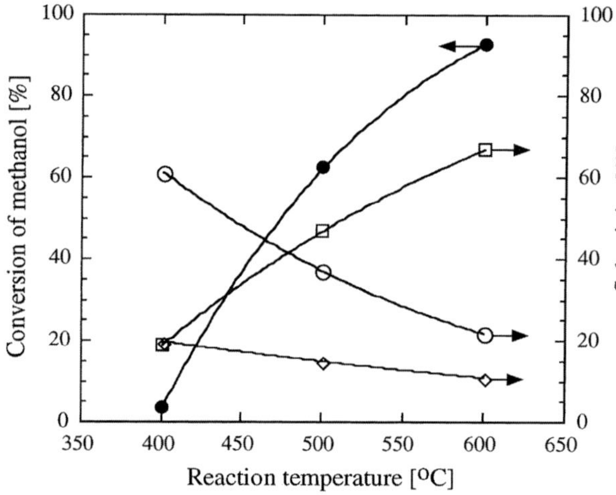

FIGURE 6.8 Temperature dependence of methanol oxidation on a quartz surface. A helium flow (4.92 ml/s) containing 1% methanol and 3.73% oxygen was passed through a crushed quartz bed (0.10 g); (●) methanol conversion, (○) formaldehyde formation selectivity, (□) CO selectivity, and (◇) CO_2 selectivity [146].

the energy of formation of CO_2 in the oxidation of methanol on this surface was −4.2 kJ/mol. Thus, CO is poorly oxidized on a quartz surface, whereas CO_2 is apparently formed from other active species during the dehydrogenation of formaldehyde.

The methanol conversion increases with the oxygen concentration, obeying, like formaldehyde conversion, the first-order law in the reactant concentration, whereas the overall sequence of the formation of the products is described by the scheme:

$$CH_3OH \xrightarrow{k_1} HCHO \xrightarrow{k_2} \{CO_a\} \rightarrow CO_2, \quad \{CO_a\} \uparrow CO$$

The energies of activation of the decomposition of methanol and formaldehyde were found to be 112 and 27.4 kJ/mol, respectively [146].

According to [135], at 600–700 °C in the presence of silica gel, the formaldehyde formed by the oxidation of methane is oxidized mostly in a homogeneous reaction (90%); i.e., the formation of CO in the presence of silica gel proceeds via homogeneous transformations of formaldehyde. In the presence of alumina, formaldehyde is oxidized mostly on its surface.

The results of kinetic simulations [138] of the catalytic conversion of methanol and formaldehyde in the presence of oxygen can be considered as the limiting case of influence of a surface on methane oxidation. These results, closely consistent with the available experimental data, show that the conversion of methanol on the surface of MoO_3/SiO_2 deposited catalyst begins to occur at 500 K and proceeds to a full completion at 700 K, with a consecutive formation of HCHO and then CO. The conversion of formaldehyde on this catalyst starts at 600 K and comes to a full completion at 900 K, with the respective temperatures for V_2O_5/SiO_2 catalyst being 500 K and 700 K.

EFFECT OF CATALYSTS ON THE GAS-PHASE DMTM PROCESS

One of the first comparative studies of the DMTM process at pressures above 100 atm in an empty copper reactor (Appendix I) and in the presence of catalysts (Cu, Ag, nickel–chromium steel, glass) was performed in [87]. A comparison of the results revealed no fundamental differences, although the authors pointed to a poor reproducibility of the experimental data, with the scatter in the yield reaching 50% even under similar conditions. As another possible reason for this, the authors mentioned the fact that the reactor walls are made of copper, which can act a catalyst. In a subsequent paper [147], with a wider variety of catalysts (Cu, Ag, Zn, Ni, and Monel), the authors could only state that nickel is more effective compared to other tested catalysts and that massive catalysts are more efficient than wire meshes. The latter, most likely, just demonstrates that the methanol yield increases with decreasing catalyst surface.

Based on their experimental studies, the authors of [55] concluded that the catalytic process is more efficient. However, even the best result they reported (a total selectivity of formation of oxygenates of 57% at a conversion of 2% on an Ag_2O/Cr_2O_3 catalyst at $P = 50$ atm) does not exceed the corresponding values for the gas-phase reaction. Of course, catalysts substantially, by 100–125 °C, reduce the temperature of the process [56], which produces a positive effect, at least, by preventing a further conversion of the oxygenates formed. Note, however, that, while in the majority of the experiments in the gas phase, no formation of CO_2 was detected [148], the CO_2 yield in the presence of a catalyst (metal oxides) even exceeded that of CO, indicative of a key role of the catalyst in the formation of deep-oxidation products.

A systematic study of the effect of catalysts on the DMTM process at $P = 30$ atm, $T = 250$–400 °C, $[O_2] = 3.9$–6%, and $t_r = 17$–50 s in one [140] of a series of works in which a very high selectivity of methanol formation was reported showed that certain catalysts (Co/Al_2O_3, hopcalite, TiO_2) promote the oxidation of methane mainly to the deep-oxidation products, CO and CO_2. Other catalysts (SnO_2, Cu/SiO_2) provide a high selectivity of methanol formation. However, the studied catalysts demonstrated no significant advantages in enhancing the selectivity of methanol formation or in lowering the reaction temperature.

An interesting feature of the DMTM process in a quartz reactor filled with quartz pieces was observed in [149]. In contrast to typical gas-phase experiments, the selectivity of methanol and CO formation at a low methane conversion (short reaction time) was very low, almost zero, with ethylene being the main product. However, with increasing methane conversion, the selectivity of ethylene formation decreased rapidly and other products appeared. The assumption made by the authors that ethylene is a primary product whereas CO and methanol are secondary products is inconsistent with the existing ideas about the kinetics of the gas-phase DMTM process [5]. One can assume that, under these conditions at a high surface-to-volume ratio and, consequently, a high rate of formation of methyl radicals, the initial stage is dominated by the oxidative condensation of methane to ethylene. As ethylene accumulates, the reaction transforms into the gas-phase process of oxidation of methane and ethylene to CO and CH_3OH.

The influence of a number of heterogeneous catalysts ($SrCO_3$, $BaCO_3$, and 7% Li/MgO), active in the oxidative condensation of methane at low pressures, on the process at 500–750 °C was studied in [80]. However, at pressures above 3 atm, no contribution from heterogeneous activation was detected, in agreement with the estimates of the pressure range of transition to branched-chain oxidation obtained in [94]. Carrying out the process in a reactor filled with a catalyst under the same conditions gives the same ratio between the condensation products and oxygenates, but a lower methane conversion compared to that measured in an empty reactor. The only difference between the results obtained in the reactor filled with the investigated catalysts and the empty reactor was an increase in the CO_2 yield at the expense of a decrease in the CO yield, which can be an argument in favour of a predominantly heterogeneous oxidation of CO under these conditions. These results are consistent with the available data on an increase in the rate of recombination of methyl radicals in the presence of an inert surface.

Experiments [81] in a flow quartz reactor at $P = 62$ atm, $T = 550\,°C$, $[O_2]_0 = 14\%$, $t_r = 0.6$s showed no significant effect of a catalyst (Sm_2O_3) on the partial oxidation of methane, except for a slight increase in the degree of conversion compared to the case where the reactor was filled with just quartz pieces. However, at a lower pressure, $P = 31$ atm, the effect of the catalyst on the methane conversion and yield of the target product becomes more pronounced.

The authors of [150] conducted a comparative analysis of the DMTM process in a Pyrex reactor at $P = 50$ atm, $T = 433\,°C$, $[O_2] = 4\%$, and a reaction time of $t_r \approx 5$ s and the S/V ratio ranging from 10 cm^{-1} for the empty reactor to 32.3 and 300 cm^{-1} when the reactor was filled with Pyrex balls of different diameters. When the S/V ratio was changed from 10 to 32.3 cm^{-1}, the methane conversion and selectivity of formation of the products did not change significantly. However, at $S/V = 300$ cm^{-1}, a complete conversion of methane under the same conditions could be achieved only at a higher temperature, 481 °C. Figure 6.9 displays the temperature dependence of the methane conversion and selectivity of methanol formation in the absence and presence of a catalyst in a Pyrex reactor, showing a rapid decrease in methanol selectivity with increasing conversion.

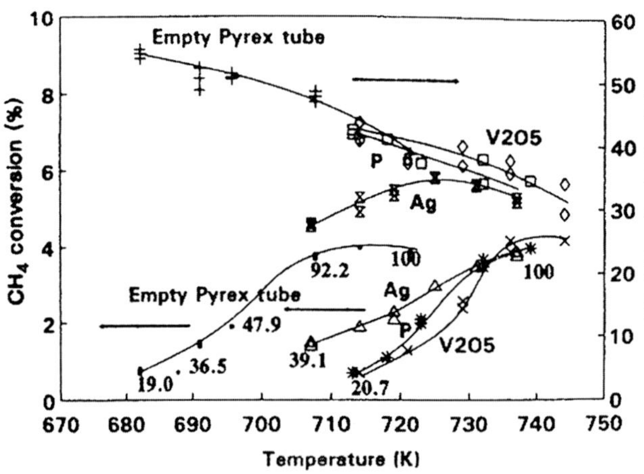

FIGURE 6.9 Comparison of the methane conversion and selectivity of methanol formation in the homogeneous and catalytic reactions. $P = 50$ atm, $[O_2] = 4.35\%$, and $t_r = 2$ s in the empty reactor and 4.5 s in the presence of the catalyst. The reactor was filled with Pyrex spheres (P), V_2O_5, and Ag/α-alumina. The numbers indicate the oxygen conversion [150].

The data presented in Fig. 6.9 show that, upon introduction of an additional Pyrex surface or a catalyst into the reactor, the same methane conversion can only be achieved by significantly increasing the temperature. In the opinion of the authors, the fact that the introduction of Pyrex spheres increases the effective activation energy to 69 kcal/mol, as compared to 56 kcal/mol for the empty reactor, clearly shows that the Pyrex surface produces an inhibitory effect.

The experiments performed in [151] at pressures of ~50 atm and a reaction time of less than 1 s were aimed at a comparative study of the characteristics of the DMTM process in an empty quartz flow reactor and in the same reactor filled with an inert material or various catalysts, such as MoO_3: UO_2, Fe/sodalite, and SnO_2, which, according to a number of patents, provide a high (up to 75%) selectivity of methanol formation. In fact, none of these catalysts showed a selectivity or positive effect superior to those typical of the gas-phase reaction, in contrast to the claims in the aforementioned patents. In all cases, the catalysts had a negative effect on the selectivity of methanol formation. The authors of [151] suggest that this is because metal oxides promote the oxidation of methane to CO_2.

The effect of the quartz reactor diameter and the S/V ratio on the selectivity of methanol formation in the DMTM process was also investigated [98]. In reactors with diameters of 11 and 19 mm, the selectivity was almost the same and decreased monotonically with increasing oxygen concentration (Fig. 6.10). However, the process in a 30-mm-diameter reactor featured a well-pronounced maximum in the selectivity at an oxygen concentration of 4–6%. When the 30-mm reactor was filled with quartz pieces, the behaviour of the reaction in it became similar that in the reactors with smaller diameters. Although the explanation of the behaviour of the reaction proposed by the author is not especially convincing, a significant influence of the surface on the reaction under these conditions is beyond doubt.

The experiments described in [105] showed that, even at atmospheric pressure, the only catalyst that provided characteristics substantially superior to those of the gas-phase process was 30% H_3BO_3/Al_2O_3. At higher pressures, 3 and 50 atm, the best results were demonstrated by the gas-phase oxidation without a catalyst.

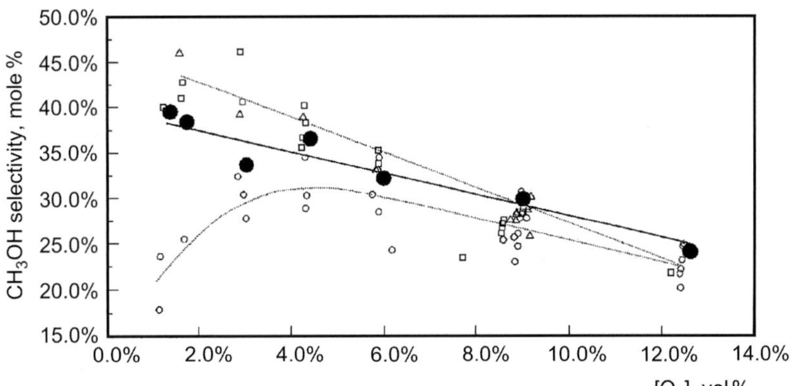

FIGURE 6.10 Effect of the reactor diameter on the selectivity of methanol formation at $P = 91$ atm, $T = 427\,°C$, $t_r = 35$ s: (□) d = 11 mm; (△) d = 19 mm; (○) d = 30 mm; and (●) d = 30 mm, the reactor is filled with quartz pieces [98].

Studying the mechanism of the partial oxidation of CH_4 by molecular oxygen on catalysts, in particular $Fe_2(MoO_4)_3$, clarified the main reasons for the low efficiency of catalysts in the production of methanol as opposed to formaldehyde production [152]. Methane is activated on surface oxygen-containing sites, which are formed by the dissociative adsorption of gaseous oxygen or from lattice oxygen supplied by diffusion. The use of oxygen isotopes revealed that the isotope exchange between $^{16}O_2$ and $^{18}O_2$ proceeds much more rapidly than the conversion of methane. This is indicative of a fast dissociative adsorption of O_2 on the catalyst surface. The possible intermediates of methane activation, such as CH_3(ads) or CH_3O(ads), react with activated surface oxygen O(ads) rapidly to form HCHO or COx. According to the authors, this is one of the main reasons why the formation of CH_3OH, which requires the protonation of CH_3O(ads), proceeds so slowly. For the selective formation of CH_3OH, it is necessary to provide a special structure of the active site of the catalyst, which should have properties of a Brønsted acid site capable of protonating CH_3O(ads) and simultaneously should protect this intermediate from further oxidation by neighbouring O(ads). In this case, the activation of methane requires a high temperature, >500 °C, at which formed CH_3OH will quickly decompose to HCHO or COx. As is well known, the degree of conversion of CH_3OH on $Fe_2(MoO_4)_3$ under the conditions typical of the partial oxidation of CH_4 is always 100%. Silica-supported MoO_3 and V_2O_5 are also efficient catalysts for CH_3OH oxidation. Thus, as pointed out by the authors of [152], there is a need to develop new catalytic systems capable of activating methane at lower temperatures.

One of these new areas of catalytic conversion of methane to methanol is the so-called reductive activation, in which methane is oxidized with a H_2-O_2 mixture. When adsorbed on an appropriate catalyst, H_2 can act as a donor of electrons and protons. There is a number of works on the selective oxidation of alkanes, alkenes, and benzene by H_2-O_2 mixtures. According to the ideas put forward in [152], CH_4 oxidation in the presence of H_2 can be described by a simple scheme

$$\begin{gathered} O_2 + H_2\ (\text{reductant}) \xrightarrow{\text{catalyst}} O^*\ (\text{active oxygen}) \\ <500\ ^\circ C \\ CH_4 + O^* \rightarrow CH_3OH \end{gathered} \tag{6.6}$$

However, it is possible that the oxidation involves OH radicals. At $T > 330\ ^\circ C$ and atmospheric pressure, a $CH_4-O_2-H_2$ mixture is converted on the $Fe_{0.5}Al_{0.5}O_4$ catalyst into methanol with a selectivity of 28% at a conversion of 0.4–1.5%. In situ IR spectra showed the presence of absorption bands of surface peroxides at a frequency of 895 cm^{-1}. At $T > 530\ ^\circ C$, in the absence of hydrogen, only small amounts of formaldehyde were formed. It was assumed that the active site was a tetrahedrally coordinated Fe (III) ion, on which the following reactions occurred:

$$\begin{gathered} O_2^{2-} + CH_4 \rightarrow CH_3O^- + OH^-, \\ CH_3{-}O{-}Fe + OH^- \rightarrow CH_3OH + FeO^-. \end{gathered} \tag{6.7}$$

Some Fe-containing catalysts provided the highest selectivity of the partial oxidation of methane to methanol. A positive effect of the presence of H_2 on the conversion of CH_4 was observed for the $FePO_4$ and $FeAsO_4$ catalysts, especially $FePO_4$. In the latter case, in the presence of hydrogen, CH_3OH was the main product, whereas in the absence, no methanol was formed at all. The presence of H_2 lowers the temperature at which CH_4 conversion on this catalyst begins and dramatically increases its rate. Most of the H_2 fed together with O_2 remains in the effluent gas from the reactor. In [152], these results were interpreted as indicating the formation of new active oxygen-containing site in the presence of H_2, in accordance with Scheme (6.6), which provides the partial oxidation of CH_4 to CH_3OH at lower temperatures. Although many other metal phosphates were tested, none of them gave a positive effect in increasing the CH_4 conversion or CH_3OH yield when H_2–O_2 mixtures were used for oxidation, results indicative of iron-containing catalyst sites playing a unique role in the formation of CH_3OH. The record selectivity of methanol formation provided by $FePO_4/SiO_2$ was attributed to the formation five-coordinated Fe, as indicated by Mössbauer spectra.

The authors of one of a few studies [86], performed under similar conditions at a high pressure ($P = 50$ atm), claimed both higher selectivity and yield of methanol formation on a Mo–V–Cr–Bi–Si oxide catalyst as compared to the homogeneous reaction. At 490 °C and $[CH_4]/[O_2] = 10$, the total selectivity of methanol and formaldehyde formation on one of the catalyst samples reached 47.6%, with their total yield being 3.39%, whereas in the absence of the catalyst, the respective values were 29.4% and 2.79%. At 460 °C and $[CH_4]/[O_2] = 10$, the total catalytic selectivity of methanol and formaldehyde formation on the same sample reached 70.5% (4.69% yield), as compared to 40.5% (3.68% yield) for the gas-phase reaction. The authors of [153], with reference to their earlier works, reported even an 80% selectivity of methanol formation at a 10% methane conversion on a similar Mo–V–Cr–Bi–Ox/SiO_2 oxide catalyst. The importance of an optimal positioning of the catalyst with respect to the overall reaction profile was discussed in a later work by the same authors [144]. Along with paying attention to interesting results reported in this work, it is worthwhile to note that the previously declared value of the selectivity was reduced. Therefore, it is necessary to perform a thorough analysis and independently reproduce these results.

The same is true regarding no less sensational results of [154], where the DMTM process in the presence NO (0–2.92%) featured a selectivity of formation of C_1 oxygenates of 40% at a very high methane conversion (40%, i.e., 16% yield) on a V_2O_5/SiO_2 catalyst with a low surface area. The conditions used in this study ($[N_2]:[CH_4]:[O_2] = 4:2:1$, atmospheric pressure, and $T = 530$–750 °C) suggest that formaldehyde is mostly formed.

As mentioned above, one of the possible reasons for such high characteristics may be an essentially heterogeneous mechanism of the process under the conditions set in these works, radically different from the mechanism of the gas-phase DMTM process. However, the reproducibility of these experiments and the possibility of scaling them to provide the conversion of gas volumes interesting for practice require special scrutiny.

A more obvious application of heterogeneous catalysis to the DMTM process is to convert part of the byproducts or highly reactive products into methanol, a more valuable and stable product. An interesting work in this respect is [68], where an increase in the methanol yield from the process promoted by 0.5% NO was achieved by passing the products of the gas-phase reaction with a temperature of 250 °C over a Cu–ZnO/Al_2O_3 catalyst. In the authors' opinion, this occurred due to the catalytic hydrogenation of CH_2O:

$$CH_2O\text{–}H_2 \rightarrow CH_3OH \qquad (6.8)$$

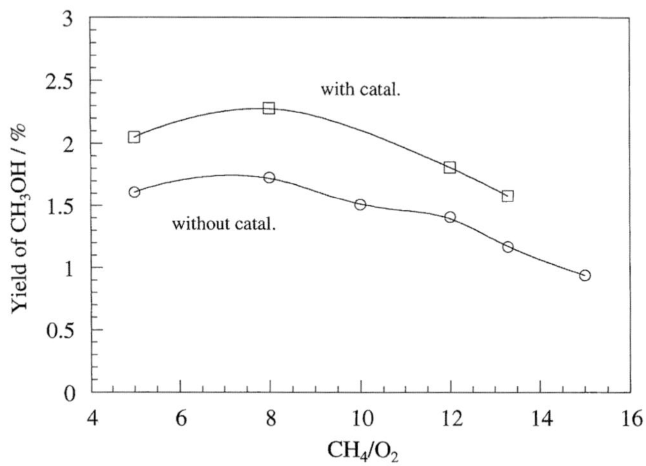

FIGURE 6.11 Dependence of the methanol yield on the $[CH_4]/[O_2]$ ratio in the (○) absence and (□) presence of the Cu–ZnO/Al_2O_3 catalyst at $P = 4$ atm and $T = 575\,°C$ [68].

Hydrogen required for this was provided, at least in part, by the reaction

$$CO + H_2O \rightarrow CO_2 + H_2 \quad (6.9)$$

Hydrogen was also assumed to come from the catalytic decomposition of formaldehyde:

$$CH_2O \rightarrow CO + H_2 \quad (6.10)$$

As a result, in the presence of the catalyst, the maximum methanol yield at $P = 4$ atm and $[CH_4]/[O_2] = 8$ increased by about 30% (Fig. 6.11).

SHORT-TIME CATALYSIS AND THE POSSIBILITY OF CONTROLLING THE DMTM PROCESS

In general, the possibility of high-temperature catalysis of the oxidative conversion of alkanes is not limited to a trivial increase in the rate of generation of active sites, so the catalysis of these processes cannot be fully replaced, for example, by physical methods of their activation. For essentially thermodynamically nonequilibrium processes of oxidative conversion of alkanes, because of their fundamentally nonlinear character due to numerous cross-reactions involving stable and unstable intermediates in the gas phase and on the surface, there exists a principle possibility for directing the evolution of the system in different ways to different end stationary states by relatively weak impacts.

The existence of multiple pathways of occurrence of such a complex reaction as the oxidation of hydrocarbons, and, accordingly, of the multiplicity of the final states, which are determined not only by the properties of the chemical system itself, but also by a variety of external factors, is the most characteristic feature of nonlinear systems and nonlinear approach in general. Note also that, while far from the point of critical transition to another

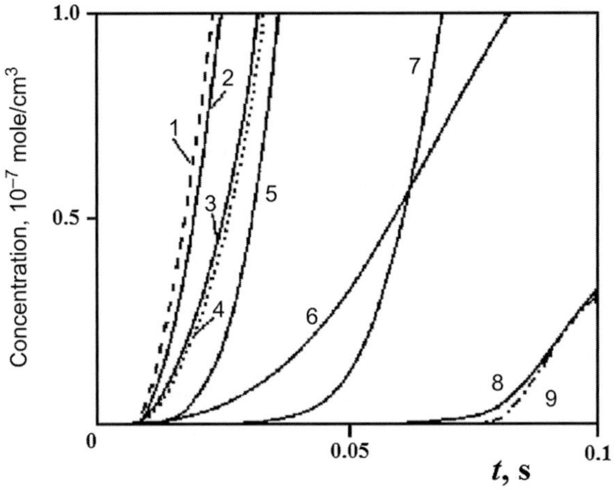

FIGURE 6.12 Calculated kinetic curves for the initial stage of the partial oxidation of methane at $T = 783$ K, $P = 85$ atm, and $[CH_4]/[O_2] = 21$: (1) CH_2O, (2) H_2O, (3) H_2O_2, (4) CH_3OH, (5) CO, (6) H_2 (7) CO_2, (8) C_2H_6, and (9) C_2H_5OH [70].

stationary state (bifurcation point), a limited external impact changes only the quantitative parameters of the system, near a bifurcation point, even a small external impact can dramatically change the properties of the system and the direction of its subsequent evolution. Nowadays, studying the behaviour of nonlinear chemical systems has become one of the topical fields. However, it is studies of the gas-phase oxidation of hydrocarbons in the first half of the last century that led to the discovery of a variety of nonlinear chemical processes, such as oscillatory oxidation, multiple ignition, cool flames, temperature hysteresis of the system parameters, etc. [13].

The progress in computer modeling and numerical methods has made it possible to perform not only qualitative but also to a certain extent quantitative analysis of complex nonlinear chemical systems. Computer modeling of complex processes of hydrocarbon oxidation described by mechanisms composed of hundreds of elementary chemical reactions provides a pattern of the formation and interaction of many intermediate products, thereby opening up entirely new prospects of controlling the kinetics of complex chemical processes.

Figure 6.12 presents the results of simulations of the kinetic characteristics of the formation of the products in the initial stage of the DMTM process. As can be seen, the processes of formation of various species are significantly separated in time. This offers the potential possibility to influence selectively on individual components of the reaction mixture. The time–space evolution of the process, for example in a fast flow, makes it possible, in principle, to guide the process in a more desirable direction to another stationary state by producing influence on individual species or system parameters. The possibility of such targeted catalytic influence on rapid gas-phase processes has been repeatedly demonstrated in recent years.

For example, it was shown [129] that the introduction of a single ~1-mm-diameter grain of Pt/ZrO_2 catalyst at 850 °C into a methane–oxygen flow with a ratio of $[CH_4]/[O_2] = 2$ (corresponding to the maximum reaction rate) affects not only the conversion of reactants, temperature profile near the catalyst, and the yield of the products, but also the characteristics of the gas-phase reactions upstream of the catalyst. At temperatures at which the conversion of the reactants in the gas-phase reaction practically does not occur, the introduction of a catalyst not only dramatically increases the conversion, but also shifts the zone of maximum heating (maximum rate of the gas phase reaction) a few millimetres upstream from the surface of the catalyst. In similar experiments with a thin catalyst layer, depending on the ratio of the layer thickness to the flow rate, a transition from complete oxidation products at low flow rates to typical incomplete gas-phase oxidation products, such as CO, H_2, and C_2H_6 (in small concentrations), at higher flow rates is observed.

Studies of short-term catalytic impacts, for milliseconds, on the initial stages of gas-phase reactions, which have been intensely developing in recent years, offer attractive prospects for controlling fast exothermic processes of gas-phase oxidation of alkanes. Changing the duration and nature of the catalytic impact and the fraction of gas involved in direct contact with the catalyst, as well as using other techniques, enables to control the temperature of the ensuing gas-phase reaction and the yield of the products. In particular, at a time of contact with Pt catalyst of ~1 ms, there was a significant increase, up to 20%, in the yield of C_2 hydrocarbons. The possibility of obtaining different products in this way, such as syngas, olefins, and oxygenates, by varying the conditions was demonstrated in [155]. For example, experiments [61] in a quartz microreactor with a 0.5-mm O-gap at $T = 1173$ K, $[O_2]/[CH_4] = 2$, and a reaction time of 20 ms demonstrated a high selectivity of formaldehyde formation, up to 82%, with the remaining 18% accounted for by C_2H_6. However, the methane conversion (0.85%) and formaldehyde yield were very low. With increasing methane conversion, the selectivity decreased rapidly, with the maximum yield of CH_2O not exceeding 2.4% at $T = 1223$ K, $[O_2]/[CH_4] = 8$, and $\tau = 60$ ms.

In the case of short-term catalysis, the basic processes of formation of the products occur in the gas phase after the interaction with the catalyst, which removes many of the issues related to changes in its properties under the influence of the products, for example, because of their deposition on the surface. High gas flow rates (~1 m/s) and short contact times (~1 ms) enable to produce a few kilograms of products per day per less than 1 g of catalyst [155]. Short-term catalysis makes it possible to realize complex combinations of temperature and concentration profiles during the reaction, as well as various versions of the relationship between the areas of catalytic and gas phase processes.

As a particular example of the possibility of controlling the partial oxidation of methane to oxygenates by directed catalysis, it is worthwhile to point to the potential consequences of a short-term heterogeneous impact on the concentration of hydrogen peroxide at the moment immediately before the end of the initial phase of the reaction. According to the results of kinetic simulations, by the time of complete conversion of oxygen and the end of the active phase of the reaction, up to 20% of the initial oxygen passes into hydrogen peroxide, the maximum concentration of which is reached at this moment (Fig. 5.2). Besides the fact that this oxygen is excluded from the process of formation of the target products, the subsequent decomposition of hydrogen peroxide is accompanied by the formation of hydroxyl radicals, which actively interact with the reaction products and lead to a marked decrease in the concentration of methanol and formaldehyde, especially in comparison with their maximum values. Indirect experimental evidence for a rapid decay of hydrogen peroxide after the reaction, along with its absence in the final products, is oxygen consistently detected in these products, which is probably formed by hydrogen peroxide decay. Theoretical calculations clearly rule out the possibility of incomplete conversion of the initial oxygen during the oxidation [5].

Experiments and kinetic modeling of methanol oxidation in supercritical conditions ($P = 246$ atm), roughly corresponding to the conditions of the DMTM process and closely described by the gas-phase kinetic model [156], showed that the main channel of methanol decomposition under these conditions is precisely the interaction with hydroxyl radicals arising from H_2O_2 decay and, to a lesser extent, with $HO_2^{\bullet}$ radicals. This means that, under these conditions, the formation and subsequent decomposition of hydrogen peroxide determines the methanol decomposition rate. The heterogeneous decomposition of peroxide in the initial phase of the reaction, before complete oxygen conversion, would return hydroxyl radicals into the branched-chain oxidation, thereby increasing its rate and the yield of the products. At the same time, the decomposition of the target products due to their interaction with hydrogen peroxide in the final stage of the reaction, after complete oxygen conversion, would also decrease significantly.

A key condition for an effective application of directed catalysis of oxidative conversion to hydrocarbons is an adequate quantitative description of the kinetics of these processes. This problem as a whole is not only extraordinarily difficult, but hardly feasible at all. First, a more or less adequate modeling of even the yield of the products of the partial oxidation of methane and oxidative coupling of methane with the participation of two lower alkanes and their derivatives, i.e., at the level of $C_1{-}C_2$ hydrocarbon chemistry, requires the a kinetic mechanism composed of hundreds of elementary reactions involving dozens of intermediate products. Given the sensitivity of the characteristics of a nonlinear system to the values of its constituent kinetic parameters, an adequate description of its behaviour requires a very high accuracy of these parameters. However, part of the rate constants of chemical reactions currently used for modeling the processes of oxidation and combustion of hydrocarbons are estimates to a factor of severalfold. Of course, should the problem of conversion of hydrocarbons be treated as one of the most important scientific problems with allocation of adequate financial resources, as was the case with atmospheric chemistry in a certain period, the situation may be somewhat improved, but not dramatically. In addition, for fast nonequilibrium processes, the concept of the rate constant, based on the existence of an equilibrium distribution over all states of the reacting molecules, becomes vague. Nevertheless, even the current level of the kinetic modeling of the oxidative conversion of hydrocarbons makes it possible to outline interesting prospects of catalytic influence on them. In addition, since the main effect of the presence of the catalyst is to change the composition of the gas mixture near the surface, the modeling of a short-time catalytic influence requires a scrupulous account of not only the relevant homogeneous–heterogeneous chemical processes, but also the actual processes of mass and heat transfer near the surface. Consideration of these factors gradually becomes part of the kinetic modeling.

Kinetic modeling results suggest that the known stationary states of the gas-phase partial oxidation of methane are apparently quite stable, since variation of even key rate constants within kinetically acceptable limits, for example, those of chain-branching steps, does not produce major changes in the behaviour of the system. However, systematic studies of the existence of other stationary states are needed. That such states may exist is suggested by the available literature data on abnormally high yields of target products. Although subsequent attempts to reproduce them failed, the situation requires careful analysis. It is also necessary to scrupulously examine all evidence concerning an increase in the product yield before the onset of the oscillatory mode, i.e., at the border of the stable equilibrium state.

CHAPTER

7

Role of Pressure in the DMTM Process

High pressure is an essential factor in the DMTM process, so it is necessary to analyze in detail the mechanisms through which it operates. There are several mechanisms [127], with different degrees of influence depending on a variety of conditions, such as the reaction time t_r, size of the reactor, and reactor surface material.

The lowest yield of methanol in flow reactors at high pressures, 50–300 atm, was observed in a copper reactor 10 mm in diameter [51,95] (Fig. 7.1). At the same pressures in stainless steel reactors and reactors of larger diameter (25 mm), the yield of methanol was significantly higher [69,76]. In stainless steel reactors of about the same diameter but at different reaction

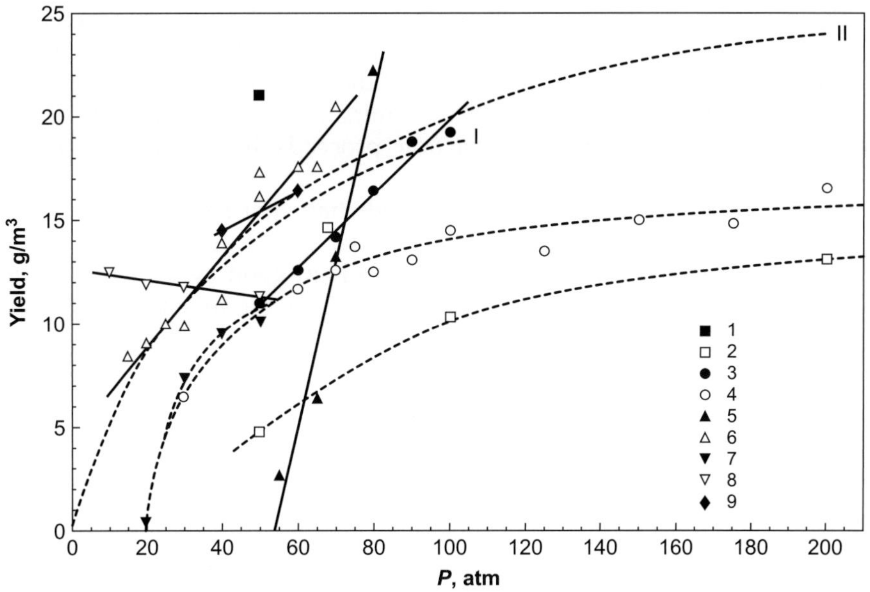

FIGURE 7.1 Pressure dependence of the methanol yield in the partial oxidation of methane according to a number of studies. (I) calculated by the kinetic model described in [2] at $T = 480\,°C$, $[O_2] = 3.5\%$ and a reactor diameter of 10 mm; (II) calculated by the equation $[MeOH] = [MeOH]_\infty/(1 + P_{0.5}/P)$. References 1–9 are given in Table 7.1, the experimental conditions are specified in Appendices I and II and Table 7.1.

Direct Methane to Methanol
http://dx.doi.org/10.1016/B978-0-444-63253-1.00007-6

times, less than 1 s [76] and tens of seconds [69], the methanol yield was higher at the shorter time. A comparison of published data shows (Fig. 7.1) that the reactor surface material (stainless steel and quartz [91]; stainless steel, quartz, and Pyrex [45]) strongly influences the pressure dependence of the methanol yield [91] and the selectivity of its formation [45]. At high pressures, the influence of the reactor material is insignificant; however, starting from a certain pressure, the methanol yield in the stainless steel reactor decreases sharply with the pressure. For the quartz reactor this pressure dependence is much weaker. As demonstrated in [45], the different influence of these materials is not due to the different decay rate of methanol formed on the respective surfaces, but due to the influence of the surface on the kinetics of methane oxidation. A relatively weak effect of pressure on the reaction kinetics was also observed in alumina (Alsint) reactors [67].

STATIONARY MODE OF BRANCHED-CHAIN REACTION

In general, the oxidation of methane at high pressure proceeds through the branched-chain mechanism. Modeling of the reaction dynamics for a $[CH_4]$:$[O_2] = 9{:}1$ mixture at 600–750 K and pressures of 1–100 atm [94] showed that a small change in the pressure, within a few atmospheres, causes a sharp reduction in the reaction time t_r by three to four orders of magnitude (Fig. 3.2). This can be explained only by a change in the mechanism of the process, more specifically, by the transition from a radical-chain process, the rate of which is controlled by a highly endothermic, slow step of homogeneous initiation, to a branched-chain process. In this mode, the oxidation rate is determined by chain-branching reactions, the rates of which are several orders of magnitude higher than that of the homogeneous initiation of radicals. Note, however, that the initial exponential growth of the concentration of radicals in the system is quickly terminated by increasing rate of their mutual quadratic recombination, so that the reaction proceeds in a quasistationary, but still branched-chain mode (Fig. 5.1).

The pressure of critical transition to the branched-chain quasistationary mode depends on other experimental parameters, typically constituting a few atmospheres. The calculated pressure dependence of the reaction time (Fig. 3.2) is in good agreement with the experimental data. Thus, the pressure determines, above all, the mode (chain or branched-chain) of this radical reaction.

EFFECT OF PRESSURE ON THE RATE OF BRANCHED-CHAIN QUASISTATIONARY REACTION AND THE METHANOL YIELD

According to the results of the kinetic modeling, after an abrupt transition of the reaction into the branched-chain quasistationary mode, the further rise of pressure is accompanied by the increase of the process rate, not so fast, but, nevertheless, monotonic (Fig. 3.2). The rise in pressure accelerates the oxidation of methane, reducing the time and/or temperature of the reaction. According to the experimental data from [54], the pressure dependence of the ignition delay time for an 8.8:1.2 methane–air mixture within 60–113 atm is described by the expression $\tau_{ig} \sim P^{-1.4}$, which is well consistent with the results of the kinetic modeling.

For example, the dependence displayed in Fig. 3.2 for a 9:1 methane–oxygen gives the expression $\tau_r \sim P^{-1.2}$ valid within 20–100 atm.

At a constant residence time of the reaction mixture in the reactor, an increase in the pressure results in a decrease in the temperature of onset of oxidation and the temperature of complete oxygen conversion. As follows from Fig. 3.1, which shows the data from [45,68,91], obtained in similar-sized flow reactors, 4–7 mm in diameter, the increase in pressure from 1 to 80 atm lowers the temperature of onset of oxygen conversion from more than 650 to 375 °C, and the temperature of complete oxygen conversion from more than 700 to ~400 °C. To some extent, the effect of increasing pressure is similar to the action of a promoter additive on the reaction at low pressure. For example, at a pressure of 5 atm, the introduction of 0.5% NO reduces the reaction temperature by ~100 °C [158]. However, with increasing pressure, the promoting effect of nitric oxide declines, being hardly noticeable against the background of the rapid branched-chain reaction at pressures above 10 atm. The accelerating effect of further increasing the pressure similarly decreases with rising pressure and, being almost imperceptible at pressures above 60 atm (Figs 3.1 and 3.2).

The effect of pressure on the gas-phase process can manifest itself, in particular, through the pressure dependence of the rate constants of the elementary reactions comprising the mechanism of the process. However, at pressures above 10 atm, most of the elementary reactions of the mechanism of methane partial oxidation occur almost in the high-pressure-limit mode; i.e., only slightly depend on the pressure.

The authors of [79] hypothesized on the possible manifestation of the 'cage effect' in the oxidation of methane at pressures above 100 atm; however, the experiments carried out in static reactors at pressures of several thousand atmospheres [46,48] did not reveal any signs of its occurrence under the indicated conditions, that is, did not support this hypothesis.

Kinetic analysis shows that the growth of the partial pressure of the reactants should lead not only to an increase in the methane conversion rate, but also to an enhancement of the role of nonlinear gas-phase radical reactions. A consequence may be an increase in the selectivity of methanol formation. Figure 7.1 and Table 7.1 show the most reliable experimental data on the pressure dependence of the methanol yield. All the experiments were performed under essentially similar conditions, optimal for achieving a high methanol yield. Only the experiments in which the oxygen conversion was close to 100% were taken into account. Comparison was conducted for the methanol yield, because it is the most reliable experimentally measured quantity. Due to a low degree of methane conversion, the selectivity of methanol formation is usually measured with great uncertainty, let alone that, in many practice-oriented studies, it was not determined at all. For the works [45] and [67], which report only the selectivity, we estimated the methanol yield based on the relationship $\Delta[CH_4]/\Delta[O_2] \approx 1$. This makes it possible to evaluate the methanol yield Y_{MeOH} as

$$Y_{MeOH} = Q\,x\,S_{MeOH} \tag{7.1}$$

where Q is the flow rate of methane, x is the methane conversion, approximately equal to the initial oxygen concentration, and S_{MeOH} is the selectivity of methanol formation.

The initial concentration of oxygen in the experiments presented in Fig. 7.1 ranged from 2.5% to 3.6%. Strictly speaking, the methane conversion and, therefore, the methanol yield depend on the initial concentration of oxygen in the mixture, being proportional to it at low O_2 concentrations; note that, in Fig. 7.1, the experimental data are presented in their original

TABLE 7.1 References and Experimental Conditions for Fig. 7.1

No.	Reactor Diameter, mm	Reactor Surface Material	Reaction Time t_r, s	Reference
1	3	Steel	5	[18]
2	~10	Copper	1.8	[95]
3	25	Stainless steel	<1	[76]
4*	25	Stainless steel	~30–40	[69]
5	10	Stainless steel	~2	[91]
6	7	Quartz	~2	[91]
7**	4	Stainless steel	~100	[45]
8**	4	Quartz	~100	[45]
9**	20	Alumina	~1	[67]

* *The values obtained at the same pressure are averaged.*
** *Estimated based on the oxygen concentration and the selectivity of methanol formation.*

form, without any reduction to the same concentration of oxygen. This increases the discrepancy between the results by no more than 30%, which in this case is irrelevant. The reaction temperature in these studies was varied depending on the other conditions, generally ranging within 410–500 °C. The reaction time varied from about 1 to 100 s. Reactors with a stainless steel, copper, quartz, Pyrex, or alumina (Alsint) surface had a diameter of 4–25 mm.

The presented data, with account of small differences in the conversion and the remarks given below on the impact of the size and surface material of the reactor, are in good agreement with each other and with the results of kinetic simulations of the process. Given the differences in the initial oxygen concentrations, it is necessary to admit that the results obtained in [76] and [91] at pressures above 70 atm are virtually identical in spite of the different diameters of the reactors (25 and 10 mm), different flow modes (laminar and turbulent, respectively), and a small difference in the reaction time t_r. The lower yield of methanol in [69] can be attributed to a much larger residence time of the gas mixture in the reactor.

Figure 7.1 also shows the results of calculations by the gas-phase model described in [2] and [89] (Fig. 7.1, curve I). Despite the apparent incomplete accounting of heterogeneous processes on the surface of the reactor, the resulting dependence satisfactorily describes the experimental pressure dependence of the methanol yield.

ROLE OF DIFFUSION OF THE REACTANTS TO THE REACTOR SURFACE

At pressures of 10–100 atm, one of the main reasons for the influence of pressure on the partial oxidation of methane is, apparently, the pressure dependence of the rate of diffusion of the reaction mixture components to the surface of the reactor. This is particularly important for laboratory reactors with a small diameter at low gas flow velocities. The magnitude of pressure influence is determined by the design of the reactor (diameter, surface material, and residence time of the reactants in the reactor) and by the partial pressure of the inert diluent.

Given that, in most laboratory experiments the gas flow was laminar, the diffusion time can be estimated using the molecular diffusion constant. The only exceptions are the experiments on the pilot plant described in [76], the flow in which was turbulent.

For a laminar flow in a cylindrical reactor, the time of diffusion of the mixture components to the reactor walls can be estimated by the known equation

$$\tau = d^2/\mu_1 D \tag{7.2}$$

Here, d is the diameter of the reactor, μ_1 is the first root of the Bessel function, and D is the diffusion coefficient, defined as

$$D = D_0(1/P)(T/273)^{1.5} \tag{7.3}$$

where D_0 is the diffusion coefficient under normal conditions, P is the total pressure, and T is the reaction temperature in K. For most of the experiments conducted on laboratory setups at pressures of about 50 atm, the value of τ constitutes a few seconds, being less or equal to the residence time of the mixture in the reactor. Therefore, at pressures below 50 atm, one can expect an intense interaction of the reaction mixture with the surface and a strong influence of the surface on the yield of the oxidation products, as evidenced by the aforementioned data on the effect of the reactor surface material on the DMTM process.

That the rate of diffusion of active species or oxygen to the reactor surface decreases with increasing pressure was also confirmed in [91], where it was demonstrated that, at a fixed CH_4:O_2 ratio, the oxidation of methane diluted with nitrogen in a proportion of CH_4:N_2 = 1:3 is characterized by a higher yield of methanol than that obtained at the same partial pressure of methane, but a lower total pressure.

Increasing the diffusion rate, in addition to raising the rate of loss of free radicals involved in the branched-chain process may increase the contribution from catalytic reactions at the reactor surface (especially, copper, iron, or nickel) with the predominant formation of the products of deep oxidation of methane, such as carbon oxides and water.

Experiments in [129] demonstrated that a supported platinum catalyst not only accelerates the reaction between methane and oxygen, but also increases the fraction of deep oxidation products, i.e., reduces the selectivity of gas-phase processes, including those leading to the formation of methanol. Since the partial oxidation of methane to methanol requires maintaining a very low concentration of O_2, the gas-phase partial oxidation into oxygenates and the surface catalytic reaction of deep oxidation in fact compete with each other. At high pressure, the rate of the catalytic reaction is determined by the diffusion of oxygen, a deficient reagent, to the surface. The decline of the competition between the surface reactions and gas-phase partial oxidation for deficient oxygen can be described by a simple expression:

$$[CH_3OH] = [CH_3OH]_\infty/(1 + P_{0.5}/P) \tag{7.4}$$

where $[CH_3OH]_\infty$ is the methanol yield at infinitely high pressure, P is the total pressure, and $P_{0.5}$ is the pressure at which the yield is half of $[CH_3OH]_\infty$. The corresponding curve plotted for $[CH_3OH]_\infty = 30\ g/m^3$ and $P_{0.5} = 50$ atm is shown in Fig. 7.1 (curve II). As can be seen, this simple equation satisfactorily describes most part of the experimental dependence.

Thus, summing up, we can name a few reasons why the pressure influences the yield of methanol in the partial oxidation of methane.

- At pressures of several atmospheres, the critical transition of the oxidation process into the branched-chain quasistationary mode manifests itself as a sharp increase in the reaction rate by several orders of magnitude.
- With rising pressure, the elementary rate constants of mono- and trimolecular reaction increase, tending to the high-pressure limit.
- As the partial pressure of the reactants increases, so does the contribution of the nonlinear gas-phase reactions leading to the formation of methanol.
- Decreasing the diffusion rate with increasing pressure reduces the contribution from the heterogeneous processes responsible for the formation of mainly deep oxidation products.

CHAPTER

8

Thermokinetic Phenomena in Partial Oxidation of Methane

EXPERIMENTALLY OBSERVED THERMOKINETIC PHENOMENA IN THE PARTIAL OXIDATION OF METHANE

A nonlinear dependence of the contribution of the most important elementary gas-phase reactions that determine the course of the DMTM process on the concentration of initial reagents, a strong temperature dependence of the key stages, and a complex interaction of the homogeneous and heterogeneous reactions give rise to a variety of thermokinetic effects in this highly exothermic system. The importance of taking into account the mutual influence of the chemical and thermal processes in the system was emphasized in [88]. The experimental data and modeling results presented in this work for a Continuous Stirred Tank Reactor (CSTR) demonstrated that this system is very sensitive to changes not only in the kinetic but also in the thermal parameters. This most clearly manifests itself in a sharp, almost stepwise change in the conversion of the reagents upon variation of the reactor temperature and in existence of a temperature hysteresis of the reaction, i.e., different patterns of the temperature dependence of the methane conversion during decreasing and increasing the temperature of the reactor walls (Fig. 8.1). The maximum methanol yield is reached on the branch of the hysteresis curve corresponding to the lowering temperature of the reactor walls, which may be an argument in favour of conducting the process at the lowest possible temperature to produce the maximum yield of methanol. The methanol yield decreases with increasing temperature mainly due to an increase in the yield of carbon monoxide.

A temperature hysteresis of the reaction in a CSTR was observed experimentally in the subsequent work of the same group [101] as well. Figure 8.2 shows the obtained dependences of the degrees of conversion of methane and oxygen, whereas Fig. 8.3 displays the selectivity of methanol formation for the rising and lowering temperature of the reactor walls.

Both figures demonstrate a clear hysteresis. The onset of the reaction under these conditions occurred at ~400 °C, while its termination in the lowering temperature mode, at 385 °C; i.e., the temperature hysteresis was about 15 °C. The methane conversion under these conditions was ~8%, whereas the maximum yield of methanol, 1.9%, was achieved at the

Direct Methane to Methanol
http://dx.doi.org/10.1016/B978-0-444-63253-1.00008-8

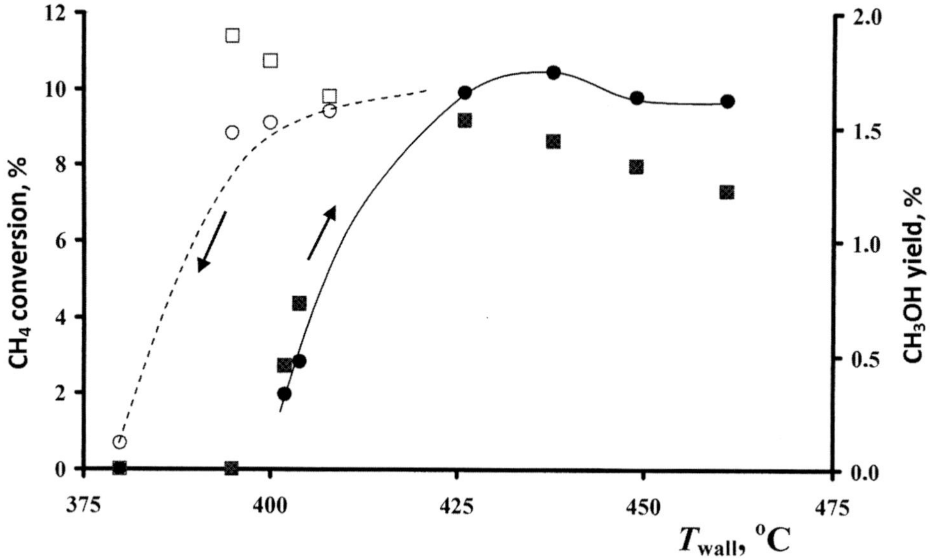

FIGURE 8.1 Dependence of the (●, ○) CH_4 conversion and (■, □) CH_3OH yield for the rising (solid line) and lowering (dashed line) temperature of the reactor walls at $P = 30$ atm, $[O_2] = 9.5\%$, $t_r = 0.2$ s. *Based on data of [88].*

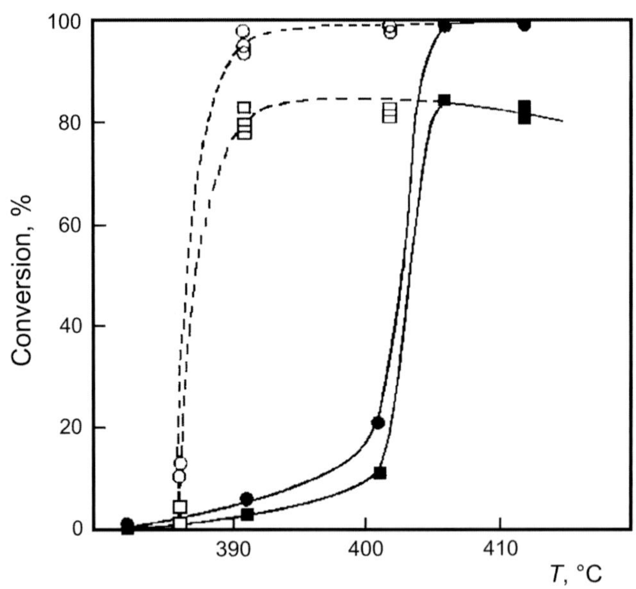

FIGURE 8.2 Dependence of the conversion of (■, □) methane (×10) and (●, ○) oxygen in a CSTR for the rising (solid lines) and lowering (dashed lines) temperature of the reactor walls at $P = 30$ atm, $[O_2] = 9.5\%$, $t_r = 30$ s [101].

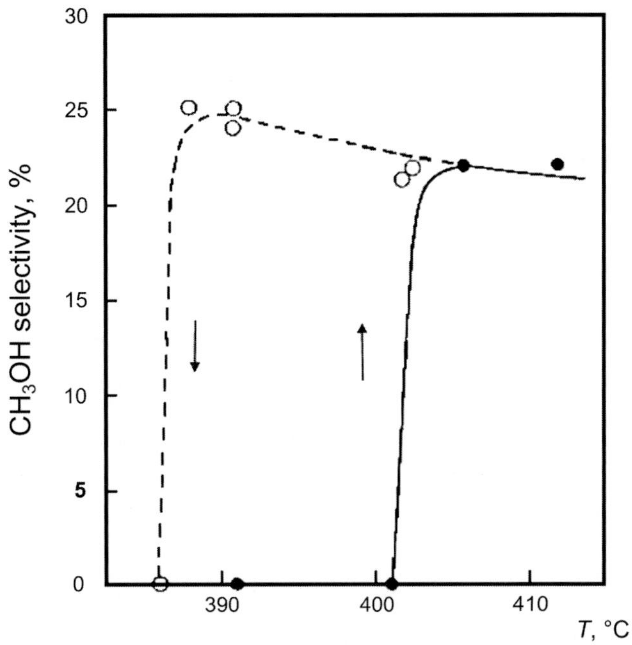

FIGURE 8.3 Dependence of the selectivity of methanol formation in a CSTR for the rising (solid line) and lowering (dashed line) temperature of the reactor walls at $P = 30$ atm, $[O_2] = 9.5\%$, $t_r = 30$ s [101].

lowest temperatures, attainable only in the lowering temperature mode. At an oxygen concentration of 9.5%, a thermal instability of the reaction was observed; however, the amplitude of these oscillations, with a period of ~6 min, was irregular.

The temperature hysteresis increases with the initial oxygen concentration (Fig. 8.4) and decreases with increasing residence time of the mixture in the reactor (Fig. 8.5).

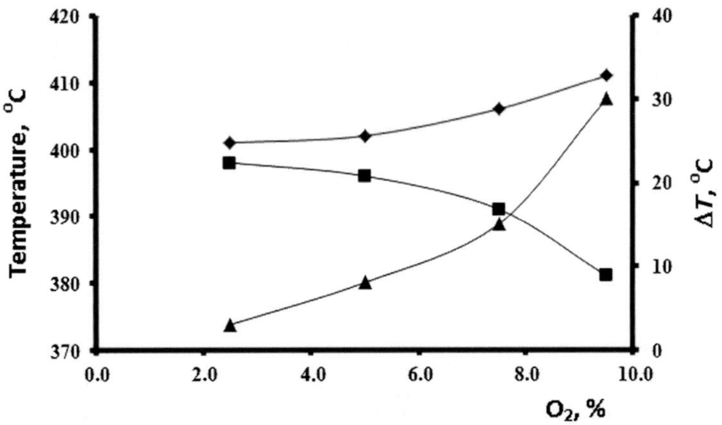

FIGURE 8.4 Dependence of the temperature of (◆) onset and (■) termination of the reaction and (▲) its temperature hysteresis on the initial concentration of oxygen in the CSTR at $P = 30$ atm and $t_r = 20$ s. *Based on data of [101].* (For colour version of this figure, the reader is referred to the online version of this book.)

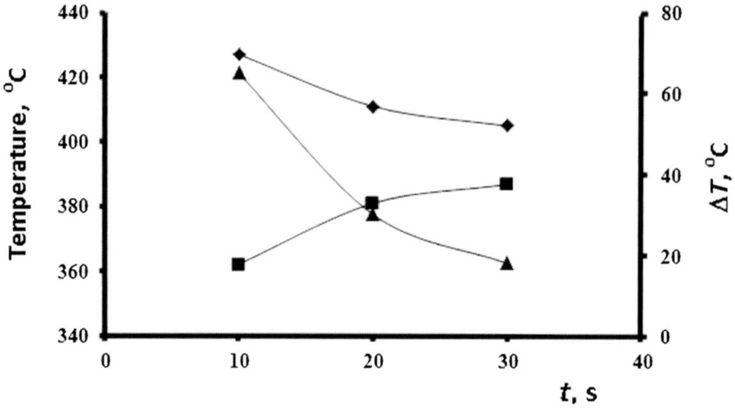

FIGURE 8.5 Dependence of the temperature of (◆) onset and (■) termination of the reaction and (▲) its temperature hysteresis on the residence time of the mixture in the CSTR at $P = 30$ atm and $[O_2] = 9.5\%$. *Based on data of [101].* (For colour version of this figure, the reader is referred to the online version of this book.)

The authors of [104] attempted a qualitative and even semiquantitative analysis of the thermokinetic interactions and the nonlinear nature of this highly exothermic system. The results obtained from nonisothermal modeling of the process are qualitatively consistent with the above experimental results. A detailed mathematical description of the kinetics and processes of heat and heat transfer in a CSTR made it possible to reproduce the appearance of temperature hysteresis and to demonstrate its dependence on the oxygen concentration, reaction time, and temperature of the reactor walls. Figure 8.6 shows the calculated self-heating

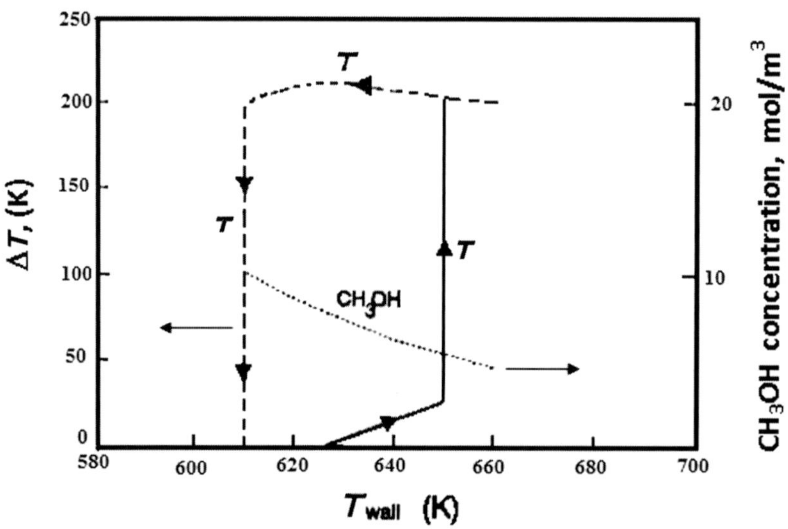

FIGURE 8.6 Simulation of the self-heating of the reaction ΔT and the methanol concentration on T_{wall} at $[O_2] = 9.5\%$, $P = 50$ atm, and $t_r = 10$ s. The arrows show temperature variation [104].

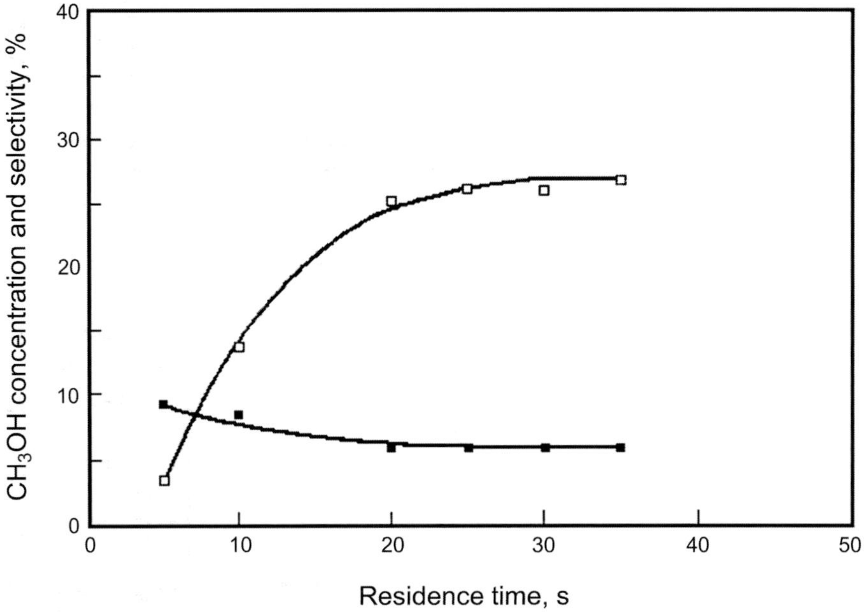

FIGURE 8.7 Simulation of the (■) methane conversion and (□) maximum selectivity of methanol formation on the residence time of the reactants in the reactor at $[O_2] = 9.5\%$ and $P = 50$ atm [104].

in the reaction at $t_r = 10$ s for the rising and lowering temperature of the reactor walls. We see that, as in the experiments presented in Figs 8.1–8.3, the highest concentration of methanol is achieved at the minimum temperature realized in the lowering temperature mode, the temperature at which no methanol is formed in the rising temperature mode.

With increasing reaction time, the calculated maximum selectivity of methanol formation increases, tending to 30% (Fig. 8.7), and the temperature hysteresis decreases, completely disappearing at $t_r = 30$ s. The methane conversion decreases slightly with increasing reaction time, reaching ~6%.

Figure 8.8 shows the calculated selectivity of methanol formation on the reactor wall temperature at different oxygen concentrations. At an oxygen concentration of 2.5%, decreasing the temperature of the reactor walls results in the onset of an oscillatory mode.

In the oscillatory mode, the calculated methanol concentration changes by several orders of magnitude during the oscillation period. Given that the oscillation period lasts for several minutes, the authors discuss the hypothetical possibility of controlling the methanol yield by adjusting the parameters of the process. Figure 8.9 displays the calculated methanol concentration in the oscillatory mode at different reactor wall temperatures T_{wall}. The oscillation period is ~90 s. The amplitude of the corresponding temperature oscillations is ~140 °C. For unknown reasons, methanol concentration oscillations have two different periods, with the concentration ranging from 1.5 to 9.5 mol/m^3.

Figure 8.10 shows the corresponding oxygen concentration oscillations. In contrast, the concentration of methanol, that of oxygen decreases to zero, with the oscillations being almost harmonic.

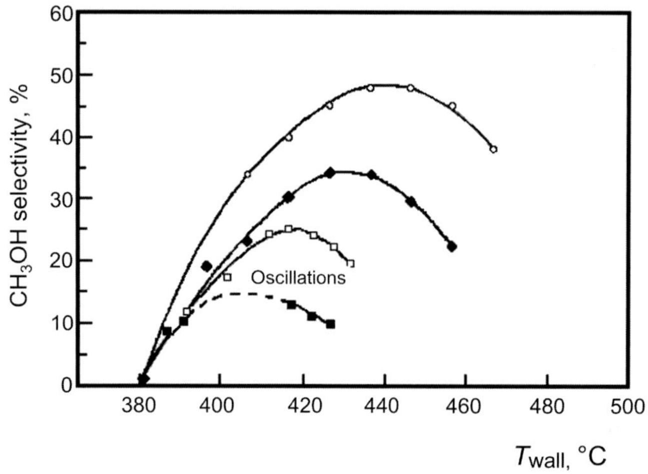

FIGURE 8.8 Simulation of the methanol formation selectivity on the temperature of the reactor walls T_{wall} at $P = 30$ atm and $[O_2] =$ (○) 9.5%, (◆) 7.5%, (□) 5%, and (■) 2.5% [104].

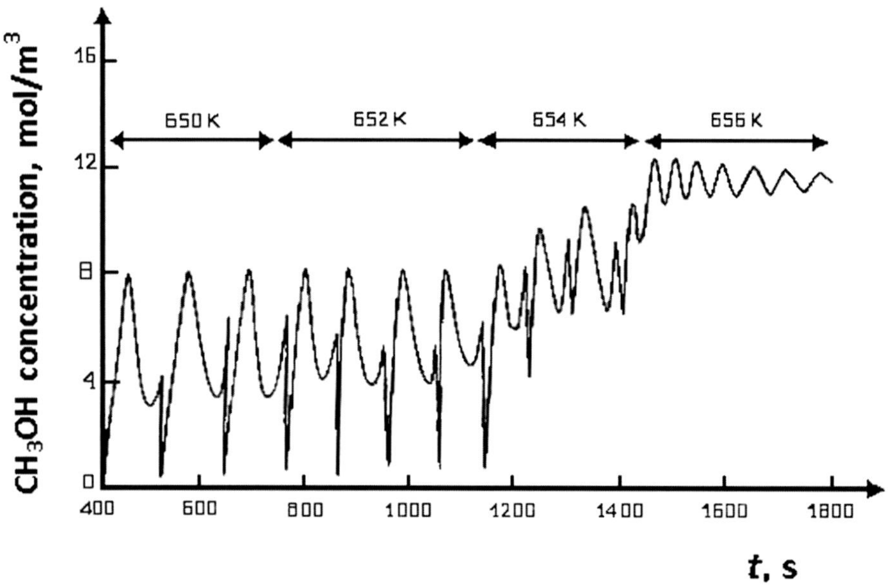

FIGURE 8.9 Simulation of time evolution of the methanol concentration in the oscillatory mode at $P = 50$ atm, $[O_2] = 9.5\%$, $t_r = 30$ s, and different temperatures of the reactor walls T_{wall} [104].

Thus, in certain conditions, under the influence of thermokinetic effects, the partial oxidation of methane in a CSTR becomes unstable. All the parameters, including the temperature and methanol yield, can vary dramatically. Figure 8.11 displays the calculated temperature evolution for the nonisothermal oxidation of methane when the temperature of the CSTR

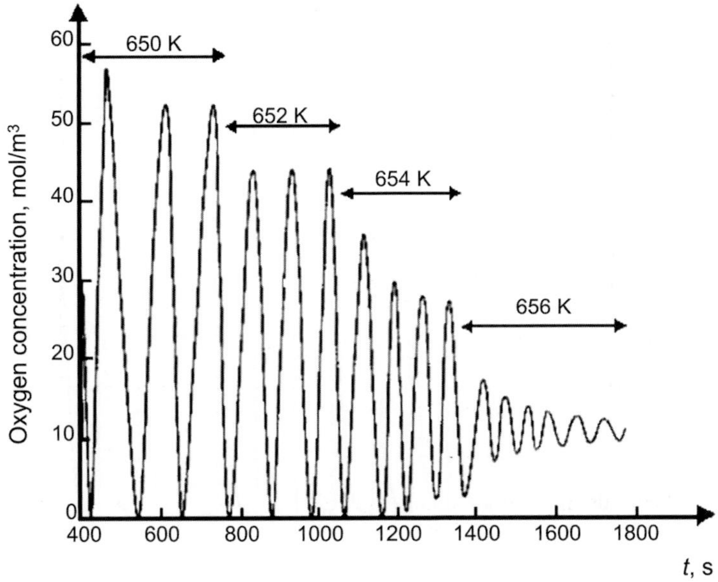

FIGURE 8.10 Simulation of time evolution of the oxygen concentration in the oscillatory mode at $P = 50$ atm, $[O_2]_0 = 9.5\%$, $t_r = 30$ s, and different temperatures of the reactor walls T_{wall} [104].

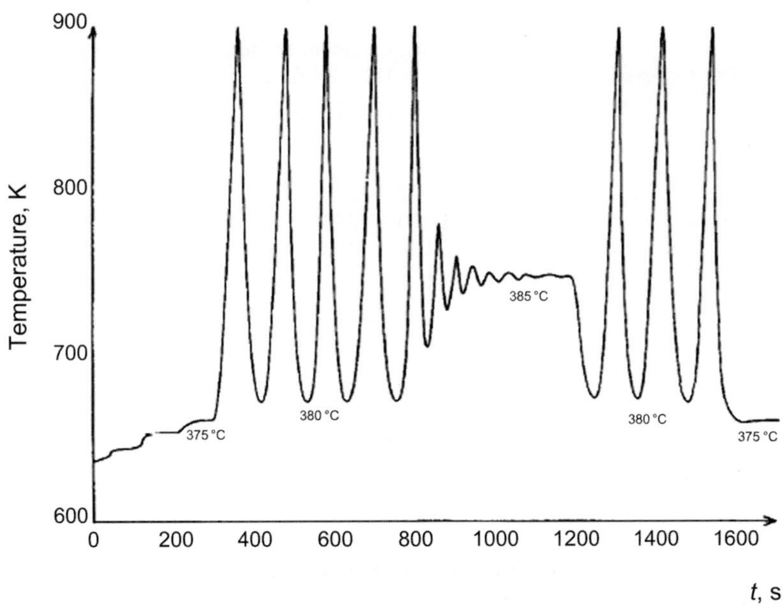

FIGURE 8.11 Simulation of temperature oscillations in a CSTR for a stepwise variation of the reactor wall temperature at $P = 50$ atm, $[O_2]_0 = 9.5\%$, and $t_r = 30$ s [99].

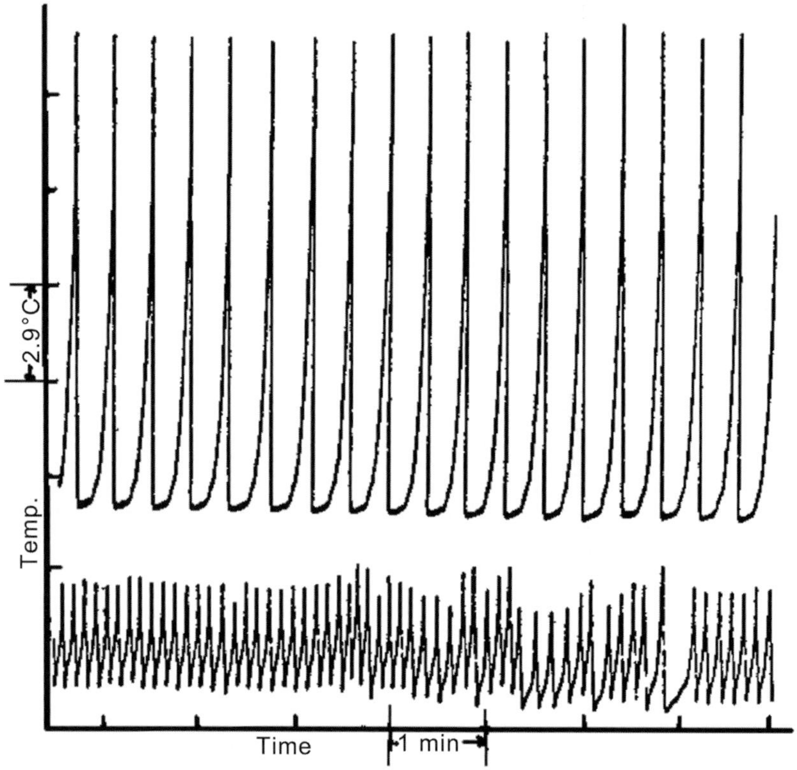

FIGURE 8.12 Temperature oscillations in the partial oxidation of methane to methanol at $P = 25$ atm, $[O_2] =$ 9.2%, and $t_r = 45$ s. The top graph corresponds to $T = 412$–$427\,°C$, an oscillation frequency of 2/min, and an amplitude of $A = 15\,°C$. The bottom graph is plotted at $T = 445$–$449\,°C$, an oscillation frequency of 8/min, and an amplitude of $A = 3\,°C$ [90].

wall changes stepwise. This is a graphic demonstration of the bifurcation behaviour of the system, which stems from the phenomena associated with the occurrence of cool flames in it.

For comparison, Fig. 8.12 shows the pattern of temperature oscillations observed in the oxidation of methane in a flow reactor at 25–35 atm and oxygen concentrations above 5% [90]. At oxygen concentrations of 8% and above and temperatures of ~410 °C, oscillations exhibited a stationary behaviour. Raising the temperature to 450 °C caused a decrease in the amplitude of the oscillations and an increase in their frequency. At 475 °C, oscillations disappeared. The amplitude and frequency of oscillations at different points along the axis of the reactor were very different. The maximum selectivity of methanol formation was achieved at the lowest temperatures, corresponding to an end of the oscillations, but it was still less than that observed at the optimal oxygen concentration ($\leq$3%).

This suggests that, in the case of existence of such oscillations, it is possible control the processes in the reactor so as to maximize the selectivity of methanol formation. This assumption is confirmed by the calculated time evolution of the methanol formation selectivity during an oscillation period (Fig. 8.13) [104]. As can be seen, at some point in time, the

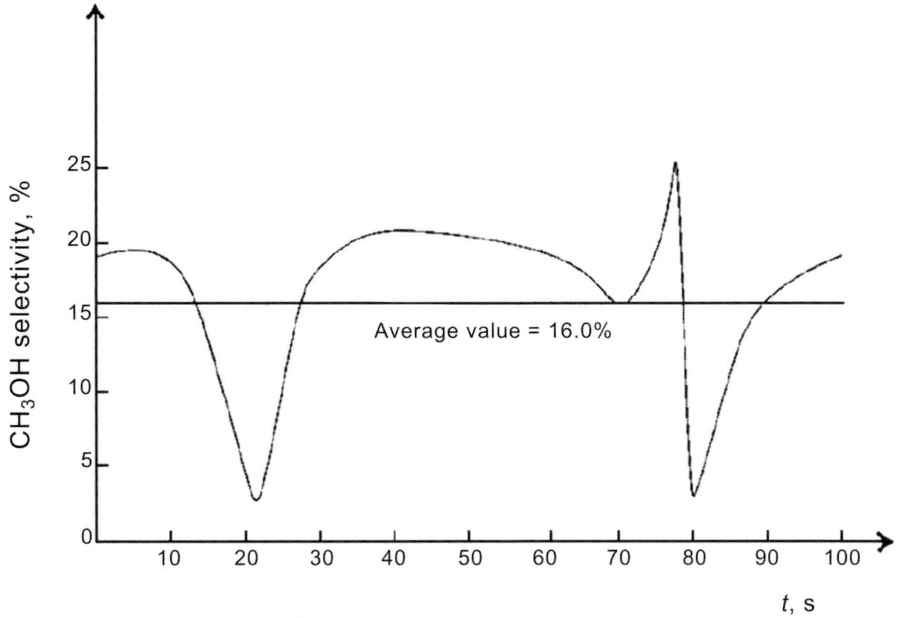

FIGURE 8.13 Simulated time evolution of the methane formation selectivity over an oscillation period at $T_{\mathrm{wall}} = 652$ K, $P = 50$ atm, $[O_2] = 9.5\%$, and $t_r = 30$ [104].

selectivity is much higher than the average value, which is close to the value of the selectivity in the steady state just before the onset of to the oscillatory mode. The excess over the selectivity in the steady state is as high as 20% and lasts for a significant fraction of the oscillation period.

Given their considerable period, the above-described oscillations under flow conditions [90] may well have a thermochemical nature, stemming from a thermal feedback through the reactor surface; at the same time, it is impossible to rule out their chemical origin, through a slow change of the state of the reactor surface under the action of the reaction.

In the oxidation of many hydrocarbons of various structures, including alkanes, purely kinetic periodic oscillations and other critical phenomena have been observed. They manifest themselves through a sharp increase in the reaction rate and flashes of light, followed by an equally sharp decrease in the oxidation rate although the reagents are far from being exhausted. Since the heating of the mixture in such flashes does not generally exceed a few tens of degrees, this phenomenon has become known as cool flames [13,159,160]. The possible occurrence of cool flames in the oxidation of methane at atmospheric pressure was first demonstrated experimentally in [161]. In [162], cold flames were observed in $2CH_4{-}O_2$ mixtures at pressures of 650−740 Torr and temperatures of $\sim$500 °C. The time evolution of the reaction mixture temperature (Fig. 8.14) features two peaks, with the heating in the first peak reaching 100−120 °C.

In [163], a region of negative temperature coefficient (NTC) of the reaction rate for methane oxidation was revealed, whereas the authors of [164] performed an experimental and kinetic-modeling study of cool-flame phenomena and NTC for the oxidation of methane

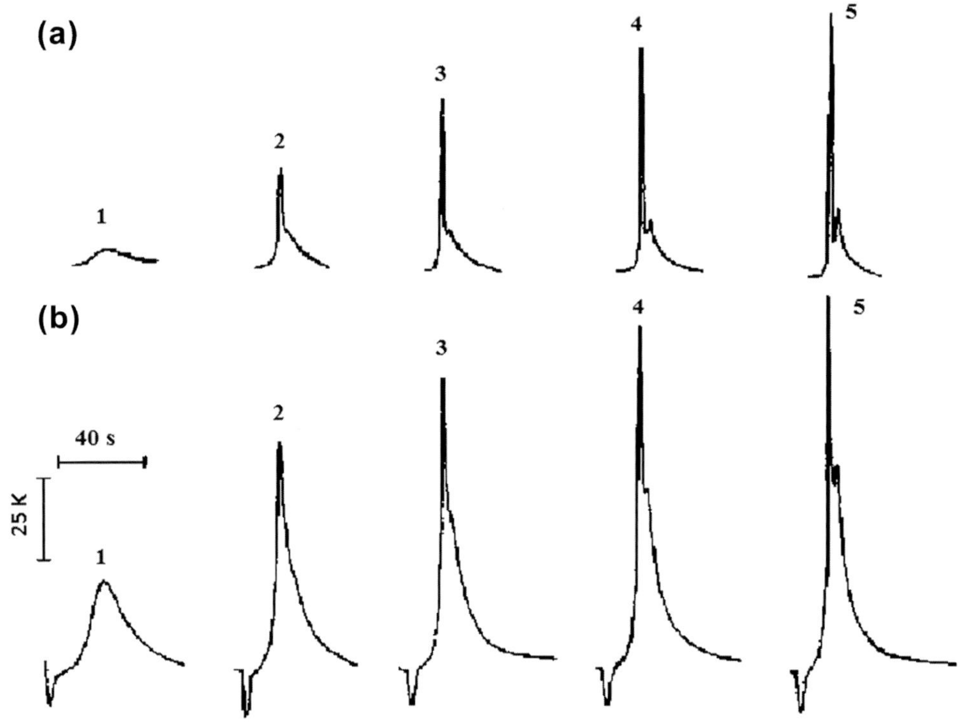

FIGURE 8.14 Kinetic curves of (a) heating and (b) emission for the oxidation of a $2CH_4-O_2$ mixture at $T = 784$ K and various pressures: P (Torr) = (1), 585, (2) 650, (3) 683, (4) 709, and (5) 727 [162].

at pressures close to atmospheric. Figure 8.15 demonstrates experimentally measured and calculated regions of cool-flame ignition for a $2CH_4-O_2$ mixture.

The appearance of cool flames is directly related to the existence of NTC region, near which the temperature rise caused by the self-heating of the reaction mixture can lead to a decrease in the rate of the process or even its complete halt. Kinetic simulations [164] demonstrated that a cool-flame flash begins as an ordinary branched-chain process, identical to the reaction under isothermal conditions, and only at a later stage becomes a thermal-chain process. The conventional explanation of the mechanism of the appearance of the NTC region is the occurrence of aforementioned reversible reaction

$$CH_3^{\cdot} + O_2 \Leftrightarrow CH_3OO^{\cdot}, \qquad (8.1)$$

the equilibrium in which shifts to the left with increasing temperature, leading to a decrease in the concentration of methylperoxy radicals $CH_3OO^{\cdot}$, which determines the chain-branching rate and, hence, the rate of conversion of methane in this process.

Although the existence of the NTC region and, respectively, cool flames in the oxidation of very rich methane–oxygen mixtures has not yet been demonstrated, this phenomenon was shown to take place for rich (up to $[C_3H_8]/[O_2] = 23$) propane–oxygen mixtures (Fig. 8.16) [165].

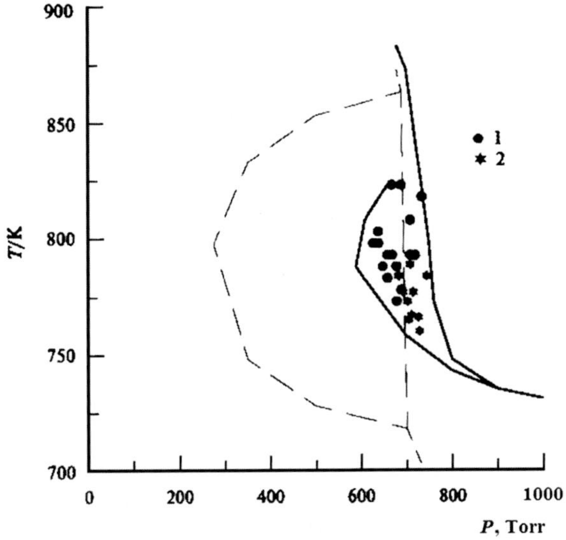

FIGURE 8.15 Cool-flame region for the oxidation of a $2CH_4$–O_2 mixture: (1) single- and (2) double-cool flame flashes. The solid and dashed lines represent, respectively, the experimentally measured and calculated boundaries of the regions of cool-flame flashes and two-stage ignition [164].

An important result of this work is that it demonstrates that the appearance of a cool flame during the oxidation of light alkanes is associated with reactions of the alkane itself and the primary alkylperoxy radical, not reactions of alkane oxidation intermediates, for example, the respective oxygenates, the oxidation of which was also observed to be accompanied by

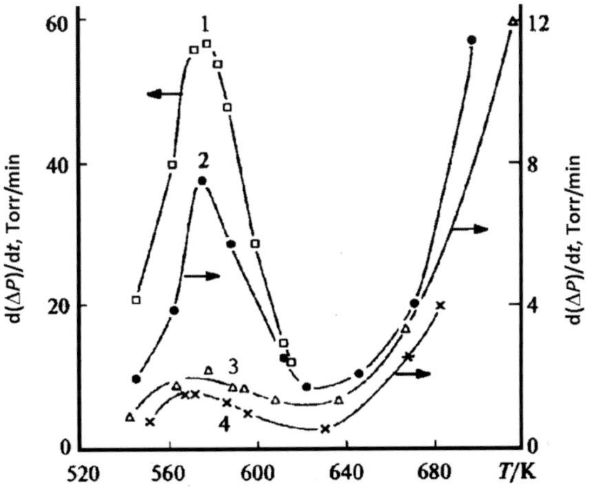

FIGURE 8.16 Experimental temperature dependence of the maximum pressure rise rate pressure $d(\Delta P)/dt$ for the oxidation of propane–oxygen mixtures at an initial pressure of $P_0 = 200$ Torr and $[C_3H_8]/[O_2] = $ (1) 1, (2) 7.5, (3) 18, and (4) 23 [165].

cool-flame phenomena. First, for the ratio of $[C_3H_8]/[O_2] = 23$, it is difficult to expect a significant role of the oxidation of secondary products in the process. Second, kinetic simulations within the framework of a model including no reactions involving oxidation products showed quite a satisfactory agreement with the available experimental data, both in the location of the NTC region and in the maximum oxidation rate.

Theoretical studies [166,167] suggest the possibility of existence of cool-flame modes of methane oxidation at high pressures. Moreover, such modes can lead to yields fundamentally different from those typical of the known modes. However, there is only one experimental observation [46] indicative of the feasibility of the cool-flame mode at high pressure. In this study, in a static reactor at a pressure of 3360 atm, oxygen concentration of 8%, and an initial temperature of 262 °C, a duel temperature spike with amplitudes of 6 and 14 °C accompanied by corresponding increases in the pressure was observed against the background of slow temperature growth. Unfortunately, this phenomenon, not studied in detail, remains an isolated event.

MECHANISM OF NONLINEAR PHENOMENA IN THE OXIDATION OF ALKANES

Kinetic modeling enabled to establish the mechanism of the nonlinear phenomena accompanying the oxidation of light alkanes. For example, simulations of the NTC for the low-temperature oxidation of propane showed that the leading species in this phenomenon are the propylperoxy radical and its isomers [165]. The simulation results obtained in [168] almost quantitatively (within 10 K) agreed with the experimentally determined temperature boundaries of the NTC region and closely reproduced the form of the temperature dependence of the induction period. An explanation of this phenomenon is based on the widely held view that contribution from the removal of $HO_2^{\bullet}$, a low-active radical, from the propylperoxy radical,

$$C_3H_7OO^{\bullet} \rightarrow C_3H_6 + HO_2^{\bullet}, \tag{8.2}$$

increases with the temperature, which, under these conditions, is equivalent to the termination of chains and reduction of the chain-branching rate. At higher temperatures, hydrogen peroxide formed by the recombination of $HO_2^{\bullet}$ becomes a source of $OH^{\bullet}$ radicals, which is equivalent to the elimination of this chain-termination channel and restoration of the normal temperature dependence of the reaction.

The NTC region for the oxidation of methane [163] is related to the temperature dependence of reversible reaction (8.1) of methyl radicals with oxygen. At elevated pressures and temperatures below 800 K, the equilibrium in this reaction, even at very low concentrations of oxygen ~1%, is almost entirely shifted to the right, towards the formation of the methylperoxy radical $CH_3OO^{\bullet}$. This leads to a situation where methyl radicals $CH_3^{\bullet}$ formed in the reaction are immediately converted into methylperoxy radicals, which participate in further transformations, providing an efficient chain branching in the system through the formation of methylhydroperoxide and its subsequent decomposition via reactions (II) and (VI) (Table 5.1). However, an increase in temperature shifts the equilibrium of reaction (8.1) to the left, causing a decrease in the concentration of methylperoxy radicals and, as a consequence, a decline in the chain-branching rate. Thus, an increase in temperature really slows down the oxidation process. At higher temperatures, near 1000 K and above, hydrogen peroxide begins to dissociate intensely via reaction (X) (Table 5.1), which provides a new growth of the chain-branching rate and the shift of the reaction from the NTC region.

The same mechanism explains the existence of two-stage ignition and cool flames observed in the oxidation of $2CH_4{-}O_2$ mixtures at pressures of 650−740 Torr and temperatures of ~800 K. The self-heating of the reaction mixture by ~100 K (Fig. 8.14) transfers it into the NTC region, where the reaction rate decreases and the process stops. Since this is accompanied by a burnout of a negligibly small fraction of the reagents, upon subsequent cooling of the mixture, it can ignite again. Such successive flashes occur until the reagents burnout enough to cause the system to leave the self-ignition region. Under flow conditions, at continuous supply of reagents, the system can switch to the mode of periodic oscillations similar to those observed during the oxidation of methane in a flow reactor [90] (Fig. 8.12). That the sensitivity of the system to all parameters is high under these conditions is explained by the process occurring near the ignition boundary. In this case, the frequency and amplitude of oscillations in addition to chemical factors are strongly influenced by heat transfer and physical conditions of supply of the reagents, i.e., in fact, these are thermokinetic oscillations.

In general, one can speak of a staged ignition of hydrocarbons [167], in which the oxidation rate and ensuing characteristics, such as the temperature, pressure, and concentrations of intermediates, vary monotonically with time. During the induction period and up to the appearance of a hot (over 1200 °C) flame, the ignition process takes place in the middle-temperature region and, accordingly, must be described by the respective kinetics. The main process providing chain branching in this region is the decomposition of peroxide compounds. Because of the difference in the bond dissociation energies of different peroxides, the temperature intervals of their intense decomposition may be slightly different. It is this sequence of poorly separated chain-branching processes, continuously overlapping on each other with increasing temperature, that ensures, according to [167], the staging behaviour of the oxidation of methane, its homologues, and other hydrocarbons. The staging is a reflection of the presence of various groups of reactions in the oxidation mechanism, which contribute individually to the chain-branching process.

For methane oxidation, the occurrence of cool flames is associated with the chain branching caused by the decomposition of methylhydroperoxide. In the oxidation of methane, the region of hydroperoxide decay is too close to the region of methylhydroperoxide decomposition to enable to clearly identify the corresponding region of so-called 'blue' flames, where chain branching is dominated by hydroperoxide decomposition. However, in the oxidation of other hydrocarbons, such as ethane, where the main role in the occurrence of cool flames is played by ethylhydroperoxide, the individual stages are more pronounced. In general, in the oxidation of more complex hydrocarbons, one can expect the appearance of a larger number of stages in the sequential changeover of the main branching agents and the respective changes in the reaction mechanism. The final stage of hot ignition, taking place only after all preparatory stages involving cold and blue flames [167], onsets at a relatively high temperature, being controlled by the main hot-flame chain-branching reaction

$$H^{\bullet} + O_2 \rightarrow OH^{\bullet} + O^{\bullet\bullet}. \quad (8.3)$$

CHAPTER

9

Promotion of the Process

The effective activation of the extremely strong C—H bond in the methane molecule, a process underlying all of its transformations, has been and remains a topical problem, attracting many researchers. It is difficult to point to any of the known methods of activation of chemical bonds that has not been applied to activating the methane molecule. Since methanol is a major primary product of the oxygen conversion of methane, a large number of studies on the chemical or physical activation of methane had a specific goal of developing an effective method for the conversion of methane to methanol. The first attempts to promote the DMTM process were made in the early 1920s, when various methods for obtaining methanol and formaldehyde from natural gas were tested, including (1) the oxidation with the use of various catalysts, (2) oxidation with ozone, (3) iodination followed by hydrolysis, (4) interaction of CO_2 with methane, (5) synthesis of C_2H_4 from methane and its subsequent oxidation, and (6) oxidation of methane in solution or with liquid oxidizers [5]. An overview of these attempts, including photochemical, electrochemical, catalytic, and electromagnetic methods can be found in [32]. Since the issue of the heterogeneous-catalytic activation DMTM was discussed in Chapter 6, this chapter examines only the homogenous chemical promotion of the DMTM, physical methods of its activation, and the occurrence of the process in supercritical (SC) conditions.

HOMOGENEOUS CHEMICAL PROMOTION OF THE PROCESS

Attempts to increase the yield of the target products of the partial oxidation of methane by means of its homogenous promotion, started in the 1920s, are still ongoing. Although a wide variety of compounds have been tested, including HCl, HBr, ozone, SO_3, and different hydrocarbons, the most promising promoters of the homogeneous oxidation of methane, actually used for this purpose in various industrial processes, remain hydrogen peroxide and nitric oxides. In most cases, the effect of promoters consists in decreasing either the temperature of the process or the induction period, but no significant impact on the product yield has been reported. Although, as shown by kinetic modeling, the initiation of the partial oxidation of methane can, in principal, produce a significant effect on the yield of the products, the failure of all known attempts can be explained, in our view, by the need to provide a very high rate of radical generation, comparable to the rate of their formation in the homogeneous branched-chain process. For example, as already

Direct Methane to Methanol
http://dx.doi.org/10.1016/B978-0-444-63253-1.00009-X

noted, an intense promotion of the oxidation of a CH_4:O_2 = 9:1 mixture at $P = 1$ atm and $T = 753$ K is expected to increase the yield of formaldehyde by almost an order of magnitude (Fig. 6.2). However, for this to occur, the promoter, like a heterogeneous catalyst, should provide an increase in the rate of primary formation of radicals by more than five orders of magnitude compared to the rate of the thermal generation of radicals by the reaction itself.

As long ago as 1931, experiments [14] on the influence of antiknock agents and promoters on the oxidation of a 3:1 methane–air mixture at 25 atm and a temperature of 300–400 °C demonstrated that small amounts of HNO_3 significantly accelerate the reaction, whereas Et_4Pb slows it down, with both these additives producing little effect on the product composition.

As promoters of methane oxidation, its heavier homologues can also be used, the effect of adding of which has been discussed above. Their small amounts significantly lower the temperature of the reaction, with little effect on the yield of the main products.

The possibility of initiating the partial oxidation of methane with ozone under flow conditions at $P \approx 1.5$ atm was considered in [169]. Although ozone significantly promotes the reaction and enhances the methanol yield, a selectivity of >42% was achieved only at a methane conversion of less than 1%, which then decreased rapidly with increasing conversion. Promotion of methanol production at high pressures, favourable for this process, requires effective methods for ozone synthesis under these conditions.

One of the most effective promoters of the oxidation of hydrocarbons, with a relatively clear mechanism of chemical action, is hydrogen peroxide. The initiation of the partial oxidation of methane with hydrogen peroxide, the effect of which consists in reducing either the temperature of the process or its induction period, was studied in [170]. Of the molecular compounds investigated in [171], H_2O_2 turned out to be the most effective in reducing the ignition delay time, twice as superior in this respect to methanol.

The authors of [171] explored the possibility of initiating the DMTM by various radicals. The actually achievable concentration of radicals in the mixture was no more than 0.1%, the introduction of which, like the introduction of ~2% different molecular promoters, reduced the ignition delay, but produced almost no effect on the product yield. The nature of introduced radicals had virtually no effect, since they are rapidly converted into each other in gas-phase radical processes.

A most extensive study of the effect of different promoters on the DMTM process was carried out in [140] at a pressure of 10 atm. The promoting effect of about 30 different compounds, including saturated, unsaturated, cyclic, and aromatic hydrocarbons, esters, alcohols, ketones, aldehydes, water, peroxides, sulphur compounds, and amines was studied. Many of the promoters significantly lowered the temperature of the process. The greatest effect was produced by diethyl ether, an additive that, even in an amount of 3.9%, reduced the temperature of the process from 402 to 225 °C. Another consequence of using some promoters was a significant increase in the yield of formaldehyde, apparently due to the corresponding decrease in the methanol yield. In some cases, an increase in the selectivity of formation of methanol and/or formaldehyde as compared to the unpromoted system was achieved; however, this requires the introduction of such a large amount of promoter that it becomes one of the main reagents, a circumstance that make this method of increasing the methanol yield economically impracticable.

Most common promoters of oxidation of hydrocarbons, widely used in industrial processes, are nitrogen oxides and nitric acid. It has long been known that the addition of nitrogen oxides increases the rate of gas-phase oxidation of methane [172,173]. Based on this effect, a large-scale process of methane oxidation to formaldehyde in the presence of 1–2% nitrogen oxides as a homogeneous catalyst was realized in Romania in 1942 [35]. A similar process was developed in the USSR [29,174].

The mechanism of influence of NO_2 on the oxidation and spontaneous combustion of hydrocarbons, primarily at low pressures, was discussed in detail in [13]. For the slow oxidation of methane, as in the case of other alkanes, addition of NO_2 was demonstrated to shorten or even eliminate (starting from a certain amount) the induction period, causing no changes in the qualitative and quantitative composition of the oxidation products. For the oxidation of a 15% CH_4–85% air mixture at $T = 480–510\,°C$ and $P = 300$ Torr in the presence of a small (1.37%) NO_2 additive, the heat-release curve featured two peaks [175], the first of which, according to the authors, is associated with the formation of formaldehyde, whereas the second, with its decomposition. This explanation is difficult to accept, because in the absence of NO_2, the formation and decomposition of formaldehyde also occur, but no double peak is observed. A double exothermic peak in the oxidation of methane in the presence of NO_2 was observed in [176] and for the oxidation of propane in [177].

As found in [178,179], the slow reaction of NO_2 with methane proceeds with $E_{eff} = 30$ kcal/mol and has a free-radical character, with its energy characteristics being entirely determined by the stage of formation of alkyl radicals by the reaction

$$RH + NO_2 \rightarrow R^{\bullet} + HNO_2 \qquad (9.1)$$

Thus, the activation energy of the primary formation of alkyl radicals in the presence of NO_2 becomes not only smaller than the activation energy of the uninitiated reaction

$$RH + O_2 \rightarrow R^{\bullet} + HO_2 \qquad (9.2)$$

with $E_{eff} \approx 50$ kcal/mol, but also lower than the activation energy of the degenerate branching involving RCHO and O_2 (endothermic by about 32 kcal/mol), which leads to a decrease in the role of degenerate chain branching. However, reaction (9.1) alone does not account for the appearance of two peaks (the second peak can be reasonably attributed the hydrocarbon oxidation reaction itself), since there seem no candidates for the role of the initial exothermic reaction.

To explain the observation of two peaks in [176], it was assumed that addition of nitrogen dioxide to a methane–oxygen mixture rapidly results in the formation CH_3NO_2 (nitromethane), which initially accumulates NO_2, and then slowly decomposes to release NO_2, a species that initiates the formation of radicals. The first maximum reflects the occurrence of the fast reaction of NO_2 with CH_4, whereas the second signifies the oxidation of methane catalyzed by CH_3NO_2. The argument in favour of this assumption was that, for mixtures of methane with NO_2, only the first peak was observed, while the second peak emerged only upon oxygen addition; according to analysis of the products, half of the NO_2 passed into nitromethane. When nitromethane was added to a methane–oxygen mixture, only the second heat-release peak was present, strictly coinciding with a sharp decline in the CH_3NO_2 concentration and with a nearly completely stop the oxidation process, with the kinetics and the product composition being identical to those found when the same amount of NO_2 was added. The effective activation energy E_{eff} was demonstrated to decrease from 46 to 25 kcal/mol when NO_2 or CH_3NO_2 was added.

It was established [13,176] that the interaction of alkyl radicals formed by reaction (9.1) with nitrogen dioxide proceeds through two channels with an almost equal probability (at 465 °C, the ratio of their contributions was estimated as 1.1, with the difference in the activation energies being of ~1 kcal/mol):

$$R^{\cdot} + NO_2 \rightarrow RNO_2 \quad (9.3)$$

$$R^{\cdot} + NO_2 \rightarrow RONO \quad (9.4)$$

Since half of the NO_2 is converted into CH_3NO_2, it means that the second half turns into CH_3ONO, and each nitrous acid molecule primarily formed in reaction (9.1) decomposes via one of the following pathways:

$$HNO_2 \rightarrow OH^{\cdot} + NO \ (Q = -50\ \text{kcal/mol}) \quad (9.5)$$

$$HNO_2 + O_2 \rightarrow HO_2^{\cdot} + NO_2 \ (Q = -33.5\ \text{kcal/mol}) \quad (9.6)$$

(the direct dissociation to $H^{\cdot}$ and NO_2 is endothermic by $Q = -80$ kcal/mol, so it is hardly possible). Although this mechanism formally explains the presence of two heat release peaks, there remains a serious contradiction between the measured (24–25 kcal/mol) and theoretically predicted (35–61.5 kcal/mol) activation energy of the process [13].

The issue of promotion of the partial oxidation of methane to oxygenates by the nitrogen oxides is of interest not only from a practical, but also from a theoretical point of view, since in addition to the promoting effect they can catalyze the formation of oxygen-containing compounds. In recent years, this subject was considered in a large group of works. In [60], it was shown that the oxidation of a $CH_4{:}O_2 = 9{:}1$ mixture at $P = 1$ atm and $T = 560$ °C in the presence of 2% NO leads to a tenfold increase in the formaldehyde yield, from 0.35% to 3.96%. The authors suggested that, in this case, not only a promoting effect of NO, but also a homogeneous catalysis, take place, since the products contained N_2O and only trace amounts of N_2. This study once again confirmed that, as in heterogeneous catalysis, the effect achieved by promoting methane oxidation to formaldehyde can be achieved by optimizing the conditions of the uninitiated gas-phase process [57] (Figs. 2.5 and 2.6), as shown in Fig. 9.1.

The idea that motivated the study [52] on the influence of NO_2 on methane oxidation is the possibility of the homogeneous inhibition of the interaction of radicals with formaldehyde formed by NO_2 molecules. It was assumed that NO_2 molecules, scavenging the radicals, prevent them from attacking formaldehyde, thereby increasing its yield, and that the respective compounds formed will easily decompose to return NO_2, so that a catalytic process takes place. The source of NO_2 was nitric acid HNO_3 (~0.26%). Indeed, addition of HNO_3 not only significantly increased the methane conversion, but also enhanced the overall selectivity and yield of organic products (from 0.7 to more than 2%), but mainly due to CH_3NO_2. At the same methane conversion, the selectivity of formation of CH_3OH and HCHO in the presence of HNO_3 changed slightly, but the total yield of oxygenates increased markedly due to the formation of CH_3NO_2. With increasing methane conversion in the presence of HNO_3, the selectivities of formation of HCHO and CH_3NO_2 increased, whereas the selectivity of CH_3OH

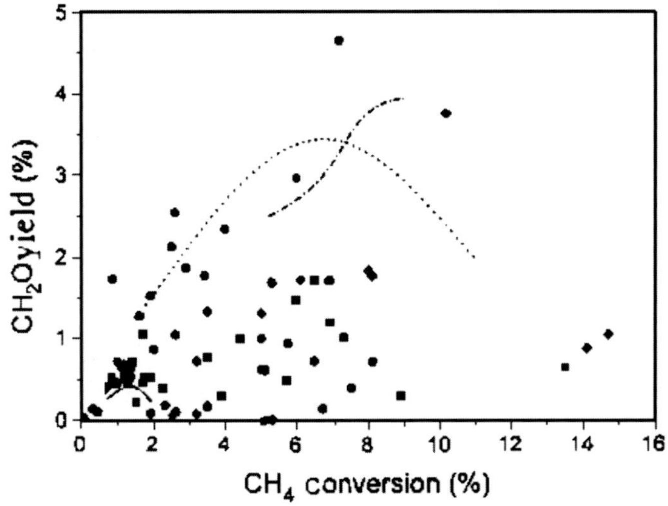

FIGURE 9.1 The production of formaldehyde in the empty quartz reactor: $P = (—)$ 1 and (·····) 5 atm [57]; (-.-.) in the presence of NO (empty reactor), and in the presence of different catalysts (symbols) [60].

formation decreased. The authors noted that the effect of NO_2 in these experiments could not be interpreted as initiation, the more so that, in all experiments with the same oxygen concentration (2.6%), the same reaction onset temperature was recorded (~410 °C). No molecular nitrogen was detected; according to estimates most of the NO_2 (40–80%) was converted into CH_3NO_2. In the authors' opinion, this suggests that CH_3NO_2 is mainly formed via the interaction of NO_2 with methyl radicals arising during the oxidation. The possibility of developing a method for the partial oxidation of methane on this basis with allowance for the conversion of CH_3NO_2 into other oxygenates and regeneration of NO_2 in the process itself or in a special cycle was also discussed.

Based on the results of kinetic modeling, the authors of [180] concluded that, in the low-temperature oxidation of methane, the promoting effect of NO is largely associated with the reduction in the contribution of the chain-termination channel involving $CH_3OO^\bullet$ radicals by converting them via the quick reaction into more active $CH_3O^\bullet$ radicals:

$$CH_3OO^\bullet + NO \rightarrow CH_3O^\bullet + NO_2 \tag{9.7}$$

In addition to accelerating the transfer of inactive $CH_3OO^\bullet$ radicals into more active $CH_3O^\bullet$ radicals, which directly lead to the formation of oxygenates, bypassing a more complex (at low temperatures and pressures) pathway through the formation and decomposition of methyl hydroperoxide CH_3OOH (Table 5.1), reaction (9.7) accelerates the conversion of NO into NO_2, thereby promoting the formation of methyl radicals $CH_3^\bullet$ in reaction (9.1). The same type of reactions of peroxide radicals is responsible for the promotion and the oxidation of NO in the presence of hydrocarbons. The latter, however, is due to [181], mainly effected by the reaction

$$HOO^\bullet + NO \rightarrow OH^\bullet + NO_2 \tag{9.8}$$

with the contribution of ROO$^•$ radical being negligibly small. Note also that considering the influence of NO on the oxidation of methane to oxygenates, it is necessary to take into account that NO simultaneously promotes the subsequent oxidation of methanol and ethanol formed in this process [182].

Thus, the effect of nitrogen oxides on the oxidation of methane is multifactorial. In addition to the promotion of the slow stage of primary formation of radicals in this system, they are capable of promoting the formation of oxygenates directly via reaction (9.7). What is more, a decrease in the rate of formation of methyl hydroperoxide reduces the rate of formation (through its decomposition) of the hydroxyl radical, the main agent destroying the oxygenates formed.

The mechanism of the promoting action of nitrogen oxides on the gas-phase oxidation of light alkanes, such as CH_4, C_2H_6, C_3H_8, and iso-C_4H_{10}, at atmospheric pressure, was thoroughly studied in [183,184]. The experiments were carried out in a quartz reactor with an inner diameter of 8 mm at a reagent flow rate of 120 ml/min ($t_r = 6$ s) and partial pressures of alkanes of 5–20 kPa, O_2 of 2.5–15 kPa, and NO of 0.5–4 kPa, diluted to a total pressure 101 kPa with helium. First of all, the promotion sharply (by ~100 °C) reduced the temperature of oxidation of C_2H_6 and C_3H_8, and even more significantly the temperature of CH_4 oxidation (Fig. 9.2).

That this is a promotion is evidenced from the absence of conversion in oxygen-free mixtures of alkanes with NO and by a sharp reduction in the effective activation energy upon NO addition, calculated from the temperature dependence of the conversion of alkanes displayed in Fig. 9.2 (Table 9.1). Note that, for the oxidation of methane, it becomes even lower than that for the oxidation of ethane and propane without NO additives. In addition, under conditions of these experiments ($P_{alk} = 20$, $P_{O2} = 10$ kPa, $P_{NO} = 2$ kPa, $P_{tot} = 101$ kPa, and $t_r = 6$ s), the alkane conversion per NO molecule was 5.0 for CH_4 (at 650 °C), 4.1 for C_2H_6 (at 550 °C), and 4.2 for C_3H_8 (at 500 °C) [184].

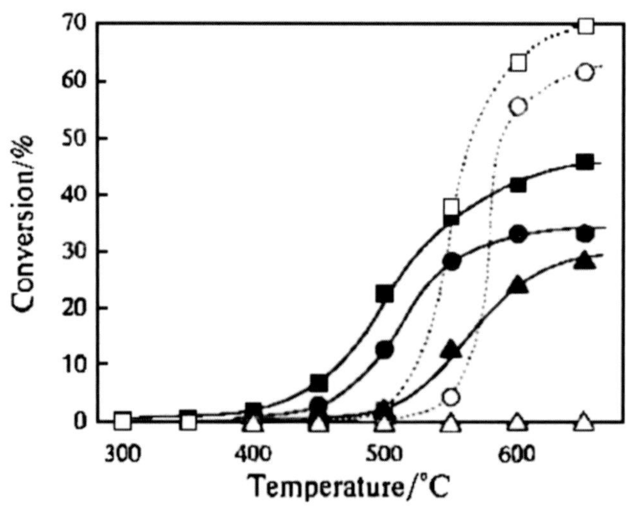

FIGURE 9.2 Effect of the addition of NO on the conversion of light alkanes: (▲, Δ) CH_4; (●, ○) C_2H_6; and (■, □) C_3H_8. Filled and unfilled symbols refer, respectively, to experiments with and without NO additives [183].

TABLE 9.1 Activation Energy of the Gas-Phase Oxidative Conversion of Light Alkanes in the Presence of NO Additives [183]

	Activation Energy, kJ/mol		
	C_3H_8	C_2H_6	CH_4
Without NO	275	361	–
With NO	107	138	208

However, the temperature of maximum conversion of the alkanes and oxygen in the presence and absence of NO additives remains almost the same (Fig. 9.3).

Under these conditions, the introduction of nitric oxide not only lowers the reaction onset temperature, but also dramatically alters the composition of the resulting products. While in the absence of NO, the main products of oxidation of C_2–C_3 alkanes are alkenes (C_2H_4 and C_3H_6), with the total amount of oxygenates being less than 9%, in the presence of NO, oxygenates, mainly aldehydes, become the main (more than 60% at 450 °C) products. These changes are particularly noticeable at low (350–500 °C) temperatures. At higher temperatures, the yield of oxygenates decreases, whereas the yield of alkenes and CO increases again.

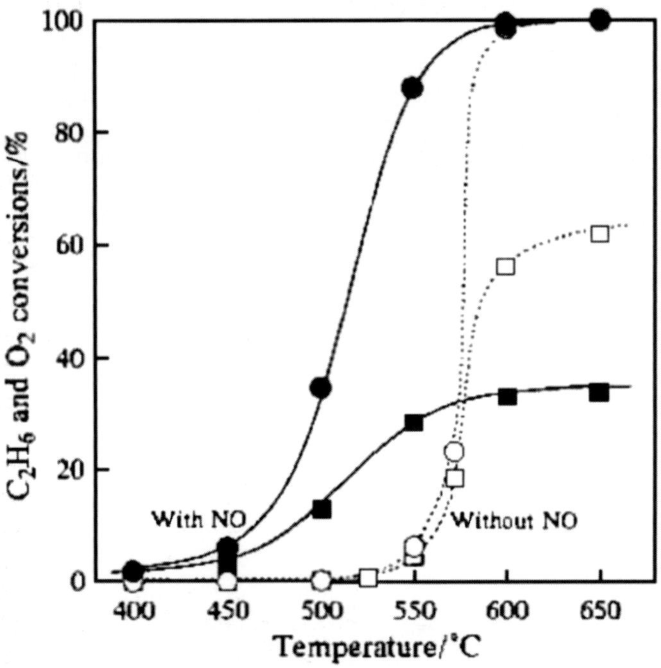

FIGURE 9.3 Effect of NO addition on the conversion of (■, □) C_2H_6 and (●, ○) O_2 ($P_{C2H6} = 20$ kPa, $P_{O2} = 10$ kPa, $P_{NO} = 2$ kPa, $P_{tot} = 101$ kPa, and $t_r = 6$). Filled and unfilled symbols refer, respectively, to experiments with and without NO additives [184].

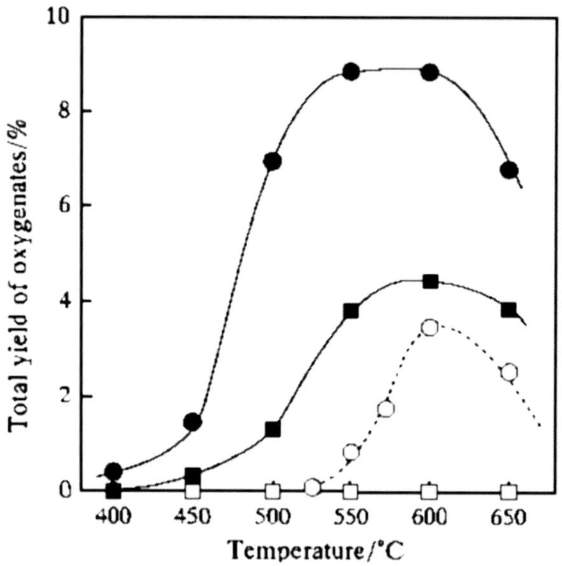

FIGURE 9.4 Effect of NO addition on the yield of the sum of oxygenates for the oxidation of (■, □) CH_4 and (●, ○) C_2H_6 ($P_{alk}=20$ kPa, $P_{O2}=10$ kPa, $P_{NO}=2$ kPa, $P_{tot}=101$ kPa, and $t_r=6$ s). Filled and unfilled symbols refer, respectively, to experiments with and without NO additives [184].

Figure 9.4 shows the temperature dependence of the yield of the sum of oxygenates for the oxidation of methane and ethane, whereas Fig. 9.5 displays the temperature dependence of the composition of ethane oxidation products.

The dependence of the yield of the products of ethane oxidation on the concentration of added NO is given in Table 9.2. With increasing NO concentration, the total yield of oxygenates increases whereas that of C_2-oxygenates decreases due to redistribution in favour of C_1-oxygenates, the result that led the authors to make an assumption on the specific acceleration

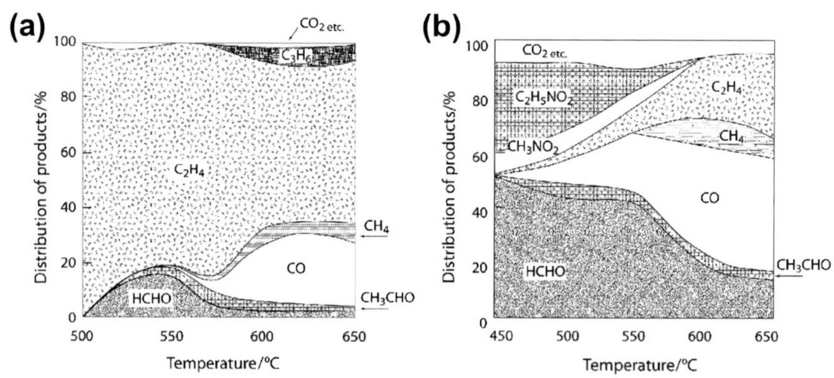

FIGURE 9.5 Effect of temperature on the distribution of products for C_2H_6 oxidation in the (a) absence and (b) presence of NO ($P_{C2H6}=20$, $P_{O2}=10$, $P_{NO}=2$, $P_{tot}=101$ kPa, and $t_r=6$ s) [184].

TABLE 9.2 Effect of NO Additives on the Yields of C_2H_4 and Oxygenates for the Oxidation of Ethane at 600 °C [184]

		Yield, %			
P_{NO}, kPa	C_2H_6 Conversion, %	C_2H_4	C_1-Oxygenates	C_2-Oxygenates	Sum of Oxygenates
0	55.8	33.7	1.5	2.0	3.5
0.5	38.8	13.0	7.3	1.3	8.5
1.0	36.8	9.8	8.0	1.2	9.2
2.0	34.1	6.2	7.9	1.2	9.1
4.0	33.8	7.1	7.9	1.0	8.9

of C—C bond cleavage. The yield of ethylene also decreases drastically, as evidenced by the decrease in the conversion of ethane. A significant reduction in the conversion of ethane in the presence of NO can be clearly seen in Fig. 9.3.

The total yield of oxygenates under optimal conditions was 6.7% for natural gas, 11.3% for ethane, 12.9% for propane, and 28.7% for isobutane [183]. In these conditions, only trace amounts of nitroalkanes and N_2 were detected. Thus, most of the introduced NO remained in the gas phase, which suggests the possibility of its subsequent separation and recycling as a catalyst for the oxidation of light alkanes. The fact that no reaction between NO and alkanes occurs in the absence of O_2 indicates that the presence of both the reagents is a necessary prerequisite, with NO_2 apparently acting as an initiator of the oxidation. All the more so that, in these experiments at low temperatures, the equilibrium

$$2NO + O_2 \Leftrightarrow 2NO_2 \quad (9.9)$$

is shifted to the right, with the $[NO_2]/[NO]$ equilibrium ratio being equal to 10 at 400 °C. Although this ratio decreases with temperature, becoming as low as 1.0 at 600 °C, let alone that this equilibrium is hardly achievable under these conditions, a specially performed experimental comparison revealed that the oxidative conversion of alkanes occurs two-to fivefold faster in the presence of NO_2 than in the presence of NO and O_2. This is a strong argument in favour of namely NO_2 being the promoter.

There is also the issue of the possible role of nitroalkanes as intermediate products, especially that small amounts of them were found in the oxidation at low temperatures. Specially designed experiments with C_1—C_3 nitroalkanes RNO_2 showed that their decomposition in the system without O_2 gives low concentration of oxygenates, primarily aldehydes, with the product composition dominated by carbon oxides and alkenes (for C_2—C_3 nitroalkanes). Addition of O_2 increases the yield of carbon oxides and simultaneously decreases the yield of oxygenates. In addition, for the C_2—C_3 nitroalkanes, the selectivity of formation of alkenes is always higher than that of oxygenates.

By contrast, similar experiments with alkyl nitrites RONO showed a high selectivity of formation of aldehydes (CH_2O and CH_3CHO), with alkenes being present in trace amounts. The monitoring of the kinetics of formation of the products by means of infrared spectroscopy showed that the products of the decomposition of methyl nitrite are formaldehyde and NO

$$CH_3ONO \rightarrow HCHO + H^{\cdot} + NO \quad (9.10)$$

The increase of the CO yield with rising temperature is apparently associated with the subsequent decomposition of formaldehyde. A similar decomposition of propyl nitrites also leads to the formation of aldehydes

$$n\text{-}C_3H_7ONO \rightarrow HCHO + C_2H_5^{\cdot} + NO \quad (9.11)$$

$$iso\text{-}C_3H_7ONO \rightarrow CH_3CHO + CH_3^{\cdot} + NO \quad (9.12)$$

The decomposition of *tert*-C_4H_9ONO predominantly yields acetone:

$$tert\text{-}C_4H_9ONO \rightarrow CH_3COCH_3 + CH_3^{\cdot} + NO \quad (9.13)$$

In the presence of oxygen, NO may be further oxidized to NO_2, resulting in the formation of C_1-C_2 alkyl nitrites and alkyl nitrates when the latter interacts with the alkyl radicals formed by reactions (9.11)–(9.13).

However, the data of [183,184] are in poor agreement with the results of another series of studies, [158,185–188], performed under very similar conditions. In this series, most of the experiments were carried out with a mixture of 55.6% CH_4, 27.7% O_2, 0.5% NOx, and 16.2% He in a quartz reactor with an inner diameter of 7 mm and a length of 200 mm at a pressure $P = 1$ atm, a temperature near 800 K, and a flow rate of 120 ml/min. The yield of the products was determined at the same methane conversion, 10%. Under these conditions, in the absence of oxygen in the mixture, the interaction of CH_4 with NO_2 produced largely formaldehyde, whereas the interaction of NO with CH_4 gave no oxidation products at all. Introduction of NO_2 into the CH_4-O_2 system, besides lowering the reaction onset temperature by more than 200 °C, resulted in a high selectivity of formaldehyde formation, 24.5%, and an even higher selectivity of methanol formation, 27.3%, although in all previous works, the methanol yield was very low if any. In addition to methanol and formaldehyde, a high yield of nitromethane CH_3NO_2 was observed, the formation of which was not mentioned previously except for [189]. Note that the final concentration of nitromethane was quantitatively consistent with the initial concentration of added nitrogen oxides (Fig. 9.6, Table 9.3). The highest selectivities of formation of methanol and formaldehyde were achieved at an initial NO_2 concentration of 0.5%, which proved to be optimal, with the selectivity of methanol formation being usually higher than formaldehyde selectivity (Fig. 9.6).

The works [158,185–188] investigated the effect of various parameters on the reaction onset temperature and the yield of the products. At pressures of 1 and 5 atm, the increase of the initial concentration of NO to 1% significantly, by ~100 °C, decreased the temperature of the process, although the maximum obtainable conversion of methane was almost the same. However, the increase of the NO concentrations to 1.5% produced practically no effect. At these pressures, the selectivity of formation of oxygenates (methanol, formaldehyde, and nitromethane) reached a maximum at an NO concentration of ~0.5%, so this concentration was adopted as optimal. The ethane formation selectivity decreased rapidly with increasing NO concentration, nearly to zero. Diluting the mixture with helium, at least up to 60%, had a moderate effect on the process. Even the CH_4/O_2 ratio, a key parameter of the

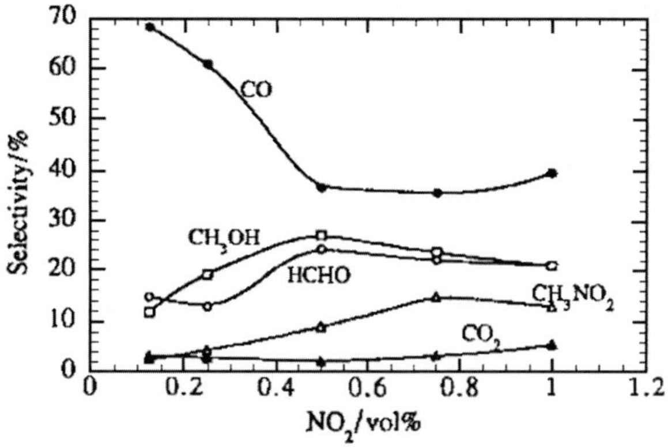

FIGURE 9.6 Selectivity of product formation as a function of NO_2 concentration at 10% CH_4 conversion (55.6% CH_4, 27.7% O_2, NO_2 and He; mixture flow rate, 120 ml/min; $P = 1$ atm): (●) CO, (□) CH_3OH, (○) HCHO, (Δ) CH_3NO_2, and (▲) CO_2 [185].

TABLE 9.3 Products of the Gas-Phase Reactions in the CH_4–O_2–NOx System under the Same Conditions as for Fig. 9.5 [186]

				Selectivity, %						
				Oxygenates				COx		
Reaction	**CH_4 Conversion, %**	**NOx, %**	***T*, K**	**Sum**	**HCHO**	**CH_3OH**	**CH_3NO_2**	**CO**	**CO_2**	**CH_3NO_2/NOx**
CH_4–NO	–	1.5	–	–	–	–	–	–	–	–
CH_4–NO_2	1.5	0.5	763	56.6	45.3	8.9	2.4	33.3	10.1	0.04
CH_4–O_2–NO	10.0	0.5	802	52.2	24.1	22.1	9.0	42.6	2.2	1.00
CH_4–O_2–NO_2	10.0	0.5	808	60.9	24.5	27.3	9.1	36.9	2.2	1.00
CH_4–O_2–NO–NO_2	10.0	0.5	806	56.6	23.8	23.8	9.0	40.5	2.9	1.00

system, although produced a maximum in the yields of methanol and formaldehyde at $CH_4/O_2 = 2$, influenced them little while being increased up to 10. Only at $CH_4/O_2 < 2$, a decrease in the methanol yield and an increase in that of formaldehyde were observed. The yield of oxygenates was also not affected either by changing severalfold the space flow rate (from 1000 to 4000 per hour) or by adding 16% water. The effect of NO_2 was generally similar to that of NO.

Thus, the authors of [158,185,186] reported consistently high values of the yields of methanol and nitromethane over a wide range of conditions. The authors' explanations why their data on the yields of methanol and nitromethane differ significantly from the results obtained in [183,184] under nearly the same conditions are not especially convincing. Since the above works still differ from each other in a number of parameters, it is difficult to establish the real dependence of the product yield on the conditions without performing a careful kinetic modeling.

As a first step in such modeling, the authors of [190] analyzed a simplified scheme of the interaction of CH_4 with NOx (in the absence of O_2); in particular, *ab initio* calculations of the activation energies and rate constants for the interaction of nitrogen oxides with methane and for other key stages were performed. According to these calculations, the activation energy of the reaction

$$CH_4 + NO \rightarrow CH_3^{\bullet} + HNO \quad (9.14)$$

is 65.6 kcal/mol, which is higher than the activation energy of 59.0 kcal/mol calculated in the same approximation for the reaction

$$CH_4 + O_2 \rightarrow CH_3^{\bullet} + HO_2^{\bullet} \quad (9.15)$$

This means that NO cannot act as an initiator of the process. In contrast, the activation energy for the reaction

$$CH_4 + NO_2 \rightarrow CH_3^{\bullet} + HNO_2 \quad (9.16)$$

turned out to be 37.6 kcal/mol, which makes it the most likely initiating step of the process. The authors assumed that, in the presence of oxygen, this reaction is accompanied by the quick reaction

$$HNO_2 + O_2 \rightarrow HO_2 + NO_2 \quad (9.17)$$

which returns NO_2.

The formation $CH_3O^{\bullet}$ in the decomposition of alkyl nitrite,

$$CH_3ONO \rightarrow CH_3O^{\bullet} + NO \quad (9.18)$$

and the known reactions of the $CH_3O^{\bullet}$ radical,

$$\begin{aligned} CH_3O^{\bullet} + CH_4 &\rightarrow CH_3^{\bullet} + CH_3OH \\ CH_3O^{\bullet} &\rightarrow H^{\bullet} + CH_2O \\ CH_3O^{\bullet} + NO_2 &\rightarrow HNO_2 + CH_2O \\ CH_3O^{\bullet} + NO &\rightarrow HNO + CH_2O, \end{aligned} \quad (9.19)$$

(for which the activation barriers were calculated) constitute a pathway for the formation of C_1-oxygenates, for which the energy barriers of all stages are below 40 kcal/mol, with reaction (9.16) being the limiting step. According to this mechanism, an increase in the NO concentration increases the yield of CH_2O and, correspondingly decreases the CH_3OH yield, in agreement with the experimental results for the CH_4–O_2–NO system.

In a subsequent work [191], the same group of researchers performed theoretical calculations of the potential barriers of the other reactions important for this system and used the obtained kinetic parameters to describe the experimentally observed selectivities of the products. The focus was on the fact that the introduction of NOx into the system, along facilitating the initiation of radicals, promotes the formation of alkoxy radicals, precursors of oxygenates.

Based on the totality of the results of the aforementioned works, it is possible to suggest the following main stages of the mechanism of the promoting and catalytic effects of nitrogen oxides on the oxidation of light alkanes:

$$\begin{aligned}
2NO + O_2 &\Leftrightarrow 2NO_2 \\
RH + NO_2 &\rightarrow R^{\bullet} + HNO_2 \\
HNO_2 + O_2 &\rightarrow HO_2^{\bullet} + NO_2 \\
R^{\bullet} + NO_2 &\Leftrightarrow RNO_2 \\
R^{\bullet} + NO_2 &\rightarrow RONO \\
R^{\bullet} + O_2 &\Leftrightarrow RO_2^{\bullet} \\
RONO &\rightarrow RO^{\bullet} + NO \\
RO_2^{\bullet} + NO &\rightarrow RO^{\bullet} + NO_2 \\
HO_2^{\bullet} + NO &\rightarrow OH^{\bullet} + NO_2 \\
OH^{\bullet} + RH &\rightarrow R^{\bullet} + H_2O \\
RO^{\bullet} + RH &\rightarrow ROH + R^{\bullet} \\
RO^{\bullet} + O_2 &\rightarrow R_1CHO + HO_2^{\bullet} \\
RO^{\bullet} + NO_2 &\rightarrow R_1CHO + HNO_2 \\
RO^{\bullet} + NO &\rightarrow R_1CHO + HNO \\
HNO + O_2 &\rightarrow HO_2^{\bullet} + NO
\end{aligned} \tag{9.20}$$

In this mechanism, NO and NO_2 act not only as promoters but also as homogeneous catalysts, since at temperatures above 600 °C, nitroalkanes formed decompose completely, releasing NO_2. According to the same mechanism, hydrocarbons themselves (and CO produced by formaldehyde oxidation) effectively promote the oxidation of NO to NO_2, even at relatively low temperatures, ~300 °C. Because of the rapid interconversion of nitrogen oxides in this system, the promoting effect is observed upon addition of any of them, though, of course, there are notable quantitative differences.

A kinetic analysis of the gas-phase partial oxidation of methane in the CH_4–O_2–NO system at atmospheric pressure within the framework of a reasonably complete kinetic scheme, composed of more than 250 elementary stages, was conducted in [187] in order to optimize the experimental conditions for this system. In all cases, the calculated values did not exceed the experimentally obtained. In our view, the main discrepancy between the experimental data and calculation results is that the kinetic calculations predict a steady decline in the selectivity of methanol formation with increasing methane conversion, while the experimental results show a linear increase up to a methane conversion of 10%. This contradiction points to the urgent need for a more careful assessment of the role of heterogeneous reactions in these experiments, carried out at low pressures. A more general kinetic model for

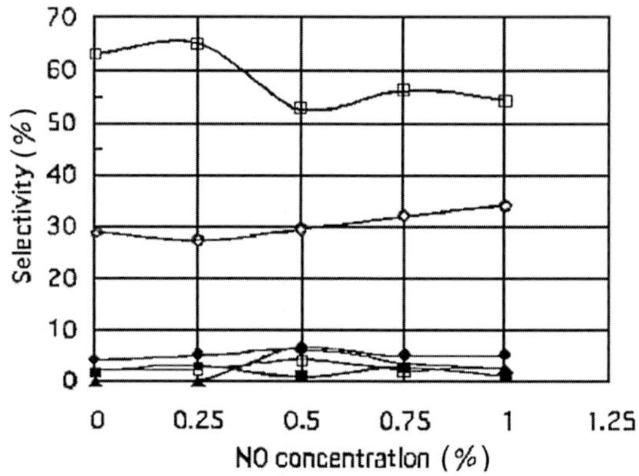

FIGURE 9.7 Selectivities of products as a function of NO concentration at the 4% conversion of CH_4. $P = 10$ atm: (○) CH_3OH, (Δ) HCHO, (□) CO, (●) CO_2, (▲) CH_3NO_2, and (■) C_2H_6 [158].

describing the low-temperature oxidation of hydrocarbons in the presence of NO was developed and analyzed in [192].

It should be emphasized that the positive effect of promotion with NOx manifests itself only at low pressure, becoming less pronounced at higher pressures [158]. Even at a pressure of 10 atm, the reduction in the reaction onset temperature (at 4% methane conversion) does not exceed 25 °C. The selectivity of formation of oxygenates is almost independent of the NOx concentration, with the selectivity of methanol formation in the absence of NOx constituting the same 30% (Fig. 9.7).

Therefore, above 10 atm, a pressure at which a maximum in the selectivity of methanol formation is observed (Fig. 9.8), the promotion becomes virtually meaningless. This is consistent with the views on the role of initiation in this reaction, which becomes insignificant upon onset of the branched-chain mode at relatively high pressure, although specific catalytic

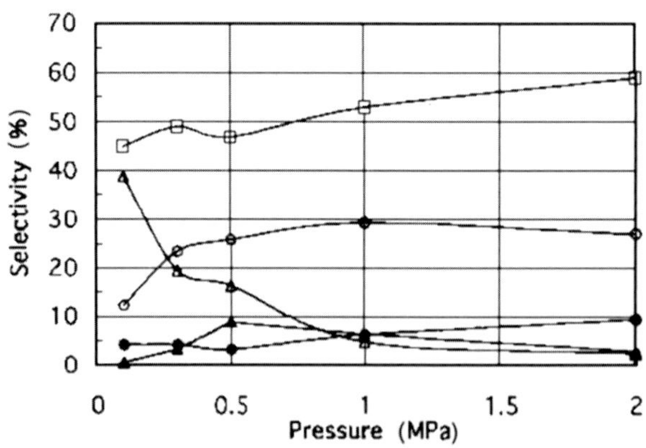

FIGURE 9.8 Selectivities of products as a function of reaction pressure at 4% conversion of CH_4 and an NO concentration of 0.5%: (○) CH_3OH, (Δ) HCHO, (□) CO, (●) CO_2, and (▲) CH_3NO_2 [158].

effects of NOx on the formation of oxygenates are likely to last even at high pressures, according to the reactions presented in scheme (9–20), but this issue requires further studies.

The promoting effect decreases rapidly with increasing NO concentration. Replacement of NO with NO_2 did not cause significant changes, in agreement with the findings based on an analysis of Scheme (9–20). It was noted that the methanol formation selectivity increases nearly linearly with the reaction time. Although the reaction time in these experiments was relatively short, only a few seconds, the reactor surface probably plays a noticeable role at the relatively low pressures used in these experiments, as admitted by the authors themselves. The same authors showed the possibility of a small increase in the methanol yield in the CH_4–O_2–NO system ($T = 550\,°C$, $P = 0.4\,MPa$) due to an almost complete conversion of formaldehyde in the presence of a Cu–ZnO/SiO_2 catalyst ($T = 250\,°C$) [68].

The possibility of the co-activation of the system with NO and a heterogeneous catalyst was investigated in [193]. Typically, the catalyst not only ensures the generation of radicals and catalyzes the formation of products, but also activates the reactions of their subsequent decomposition. To separate these functions, the authors used NO as an initiator of radical generation, an approach that made it possible to conduct the subsequent process over a V_2O_5/SiO_2 catalyst with a low specific surface area ($\sim 1\,m^2/g$). Due to a low rate of decomposition of partial oxidation products on this catalyst, a substantially higher (7%) yield of C_1-oxygenates (methanol and formaldehyde) was obtained as compared to that on catalysts with developed surface. A subsequent study by this group [154] reported an even higher total yield oxygenates, up to 16%, which is, apparently, the highest of the published values.

Another example of the combined effect of nitrogen oxides and a heterogeneous catalyst in the oxidation of methane at atmospheric pressure is given in [188]. In this paper, the authors demonstrated that, by using MgO and CaO, alkaline-earth metal oxides with a high adsorption capacity with respect to $OH^{\bullet}$ radicals, as catalysts, it is possible to increase the yield of C_1-oxygenates in the NOx-promoted process by $\sim 10\%$, evidently due to a decrease in the rate of consumption of formed oxygenates in reactions with radicals.

Although most works on the promotion of methane oxidation by nitrogen oxides at atmospheric pressure report relatively modest values of the total yield of oxygenates (4–7%), the authors of [194] claimed achieving an 11% yield for the oxidation of methane, 16% for the oxidation of methane with traces (0.7%) of ethane, and 24% for the oxidation of ethane. However, in contrast to the results of [154,193], where the main product is methanol, in this case, formaldehyde is predominantly formed. Note, however, that, in these experiments, the concentration of NO is only slightly lower than the concentration of the hydrocarbon and is almost twice the O_2 concentration, which, in our opinion, completely converts NO to NO_2 and prevents oxidation of products formed by oxygen. In addition, a high temperature (600 °C) prevents the formation of CH_3NO_2. Until these promising, but rather contradictory results receive independent confirmation, they can hardly be viewed as reliable.

One of a few experimental studies on the promoting effect of NOx oxidation of hydrocarbons at pressures much higher than atmospheric was performed in [195]. The oxidation of very rich (with a fuel equivalence ratio of $116 > \Phi > 99$), stoichiometric ($\Phi \approx 1$), and very lean mixtures ($\Phi \approx 0.04$) highly diluted with N_2 was studied at pressures of 20, 50, and 100 atm and moderate temperatures, 600–900 K. In addition, kinetic simulations were performed. The main conclusions of this work, consistent with the results of the aforementioned studies, claim, in particular, that the promoting effect of NOx arises due to a cyclic

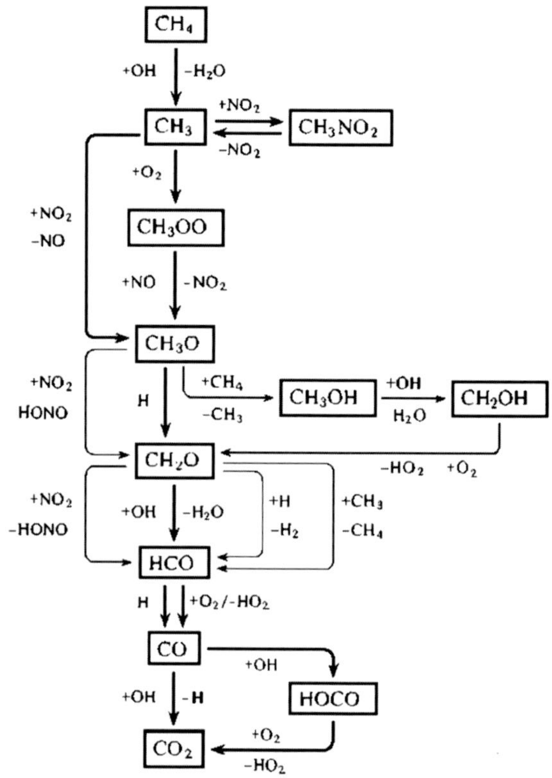

FIGURE 9.9 Main pathways of the conversion of intermediate products in the oxidation of CH_4 in the presence of NOx [195].

interconversion of NO and NO_2 during their interaction with the radicals involved in the process. Although the respective mechanism includes a sufficiently large number of different reactions, the most important of them are the transformations of $CH_3OO^•$ and $CH_3^•$ radicals to $CH_3O^•$ radicals via the interaction with NO and NO_2, respectively, as illustrated in Fig. 9.9.

An important role is played by the abstraction of H atoms from hydrocarbon molecules and radicals, especially from methane (reaction (9–16)). The side reaction of CH_3NO_2 formation leads to a depletion of NOx and may even cause a temporary reduction in the hydrocarbon conversion rate.

Although no significant reactions leading to methanol occur between nitrogen- and carbon-containing compounds, its production in the presence of NOx is facilitated by a faster formation of $CH_3O^•$ radicals. However, experiments [195] demonstrated that, at a high pressure of 100 atm, the difference between the highest concentrations of CH_3OH in the absence and presence of NOx is negligibly small; i.e., high pressure provides a sufficiently effective mechanism for the formation of methanol even in the absence of NOx.

Of the studies on the promotion of methane homologues with nitrogen oxides, along with the above-mentioned works [183,184], we would like to single out [196], where the oxidation of methane–propane mixtures to formaldehyde was investigated. It was found that, for the oxidation of methane, propane, and mixtures thereof with oxygen in the presence of nitrogen oxides under flow conditions at atmospheric pressure and a residence time of 0.1–0.2 s, the maximum yield of formaldehyde was reached at 640–700 °C. The oxidation was carried out at a constant hydrocarbon-to-air ratio of 1:2, i.e., at 13.8% O_2 in the mixture. The bulk of the experiments was carried out with an 80% CH_4–20% C_3H_8 mixture. Over the entire temperature range, the yield of formaldehyde from the methane–propane mixture (1.9 vol % at 640 °C) was higher than that for methane oxidation (0.9%), but lower than that for the oxidation of propane (2.6%). The maximum yield of CH_2O was reached at a residence time of 0.10–0.15 s, and then virtually stopped growing. The content of methanol in the liquid products of oxidation of the methane–propane mixture was higher than in the methane oxidation products, but did not exceed 30–35% of the amount of formaldehyde formed. The products contained trace amounts of acetaldehyde. The formaldehyde yield rose rapidly with increasing propane concentration in the methane–propane mixture up to about 20%, after which the effect of the propane concentration began to decline. Similarly, the relative rate of oxidation of the mixture increases rapidly with the propane concentration up to 20–25%, after which the growth of the rate slows down.

PHYSICAL METHODS FOR INITIATING THE PROCESS

First attempts to use physical methods to control the rate and selectivity of the DMTM process date back to the 1920–1930s. Even then, the effects of an electrical discharge, radiation, and ultraviolet radiation on the process have been examined. However, no significant progress in any of these areas has been made. The reasons why the DMTM process is poorly promoted relate, as discussed above, to the two main features of its mechanism: a high rate of the intrinsic generation of radicals by this branched-chain reaction and a very short length of its chains. Nevertheless, an analysis of the research conducted in this area is useful, because, under certain conditions, physical methods, as well as short-time catalysis may prove a useful tool for controlling the complex branched-chain DMTM reaction.

Physical methods of initiation are highly diverse and comprise such influences as mechanical impact (ultrasound activation, shock waves), thermal, radiation, including photochemical, and electromagnetic effects, different types of electrical discharges and plasma-chemical processes. In many cases, the impact is of mixed nature with hard-to-separate individual components. Almost all of these methods were somehow tried in activating methane, in particular, during the DMTM process.

One of the simplest methods of physical initiation of the conversion and oxidation of methane is to activate it by a heated surface. Back in 1921, the author of [197] studied the thermal condensation of methane flowing through a porcelain or quartz tube heated to 1050 °C. The yield of condensable products reached 18%. Then, a thermal diffusion column was used to study the thermal and oxidative coupling of methane [198–200]. In the thermal diffusion column, the gas flow passes through a vertical tube with cold external walls and a hot wire stretched along the axis. Due to thermal diffusion, free radicals formed on the hot surface and arising reaction products are separated in space, with the products being condensed on the cold column walls.

Using this technique, the authors of [198] achieved a 42% conversion of methane at a hot wire temperature of 1300 °C and a cold wall temperature of 30 °C. The resulting liquid products were represented mainly by aromatics, whereas the solid products, by compounds such as naphthalene. It was found [199] that, when the gas passed through a thermal diffusion column from top to bottom under the same conditions, the main product was ethylene, formed with a selectivity of up to 91.5% at a conversion of 9.4% and wire temperature of 1200 °C. The authors of [200] also obtained a wide variety of liquid and solid products in a thermal diffusion column.

The low-temperature high-pressure slow oxidation of methane to methanol in a static reactor at 40 atm initiated with a heated wire filament (Fig. 9.10) was studied in [92]. The thermal initiation let to decrease the temperature of oxidation onset to 200 °C and even lower. Oxygen conversion started after a long induction period, ~100 min, and was complete within ~600 min. The methane conversion in a CH_4:air = 4:1 mixture attained 4%. At the initial stage of the process, at a low conversion of the reagent, the methanol selectivity reached 90%, apparently due to the domination of the following sequence of reactions:

$$CH_4 \xrightarrow{T} CH_3^{\bullet} \xrightarrow{+O_2} CH_3OO^{\bullet} \xrightarrow{+CH_4} CH_3OOH \rightarrow CH_3O^{\bullet} \xrightarrow{+CH_4} CH_3OH \quad (9.21)$$

However, with increasing duration of the process, accompanied by the growth of the conversion of reactants and the concentration of the products in the reactor, the methanol selectivity dropped to ~22%, i.e., a value typical for this pressure range (~40 atm). After decreasing the pressure from 40 to 5 atm, the selectivity of methanol formation in this initiated radical—chain reaction decreases sharply to 2%, thereby confirming the importance of this parameter for the formation of methanol. Thus, reducing the process temperature by more than 200 °C due to its initiation does not show any advantage in terms of selectivity of methanol.

Acoustic activation is used to stabilize the combustion of methane [201]. There are works on ultrasonic initiation of the low-temperature oxidation of methane in water, where the initiation of chemical reactions occurs in cavitation gas-vapour bubbles. It was shown [202] that,

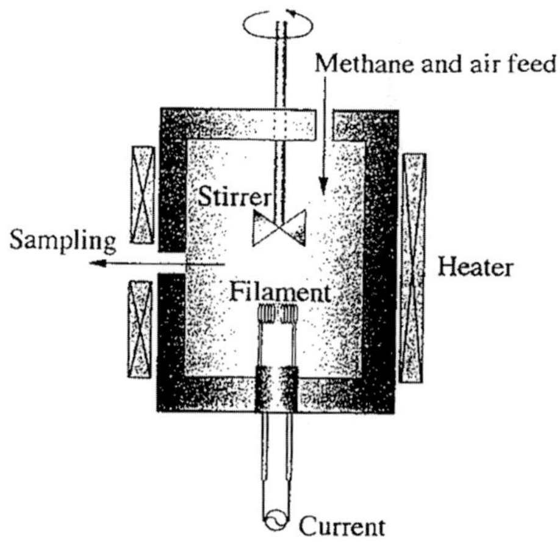

FIGURE 9.10 Static reactor with initiation with a heated filament [92].

in aqueous solutions, under the influence of ultrasound, methane is oxidized to form formaldehyde. When ethylene is added to methane, the yield of formaldehyde decreases, but acetaldehyde is formed.

Attempts to conduct the photochemical conversion of methane in water vapour at temperatures below 100 °C and atmospheric pressure were made in [203]. Since the first absorption band of paraffinic hydrocarbons is located in the Schumann ultraviolet region, starting from ~144 nm for methane, it was assumed that the conversion of methane is initiated by hydroxyl radicals formed by water vapour photolysis. The main products were methanol (with a selectivity of ~70% at 90 °C), formic acid (11%), ethanol (5%), formaldehyde (5%), acetone (4%), and acetic acid (3%). The photolysis was carried out with a 20-W low-pressure mercury lamp. Given that the absorption of UV radiation by water vapour starts at ~185 nm, which is the lower boundary of mercury lamp radiation transmitted by a quartz bulb, the photochemical conversion of methane in these conditions is unlikely to be sufficiently efficient.

In the 1980s, at the Los Alamos National Laboratory (USA), experiments on laser initiation of the DMTM process were carried [204,205]. Judging from the available results, at 450 °C, a pressure of 2.04 atm, and 11% oxygen concentration, only an increase in the rate of the process was achieved, without significant changes in the selectivity of methanol formation and the composition of the products. However, performing a kinetic analysis of the system, the authors concluded that, at elevated temperatures (530–730 °C), a pressure of 60 atm, and a $CH_4:O_2 = 2:1$ ratio, one can expect a 50% methanol selectivity at a methane conversion of up to 25%. The time scale of the process (3 ms) motivated the use of a reactor with a supersonic nozzle [205]. However, the authors apparently failed in solving the problem of rapid removal of the heat of reaction to maintain isothermal conditions postulated in the analysis.

There have been reports on attempts to create an electron-beam technology of natural gas processing based on the fragmentation of gaseous alkanes into ions and radicals under the action of high-energy electrons generated by an accelerator [206]. However, energy efficiency of technologies of such type is highly questionable.

There is a vast body of works on the microwave plasma-chemical activation of methane. Obviously, this is due to the availability and popularity of the appropriate technique, a large number of teams working in this field, and relatively well-developed physical foundations of gas-discharge processes. Unfortunately, the chemistry of hydrocarbon oxidation in plasma is still poorly studied, especially the parameters of the elementary processes of interaction of charged and excited species with hydrocarbons and carbon-containing radicals. Therefore, most of the works is purely empirical and limited to recording the general effect of different types of discharge with different parameters and different modes of generation on a methane-oxygen mixture, as well as to determining the composition of the products, with little attention given the specifics of the initiated reaction and the chemistry of the initiation process. The vast majority of works on the plasma activation of methane have studied its conversion to syngas and condensation products. The number of works on plasma-chemical conversion into oxygenates is substantially smaller. Since the general principles of the plasma-chemical conversion of methane into various products and even the results and conclusions are sufficiently similar, we will limit ourselves to an overview of works dealing directly with the activation of the plasma-chemical conversion of methane to oxygenates and of several reviews on the subject.

The oxidation of methane and ethane assisted by a barrier discharge was studied in [207]. The conversion of a methane-oxygen mixture with 7 vol % oxygen was low, with the most abundant products being carbon oxides (40–50%). The total yield of oxygenates (methanol, formaldehyde, acetaldehyde, and ethanol) was within 15%, with a significant prevalence of methanol. The fraction of oxidative coupling products (ethane, ethylene, and C_3-hydrocarbons) was 30%. The formation of resinous mass on the reactor walls was observed. The energy efficiency of formation of oxygenate did not exceed 2%.

It was demonstrated [208] that, in a methane plasma, methane dimerizes into C_2-hydrocarbons with a selectivity of above 95% at a conversion of 30–90%. However, the energy efficiency of this thermodynamically unfavourable process was only 0.2–3.3%. The yield of oxygenates in plasma-chemical initiation is usually much lower than that of C_2 products and carbon oxides due to a low chemical stability of oxygenates. Studying the oxidation of methane to methanol under the influence of a microwave discharge, the authors of [209] found that it is substantially influenced by the design of the reactor. Supply of methane downstream of the plasma zone enabled to significantly reduce the yield of C_2-hydrocarbons, but the methanol formation selectivity remained extremely low. Of the several options for organizing the process presented in Fig. 9.11, the highest selectivity to methanol was obtained for version (e), where that inner tube had three small radial holes at the tip for introducing the reagent. This design provides a better mixing of the reagents after the discharge zone, which is essential for ensuring a high methanol yield. The best results are obtained when the methane and oxygen are supplied through the inner and outer tubes, respectively. In addition, in contrast to the partial oxidation of methane at high pressure, in low-pressure plasma, an efficient formation of methanol is apparently facilitated by a low CH_4/O_2 ratio.

Of considerable interest are studies in which microwave radiation does not act directly on the gas medium, but on a solid catalyst introduced into the activation zone [210]. Pulsed microwave radiation acting on a system containing methane and a catalyst heats the active component of the catalyst. It is assumed that, in this case, due to a rapid quenching of the activation products on the radiation-unheated (cold) catalyst support, high selectivities of formation of individual products, exceeding those typical of the high-temperature catalytic or gas-phase activation of methane can be achieved. Another possible mechanism of activation in this system may be associated with gas discharges induced by microwave radiation near the catalyst surface. In these experiments, the efficiency of conversion of microwave radiation energy into the energy of chemical bonds was as high as 10%. An example of the joint action of a nonthermal plasma or a barrier discharge and a catalyst is the combined steam-oxygen reforming of methane studied in [157], where the action of a supported Ni catalyst was compared to that of the support alone.

One of the types of discharges most widely used in plasma-assisted methane conversion is a barrier discharge. Figure 9.12 shows a schematic of a setup usually used in such studies [211]. The discharge occurs in a coaxial gap of thickness 1 mm and length of 310 mm between a steel cylinder, which serves as the outer electrode, and an inner quartz cylinder with a wall thickness of 2.5 mm. The inner electrode was a metal foil tube inserted into the quartz cylinder. A sinusoidal voltage with a frequency of ~30 kHz, amplitude of up to 20 kW, and a power of 100–1000 W was applied to the electrodes. The reagent flow rate ranged from 0.1 to 4 l/min, which corresponds to a residence time of 1.5–60 s.

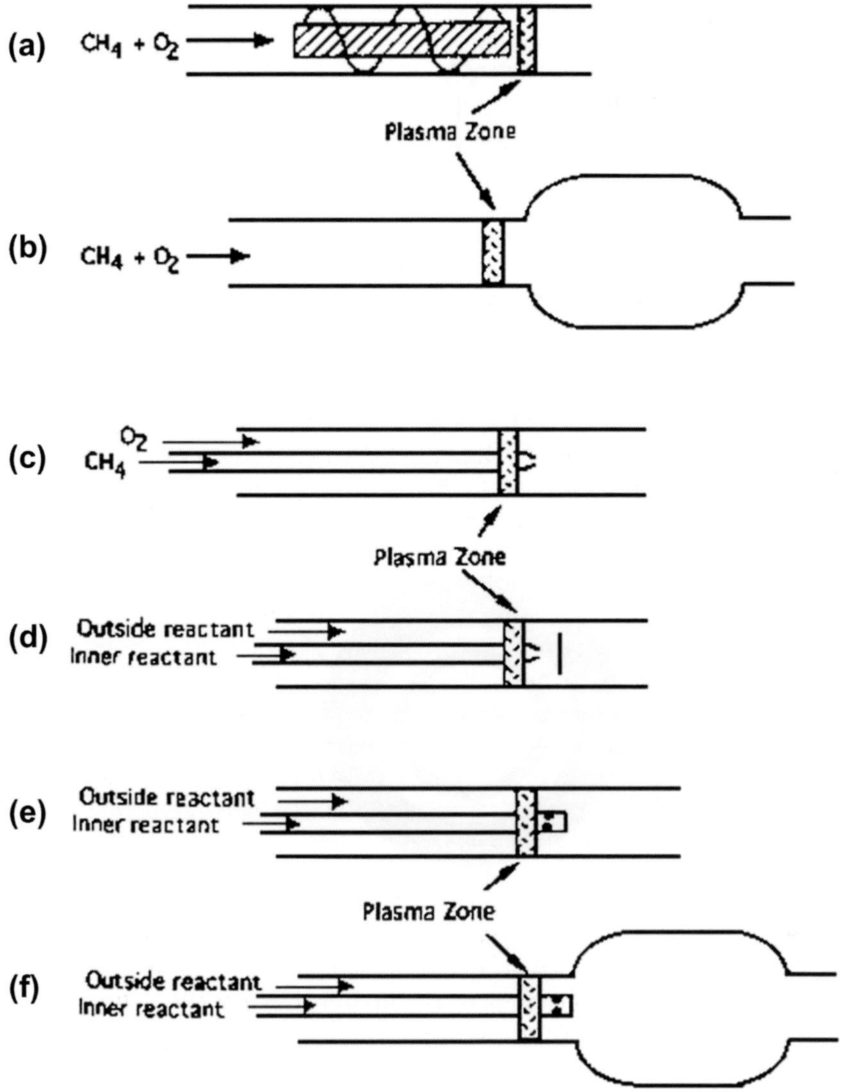

FIGURE 9.11 Versions of design of the quartz plasma reactor [209].

The dependence of the methane conversion and methanol yield on the specific discharge energy per unit volume of feed gas, varied by changing the applied voltage amplitude and the gas flow rate, is shown in Fig. 9.13.

Methane conversion increases with increasing specific energy of the discharge, with the methanol yield reaching a maximum of ~3% at a specific energy of ~6.7 kWh/m^3. The maximum selectivity of methanol formation, 30%, was achieved at its yield of 1% at a specific energy of 1.1 kWh/m^3. The difference between the impacts of changes in the discharge power and in gas flow rate shows that a decrease in the flow rate, that is, an increase in the

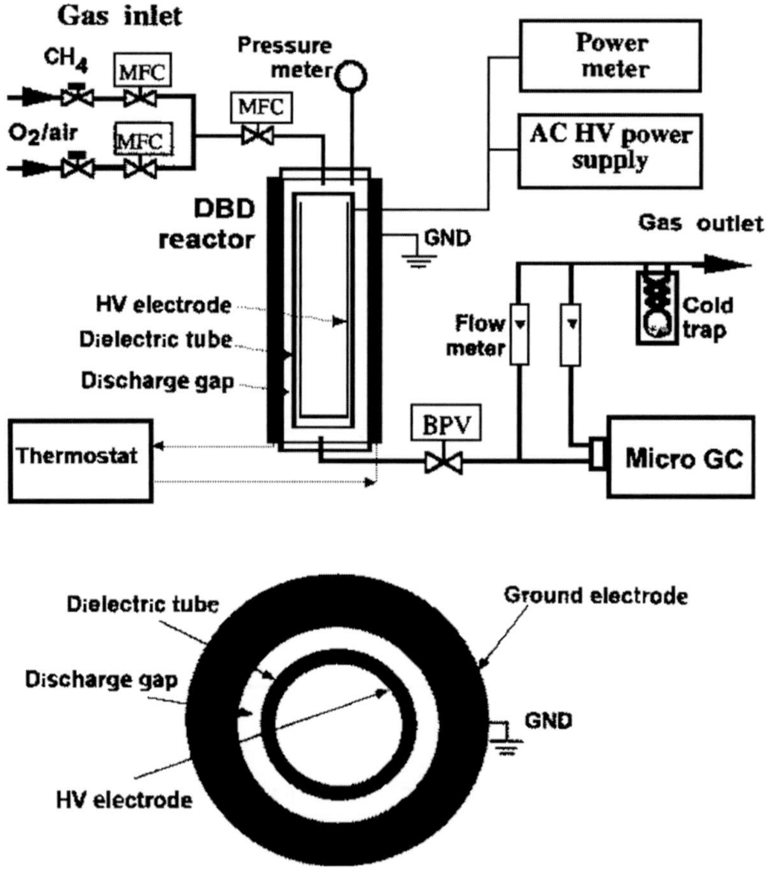

FIGURE 9.12 Dielectric barrier discharge configuration and experimental setup; MFC is the mass flow controller, and BPV is the backpressure valve [211].

residence time t_r in the discharge zone does not favour the formation of methanol. A rise in the specific energy of the discharge increases the yields of CO and C_2H_6. Changing the temperature of the reactor wall has no significant effect on the conversion of methane and the yield of the oxidation products.

The dependence of the methane conversion and methanol yield on the oxygen concentration in the feed gas is shown in Fig. 9.14. At the oxygen concentration of 30%, the methanol yield reaches its maximum, 2.9%. With increasing oxygen concentration, the yield of oxygen-containing products grows due to the reduction of the yield of C_2 products, with the yields of CO and CO_2 growing faster than that of methanol.

Replacement of oxygen with air does not lead to significant changes in the yield and selectivity of formation of the products (Fig. 9.15).

The effect of pressure on the methane conversion and methanol yield is shown in Fig. 9.16. The yield reaches a maximum at $P = 2$ atm after which it rapidly decreases, primarily due to the reduction of the methane conversion.

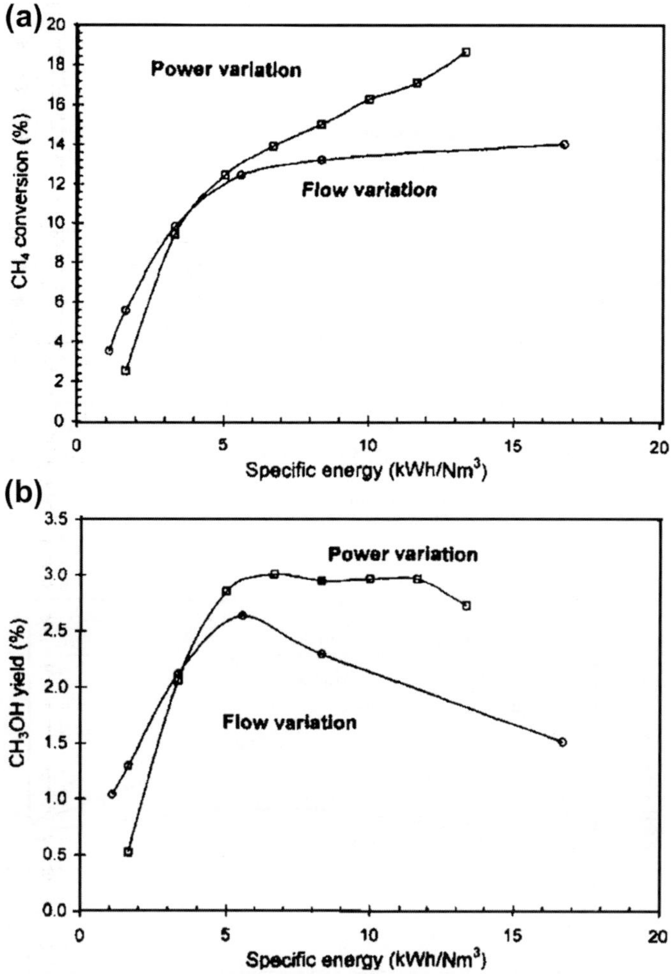

FIGURE 9.13 Influence of the specific energy input on (a) methane conversion and (b) methanol yield at $P = 2$ atm, $T_{wall} = 80\,°C$, $CH_4/O_2 = 8{:}2$, and $t_r = 6$ s. The power was varied at a flow rate of 1.0 Nl/min; the flow rate was varied at a power of 200 W [211].

Thus, a barrier discharge in a mixture of methane and oxygen makes it possible to effectively convert methane at low temperatures, less than 100 °C. In addition to methanol, a wide variety of products, including CO, CO_2, H_2O, and C_2H_4, C_2H_6, and C_3H_8, are produced. A too high intensity of the discharge reduces the selectivity and yield of methanol (Fig. 9.13), probably because of its decomposition in the discharge, accompanied by an increase in the yield of carbon oxides.

The presence of oxygen significantly speeds up the formation of radicals in methane plasma, most likely because of the reaction of methane with excited oxygen atoms:

$$O(^1D) + CH_4 \rightarrow CH_3^{\cdot} + OH^{\cdot} \tag{9.22}$$

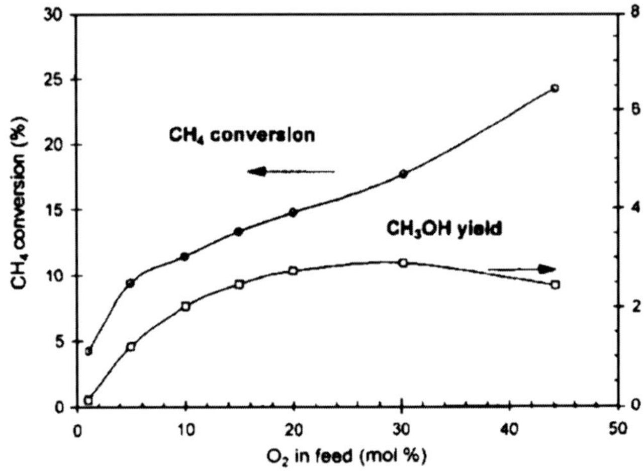

FIGURE 9.14 Influence of O_2 content on the methane conversion and methanol yield at $P=1$ atm, $T_{wall}=80\,°C$, and $E_{sp}=3.3$ kWh/m^3 [211].

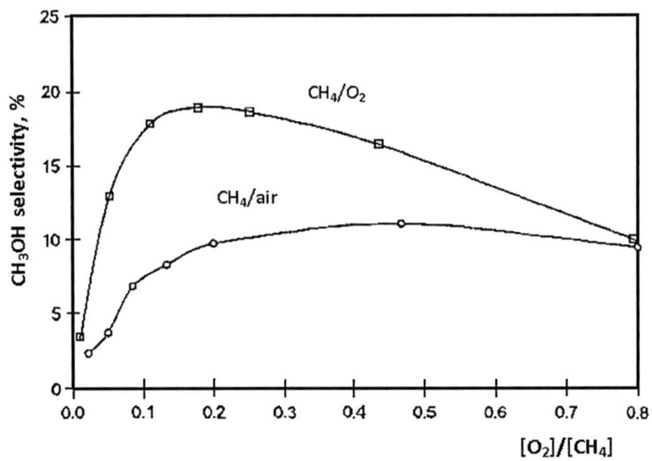

FIGURE 9.15 Comparison of methanol selectivity for the oxidation of methane with oxygen ($P=1$ atm, Power = 200 W) and air ($P=2$ atm, Power = 400 W) at $T_{wall}=80\,°C$ [211].

which has a very high rate constant, $k=1.4\times10^{-10}$ cm^3/molecule s, four times higher than the rate constant for quenching O(^{1}D) by O_2 molecules.

At low temperatures, typical of barrier-discharge conditions, reaction of methoxy radical $CH_3O^{\bullet}$ with methane, the main channel for the formation of methanol at higher temperatures, has a very low rate constant. Therefore, the mechanism of methanol formation under these conditions is likely to differ from the mechanism of its formation in the thermal partial oxidation of methane. This is evidenced by a significantly higher (~15%) oxygen concentration at which the maximum selectivity to methanol is achieved (Figs 9.14 and 9.15). Therefore,

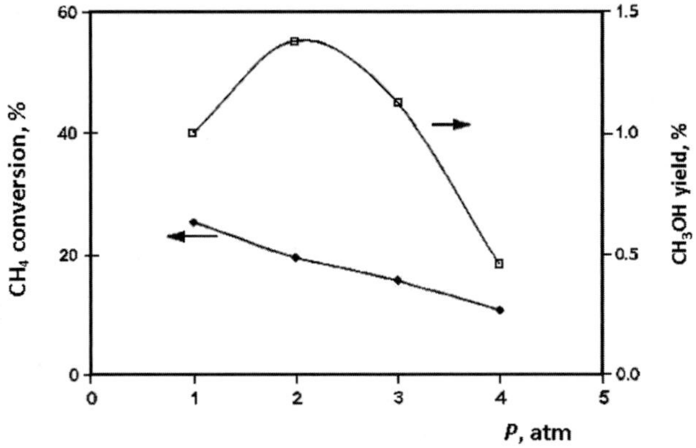

FIGURE 9.16 Effect of pressure on the (◆) methane conversion and (□) methanol yield for the oxidation of a methane–air mixture at $T_{wall} = 80\,°C$, Power = 200 W, and CH_4/air = 7:3 [211].

despite a lower methanol formation selectivity (~22%) as compared to that for the thermal partial oxidation of methane, a higher methane conversion (14%) under optimal conditions enables to achieve a significant (~3%) methanol yield. However, the specific energy consumption, ~40 eV/molecule, as estimated in [212], is too large to make this process practically feasible.

The formation of organic oxygenates during the partial oxidation of methane in a barrier electric discharge was also studied in [213]. The main liquid oxygenates were methanol, formaldehyde, methyl formate, and formic acid. A decrease in the partial pressure of oxygen caused a decline in the energy efficiency of the process. For a sequential arrangement of three reactors, with extraction of liquid products and addition of oxygen in between, the final conversion of methane reached 59% at a 35% yield of liquid organic products. In the best experiments, the energy consumed to convert one methane molecule was 51 eV. At a total selectivity to oxygenates of 56%, it was 91 eV per molecule.

The formation of oxygenates in a barrier-discharge plasma, with CO_2 acting as the oxidant, was studied in [214]. As a result, a large set of oxygenates, including acetic and propionic acids, methanol and ethanol, and a number of other compounds was obtained.

At long stay in the discharge zone, primary oxygenates formed decompose giving rise to more complex products. A study of the effects of corona- and barrier-discharge plasmas on methanol [215] has shown that this process produced more complex oxygenates, such as ethanol, propanol, ethylene glycol, and others. Great influence on the composition and yield of oxygenates was produced by the type of discharge and the conditions in the discharge zone.

To increase the yield of oxygenates in the plasma-chemical conversion of natural gas in a barrier discharge, the authors of [216] proposed an effective method for withdrawing them from the discharge zone, more specifically, stimulated condensation, with plasma ions acting apparently as nucleation centres. The aerosol formed directly in the discharge zone of the plasma-chemical reactor is used to remove the conversion products from the discharge zone and from the gas phase. At an oxygen concentration of 30%, the conversion of the initial mixture of hydrocarbons reached 28%. The resulting liquid condensate is dominated by water (50.1%), formic acid (29.7%), and methanol (11.9%). In addition, methyl formate (5.1%), acetic acid (1.6%), acetaldehyde (0.8%) and ethanol (0.8%) were found. Modeling the process within the framework of a simplified kinetic scheme demonstrated a qualitative agreement with this set of products.

An interesting method for reducing the impact of the discharge plasma on the products of oxidation of methane to methanol by using a plasma microreactor was proposed in [217]. A high-voltage (2 kV) discharge with a sinusoidal frequency of 75 kHz occurred between a twisted metal wire placed inside a Pyrex microreactor, 1 mm in inner diameter and 60 mm in length, and a massive copper block, which acted as a heat reservoir (Fig. 9.17); this type discharge is similar to a barrier discharge. The power consumption ranged from 3 to 10 W. At atmospheric pressure in the reactor and O_2/CH_4 ratios of 0, 0.2, 0.5, and 1.0, the temperature in the reactor changed from 25 to 200 °C. The idea of the experiment was that the least reactive component, methane, was activated by high-energy electrons, while the decomposition of more reactive products was minimized by a sharp quenching of the chemical processes on the cold walls of the reactor.

The selectivity to methanol for such a microreactor turned out to be much higher than that to other oxygenates. The maximum total selectivity of formation of oxygenates was 60%, whereas the methanol yield was 10% per pass through the reactor. According to the authors, optimal conditions can provide a specific energy consumption of 69 kWh per kg of methanol.

An interesting approach is to perform an electric discharge in a two-phase gas–liquid medium. One of the obvious advantages of this system compared to the gas discharge is a quick absorption of products produced in the discharge by the liquid phase, as well as a rapid dissipation of the heat released by the reaction. In addition, for a high-power pulsed discharge, a certain contribution can come from acoustic activation. The formation of methanol and formaldehyde in an electric discharge in a two-phase system in which one of the electrodes is in

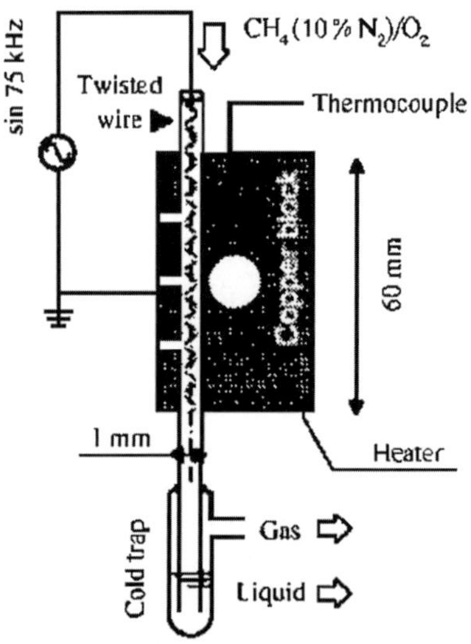

FIGURE 9.17 Micro-plasma reactor [217].

the gas phase, whereas the other, in the liquid, was described in [218]. A third largest-yield product was hydrogen peroxide, which was formed only when one of the electrodes was in the liquid phase, as opposed to methanol and formaldehyde, which were formed, although in much smaller numbers, in the absence of the liquid phase. The energy efficiency of formation of formaldehyde in this system was significantly higher for discharge treatment in the methane-water vapour system, but lower for the methane–air and methane–oxygen systems with gas-phase discharge.

The aforementioned works on the plasma-chemical initiation of methane oxidation suggest that the majority of plasma-chemical methods for hydrocarbon conversion, because of being nonselective and involving multiple cleavages and reformations of the same bonds, are characterized by a too high specific-energy consumption per converted molecule, significantly greater than the energy required for the appropriate rearrangement of chemical bonds. Therefore, the reduction of specific energy consumption remains the main requirement for the practical use of plasma-chemical technologies. Table 9.4 lists the lowest values of the specific energy consumption for the pyrolysis and oxidative conversion of methane for different types of discharges reported in experimental studies.

Below are presented some general conclusions on the characteristics and prospects of the plasma-chemical methods for the conversion of methane from the review [212]:

1. All types of discharges can be divided into two groups: nonuniform discharges (arc, spark, surface) and volume discharges (barrier, corona).
2. In nonuniform discharges, the methane conversion is higher. The energy consumption for the decomposition of methane is less than 10 eV/molecule; the degree of decomposition in arc discharges is above than 90%, whereas selectivity of formation of equilibrium products can exceed 90% (acetylene in plasma pyrolysis and hydrogen and CO in steam reforming).
3. In volume discharges, a high conversion (50%) is achieved only at a high consumption of energy in the decomposition of methane (above 40–50 eV/molecule). This process produces a wide variety of products with low selectivity.
4. The conversion of methane in mixtures with oxygen makes it possible to reduce the energy spent for methane decomposition, but its value is still higher than that characteristic of nonuniform discharges. The methane conversion in mixtures with oxygen and selectivity of formation of the products remain low.
5. A significant reduction in the energy consumption to less than 1 eV/molecule can be achieved by organizing the branched-chain process of methane conversion.
6. A promising line in the plasma-chemical conversion of methane is to organize this process in a mixture with water. In this case, the energy consumption in a gliding discharge is lower than the energy of the C–H bond, energy of methane decomposition, and enthalpy of decomposition of methane under equilibrium conditions. The hydrogen yield is much higher than the equilibrium value.

Thus, with the exception of a number of technologies for producing equilibrium products (acetylene, syngas), the energy consumption and, consequently, the costs of the conversion of natural gas by means of discharge and plasma-chemical technologies, especially with the real efficiency of the transformation of the input electricity energy into the energy of an electric discharge remain unacceptably high. A more promising approach is to use of plasma-chemical methods for producing a powerful initiating impact on nonequilibrium chemical

TABLE 9.4 Specific Energy (ε, eV/molecule) for the Pyrolysis and Oxidative Conversion of Methane and the Methane Conversion (η, %) for Different Types of Discharges [212]

	CH_4		$CH_4 + O_2$		$CH_4 + CO_2$		$CH_4 + H_2O$	
Type of Discharge	ε	η	ε	η	ε	η	ε	η
Arc DC discharge	3–5	95						
Arc + H* discharge	11–15	14						
Spark discharge	3.8	29					5.3	55
	6.1	60						
Pulsed discharge (240 Hz)			29					
Pulsed discharge (8 kHz)	8.5				15.5			
Pulsed discharge (10 kHz)	3.8	23						
Pulsed discharge plus Ni-catalyst							0.6	30
Gliding discharge							0.9	50
RF gliding discharge, $CH_4 + Ar$	5.2	33						
	11.6	75						
Continuous electron beam, 400 keV	3.2–10							
Continuous electron beam, 14 keV			19	4	15	5	24	1.6
					20	6.6		
Barrier discharge	52	20	5–7	27	38	10	14	8
	12–33				100	65		
Barrier discharge, $CH_4 + He$					156	12		
Barrier discharge + catalyst	52	40						
Corona discharge	52	20			85			
Pulsed corona discharge					24–66			
Microwave discharge + heating	5.8	90	2.8	70			21	
			3	100				
RF discharge $CH_4 + N_2$	2700	90						
Nonself-maintained discharge			4–6				7.2	15
Glow discharge			18					
Uniform pulsed glow discharge (50 Hz)	9.1							
Pulsed electron beam	100–130		0.1	100				

systems, such as branched-chain processes, possibly in combination with heterogeneous catalysis. In these cases, significant changes in the chemical behaviour of the system can arise even under the action of low-energy impacts.

PARTIAL OXIDATION OF ALKANES IN SC CONDITIONS

Given that many processes of conversion of alkanes proceed efficiently at high pressures, it is not surprising that the rapid advances in the technology and practice of supercritical (SC) fluids for a variety of chemical processes stimulated great interest in the oxidation of hydrocarbons at SC temperatures and pressures [219–221].

This interest stems from at least two circumstances. First, in approaching to and entering the SC state of the fluid, not only the physicochemical properties, but also the reactivity of many substances change. This makes it possible to perform unusual reactions by using quite common and available reagents, such as water, ammonia, CO_2, and CH_4. In addition, since many practically important gas-phase processes occur more effectively at high pressure (high densities of the reaction mixture), the transition to high densities characteristic of SC fluids is potentially interesting from the point of view of technological perspectives.

Oxidation of light alkanes in the SC conditions is also interesting because they themselves can be transferred relatively easily into the SC state (Table 9.5). Therefore, comparison of the processes of their oxidation at widely various temperatures and pressures can provide ample data for understanding the role of the transition to the SC state for the reactions of different types.

So far, a relatively limited number of works has been published on the oxidation of light alkanes under SC conditions, mostly on the oxidation of methane in supercritical water (SCW). The role of water, which under SC conditions even at moderate temperatures, below 600 °C, cannot be considered as an inert component, is of special interest. Unfortunately, only a few experimental studies have been performed on the oxidation of methane in other than water SC fluids, a comparison with which would be very interesting for understanding the role of the state of the medium in these reactions.

TABLE 9.5 Thermodynamic and Critical Parameters for C_1–C_4 Alkanes

Hydrocarbon	$\Delta H^o_{f,298}$, kJ/mol	$-\Delta H^o_{f,298}/n$, kJ/mol	S^o_{298}, J/(mol K)	$T_{critical}$, °C	$P_{crtitical}$, atm	$\rho_{critical}$, g/cm^3
CH_4	−74.85	74.85	186.19	−82.1	45.8	0.162
C_2H_6	−84.67	42.33	229.5	32.3	48.2	0.203
C_3H_8	−103.9	34.6	269.9	96.8	42.0	0.220
n-C_4H_{10}	−124.7	31.2	310.0	152.0	37.5	0.228
iso-C_4H_{10}	−131.6	32.9	294.64	134.9	36.0	0.221

The oxidation of methane in water under SC conditions, typically at pressures above 250 atm and temperatures above 400 °C, has been extensively studied in recent years. The properties of water near the critical point differ significantly from the properties of water under normal conditions. In particular, the dielectric constant, strength of hydrogen bonds and their number per water molecule are substantially lower. Many substances, both gaseous and condensed under given conditions, are completely soluble in SCW, which enables to conduct the process at much higher concentrations of reagents and without complications caused by interphase mass transfer. Since many of the properties of water in the SC state and near the point of transition into it (viscosity, dielectric constant, etc.) depend more strongly on the external parameters than those of normal water, changing the pressure and temperature of SCW enables to widely vary the reaction medium parameters so as to control the occurrence of the reaction. In particular, it was shown that the selectivity of formation of partial oxidation products from some hydrocarbons increases with increasing pressure, i.e., the SCW density [222]. An additional advantage of the oxidation of hydrocarbons in SCW is the possibility of more easily removing the heat released by the reaction.

Because of its high reactivity, SCW can act as both an oxidant and catalyst. As was shown experimentally, even in the gas phase at low temperatures, water, due to its ability to form hydrogen bonds, can act as a molecular catalyst, capable of increasing the rate of the reaction of hydroxyl radicals with acetaldehyde [223]. Such effects are even more expected in SCW.

Another feature of the process in SCW is a high probability of carbon monoxide formation. This, in turn, gives rise to subsequent reactions (6.9) of water gas. As a result, molecular hydrogen appears in the system, so hydrogenation processes can rapidly occur even without special introduction of hydrogen.

It makes sense to assume that the oxidative processes in SCW, at least at sufficiently high temperatures, occur largely as the same free-radical processes of gas-phase oxidation in the same temperature range [220], which facilitates kinetic analysis of these processes. Nevertheless, so far, there have been no direct observations of radicals in these conditions. However, Fourier-transform infrared spectroscopy, Raman, and emission spectroscopies enabled to detect intermediate molecular species in diffusion flames at high pressures [220].

At 450 °C, in the absence of oxygen, methane is quite stable in SCW. The thermolysis of methane in SCW was investigated in [224] at 600 °C, 600 atm, and a water/methane ratio of 10:1. Under these conditions, the methane conversion was 2.54% after 60 min and 3.50% after 120 min. The main products of the reaction were CO_2 and H_2. Also observed was the formation of ethylene – i.e., the process occurs via two main pathways: pyrolysis (dehydrogenation) to ethylene and hydrolysis to CO_2.

At temperatures above 450 °C and a sufficiently high O_2/CH_4 ratio, methane is oxidized in SCW mainly to CO and CO_2 [225], and even methane ignition was observed [224], although under the gas-phase conditions, presence of more than 30% water vapour prevents the ignition of methane–air mixtures of any composition [226]. In analyzing the oxidation of methane in SCW, we will confine ourselves to the partial oxidation at $O_2/CH_4 < 0.1$, a ratio at which a significant yield of methanol and other oxygenates is observed. A detailed review of the work on the oxidation of light alkanes in SCW is presented in [221].

The partial oxidation of methane to methanol in SCW was investigated in [222,224,225,227,228] (Table 9.6). As in the gas-phase oxidation, a significant yield of methanol was observed only at a large excess of methane.

TABLE 9.6 Experimental Conditions and Main Products of Methane Oxidation to Methanol in Supercritical Water Experimental Conditions and Main Products of Methane Oxidation to Methanol in Supercritical Water

P, atm	*T*, °C	$(O_2/CH_4)_0$	Main Products	References
200–350	400	0.03	CO, CH_3OH, CH_2O	[222]
300–1000	382–440	0.025–0.33	CO, CO_2, H_2, CH_3OH	[224]
250	400–450	0.03–0.05	CO, CO_2, CH_3OH, H_2	[225]
330	450	0.07–8.7	CO_2, CH_3OH	[227]
230–430	382–481	0.04–0.1	CO, CO_2, CH_3OH, CH_2O	[228]

In [227], one of the first works on the subject, along with the deep oxidation, the partial oxidation methane to methanol (at $O_2/CH_4 = 0.07–0.73$) in SCW was carried out in a static reactor at 450 °C and 330 atm in the presence of a Cr_2O_3 catalyst. It was noted that, although SCW inhibits the catalytic oxidation of methane on Cr_2O_3 (Fig. 9.18), it increases the selectivity and yield of methanol formation almost by an order of magnitude compared to the gas-phase reaction under the same conditions. The reaction order with respect to oxygen was low and negative; i.e., an increase in the concentration of oxygen leads to some reduction in the reaction rate. The reaction products contained only CO_2 and methanol. The absence of CO was explained by a substantial contribution from the reaction of water gas, whereas the formation of formaldehyde is not mentioned at all.

The issue of the formation of formaldehyde is one of the main contradictions in the available studies. According to [224], the oxidation of methane in SCW (70% H_2O and 30% CH_4) at

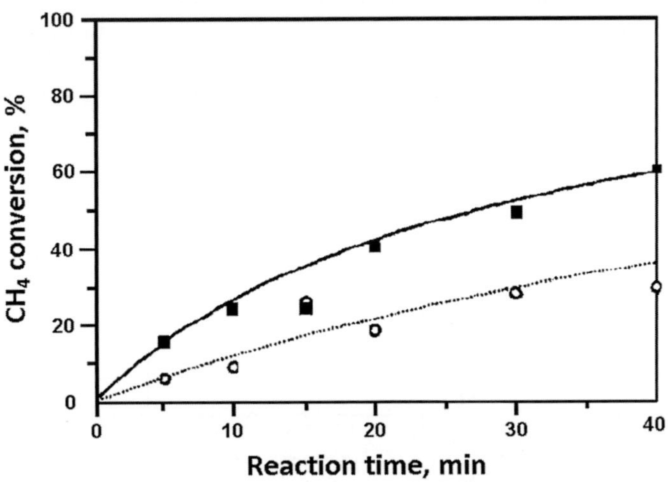

FIGURE 9.18 Comparison of the conversion of methane in the gas phase at (■) 114 atm and (○) in SCW at 439 atm and $T = 450$ °C [227].

a large excess of methane with respect to oxygen ($O_2/CH_4 = 0.025-0.33$), pressures of 300–1000 atm, and temperatures of 380–440 °C mainly produced CO, CO_2, H_2, and CH_3OH, with the yield of formaldehyde being negligibly small. On the other hand, the partial oxidation of methane under flow conditions with very similar parameters ($O_2/CH_4 = 0.03$ and 400 °C) [222] gave in all experiments, along with CO and methanol, detectable amounts of formaldehyde; small amounts of CO_2 and H_2 were also observed. The formation of formaldehyde was also observed in static conditions in [228] at temperatures of 382–481 °C and pressures of 230–430 atm, as opposed to [225], where no formaldehyde was formed in the laminar flow in an isothermal reactor at 250 atm and 400–450 °C. In [225], the formation of H_2 and in part CO_2 was attributed to the water-gas reaction, although, on the other hand, according to this work, the selectivity of CO formation grows rapidly while that of CO_2 decreases with increasing oxygen conversion.

According to the experimental data from [225], in the partial oxidation of methane in SCW to all products, including methanol, methane and oxygen are consumed in approximately equimolar proportion, as in the gas-phase oxidation.

As in the gas-phase conditions, at a large excess of methane, the reaction order with respect to methane is close to unity, whereas that with respect to oxygen is nearly zero. In [225], it was reported to be 0.3, whereas in [227], as noted above, even negative. This correlates with the results obtained in the gas-phase conditions, under which oxygen was observed to inhibit the oxidation rate. The values of activation energy obtained in [224] based on an analysis of their own data and the results of other works range within 32–34 kcal/mol (Fig. 9.19), which is about 10 kcal/mol lower than the activation energy of the gas-phase reaction. The activation energy obtained in [225] in a very narrow temperature range (25 °C), 96 kcal/mol, seems too high to be realistic.

At an oxidation temperature of 425 °C and an oxygen concentration of 5%, its total conversion is achieved within ~10 s. In [222], under almost identical conditions, a similar time of complete oxygen conversion was reported, ~60 s.

Low yields of methanol and formaldehyde are largely attributed to their rapid decomposition during the oxidation reaction; however, methanol turned out to be relatively stable in

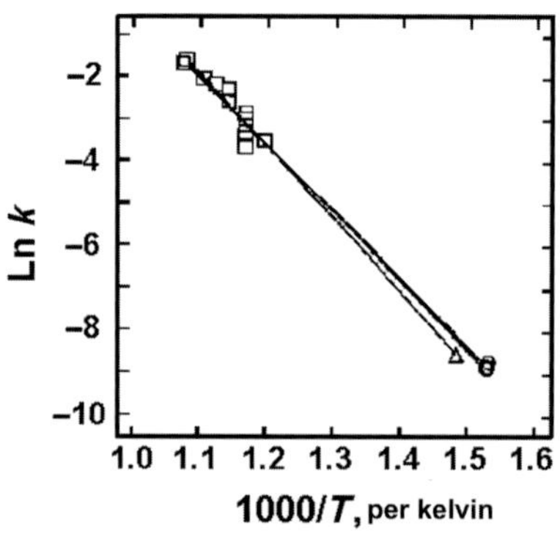

FIGURE 9.19 Temperature dependence of the rate constant k of the methane oxidation in supercritical water [224].

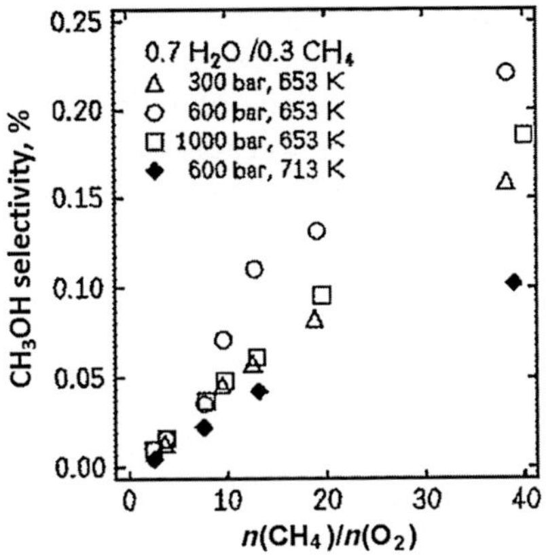

FIGURE 9.20 Dependence of the selectivity of methanol formation on the methane/oxygen ratio for the oxidation of methane in SCW (70% H_2O–30% CH_4) [224].

SCW in the absence of oxygen [224]. At that, according to this work, the selectivity to methanol did not exceed 22% (Fig. 9.20).

Increasing the density of SCW shifts the composition of methane conversion products in the direction of greater yield of incomplete oxidation products, including formaldehyde [222]. As in the gas-phase reaction, the selectivity of methanol formation increases with decreasing initial concentration of oxygen (Fig. 9.20), and degree of its conversion, with the selectivity remaining below that obtained under the gas-phase conditions. Low selectivities of methanol and, therefore, its yield were observed in almost all studies. The highest selectivity, 75%, at a very low (0.3–0.6%) oxygen conversion was obtained in [228]. The highest methanol yield, 4%, at a selectivity of ~40% and a methane conversion of ~10%, was obtained in [227]. The maximum yield of methanol was observed at short times, less than 5 min. In [225], a lower methanol yield was reported, less than 1%. At a conversion of 1–3%, the selectivity to methanol was ~35%.

According to [224], the pressure generally increases the selectivity of methanol formation, although this dependence is nonmonotonic. For example, the increase in pressure from 300 to 600 atm increases the methanol formation selectivity from ~15.8% to ~21.5%, but a further increase in pressure, up to 1000 atm, reduces the selectivity to ~18.5% (Fig. 9.20). A possible reason, according to the authors of this work, could be the emergence of conditions for methane ignition, they observed in SCW. At the same time, according to [222], although an increase in pressure significantly increases the yield of all main products, only the selectivity of formation of formaldehyde increases with the pressure, while the selectivity to methanol and CO_2 remains practically unchanged, with that to CO even slightly decreasing. Temperature produces a more pronounced effect on the selectivity: it decreases twofold temperature increase by only 60 °C (Fig. 9.20).

There is also a controversy concerning the CO/CO_2 ratio. As reported in [222], this ratio exceeds 10 at lower pressures and slightly decreases with increasing pressures, generally in agreement with the results on the oxidation under the gas-phase conditions. On the contrary, according to [224], the CO/CO_2 ratio is mostly well below unity, being slightly higher than that at the lowest O_2/CH_4 ratio, 0.025, and temperatures no higher than 380 °C.

Several attempts were made to increase the methanol yield in the oxidation of methane in SCW by introducing heterogeneous catalysts into the system. As already mentioned above, in [227], methane oxidation in SCW was carried out in the presence of Cr_2O_3. This oxide is not a very good catalyst for oxidation in SCW, because it is not stable under these conditions (chromium is leached from the catalyst [220]). There have been attempts to increase the yield of methanol from methane oxidation in SCW through introduction of other heterogeneous catalysts (platinum chloride and cobalt acetate) into the system, which, however, also did not give a significant effect [224].

As noted above, a comparison of the results on oxidation in SCW and in other SC fluids, including methane itself, would be very important for understanding the role of the fluid in the oxidation of methane under SC conditions. In addition to comparative data from [229] on the oxidation of methane in SCW and nitrogen, it is worthwhile to mention the known results on the gas-phase oxidation of alkanes at sufficiently high pressure (250 atm for methane) [3,5]. However, in most of these works, even if the parameters of state formally exceeded the critical values, the density of methane was low. From this point of view, of considerable interest is the aforementioned work [48] on the partial oxidation of a 92% CH_4–8% O_2 mixture in a static reactor at very high pressures, 1700 and 3400 atm, temperature of 270–480 °C, and reaction times from 0 to 60 min. In these experiments, the density of methane was 80 mol/l. Although, as can be estimated from the data presented in this work, one of the formally highest values of the methanol yield, up to 4%, was achieved at a methane conversion of ~8%, it is well comparable with the results obtained at pressures 10 times lower. Moreover, at a pressure of 3400 atm, the yield was 2.5 times lower (~1.6%) than that obtained at a pressure of 1700 atm and the same temperature. The composition of the products was also quite common: water, methanol, formaldehyde, CO and CO_2, as well as acetone and (in some experiments), ethanol, and formic acid.

In general, it should be emphasized that the oxidation of methane at high pressures, including SCW conditions, has not provided a yield of target products exceeding that obtained in the 'traditional' gas-phase oxidation. Some authors concluded that the results obtained under SC conditions can be satisfactorily described by the kinetic models developed for the gas-phase oxidation [156,230], suggesting that there are no fundamental differences in the mechanism of these processes. Sensitivity analysis demonstrated a significant role of reactions involving H_2O_2, $HO_2^{\bullet}$, and $OH^{\bullet}$, which is quite natural for a process in SCW. According to the simulation results [222], as in the gas-phase oxidation, a key intermediate determining the yield of partial methane oxidation products in SCW is the $CH_3O_2^{\bullet}$ radical.

Along methane conversion, the reactions of ethane in SCW were studied [224]. The mixture exposed to a pressure of 700 atm and temperatures of 500 and 550 °C for as long as an hour produced no detectable amounts of products. Conversion became noticeable at 600 °C, when gaseous H_2, CO, CO_2, CH_4, and C_2H_4 appeared; i.e., as in the case of methane, ethane conversion in SCW occurs via two parallel pathways: cracking and hydrolysis. The yields of CO_2 and CO were small, with the main gaseous product being methane. Some amounts of liquid, oily products and soot were also formed. Comparison with the reaction in nitrogen at a given temperature, 600 °C, showed that the process in SCW at 700 atm proceeds at about the same rate as in nitrogen at a pressure of 1 atm and 3.5 times more slowly than in nitrogen at a pressure of 700 atm [224].

Also, as for the oxidation of methane, ignition limits were found for the oxidation of ethane in SCW, with the region of the flameless oxidation for the water–ethane system being considerably narrower than for the water–methane system. Under the same conditions, the soot yield for the oxidation in nitrogen was more than an order of magnitude greater than that for the oxidation in SCW. Under the ignition conditions at 400 °C, only a small fraction of the ethane consumed, less than 3%, is converted to oxygenates. The remaining 97% are converted into gaseous products, such as H_2, CO, CO_2, and CH_4. Outside the ignition limits, the yield of liquid products increases even more significantly than for the oxidation of methane. At a large excess of ethane with respect to oxygen, up to 70% of the ethane turned into valuable oxidation products (alcohols, etc.), basically C_2-products, which is also related to the increase of their yield with increasing pressure in the gas-phase oxidation of ethane. Of the C_1-products, only methanol was observed in large quantities. The oxygen-free products are dominated by methane. Very small amounts of ethylene, only under the ignition conditions, were also detected. The main gaseous products of the oxidation of ethane, like methane, were CO and CO_2.

A rather detailed study of the partial oxidation of propane to oxygenates in SCW and near the transition to the SC region, both in static and in flow conditions was carried out in [231]. Apart from experiments under homogeneous conditions, the process was investigated in the presence of some oxide catalysts (Carulite 300, MnO_2, Co_2O_3, MnO_2–Co_2O_3, and MoO_3 supported on γ-Al_2O_3). Long-term experiments in static conditions lasted for up to 6 h at temperatures of 400–500 °C, pressures up to 245 atm and a propane:air:water mixture composition of 1:2.5:250. Experiments under flow conditions were carried out at temperatures of 360–420 °C, pressures of 167–280 atm, and residence times in the reactor of 0.4–7.5 s. The propane:air:water molar ratios ranged within 1:2–25:100–1000. Preliminary experiments showed that, in the absence of oxygen, propane is stable under these conditions.

Experiments under static conditions gave liquid products containing various oxygenates, such as methanol, ethanol, acetaldehyde, acetic acid, *n*-propanol, isopropanol, propionaldehyde, acetone, acrolein, propionic acid, and acrylic acid. The gas phase contained methane, ethane, and ethylene. The conversion of propane ranged from 15% to 61% at a total selectivity of up to 15% to liquid oxygenates and ~10% to C_1–C_2 hydrocarbons.

Under flow conditions, the propane conversion was varied from 30 to 100% by changing the residence time in the reactor and the oxygen concentration. In all cases, methanol was the main liquid product (as in the gas-phase oxidation), which was formed with a selectivity of 12%. In addition, acetone and acetic acid were formed, and to a lesser extent ethanol, acetaldehyde, propionaldehyde and propionic acid, and carbon oxides. The total selectivity to oxygenates reached ~15%, whereas the composition of the product was similar that observed in the static experiments.

The pressure change from 160 to 280 atm, which corresponded to a density change from 0.07 to 0.27 g/cm^3 in the temperature range covered, includes the transition to the SC state and shows some sharp, albeit slight growth in the propane conversion at about 200 atm, close to the critical pressure of water. With further increase of pressure at a constant temperature and reaction time, the conversion remains unchanged. The temperature dependence of the propane conversion (Fig. 9.21) shows a sharp increase, by almost 50%, near the critical temperature of water. With a virtually complete conversion of oxygen in the entire temperature range, the growth of the propane conversion in passing through the critical point can hardly

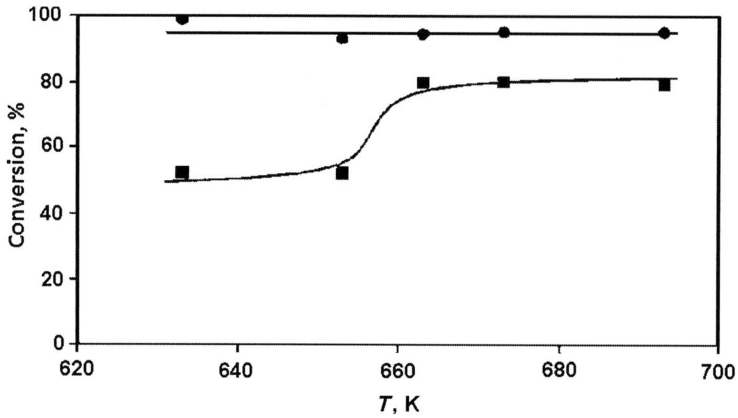

FIGURE 9.21 Temperature dependence of the conversion of (■) propane and (●) oxygen for the noncatalyzed oxidation of propane at $P = 240$ atm and propane:air:water = 1:0.5:250 [231].

be explained, as in [231], only by a high degree of homogenization of the reaction mixture. Rather, it can be assumed that, at temperatures below the critical point, part of the oxygen is converted to some intermediates, which provide an increase in the conversion of propane when the system passes through the critical point.

As the water:propane ratio increases from 250 to 1000, the propane conversion and the selectivity of methanol formation decrease linearly; i.e., as in the partial oxidation of methane, water in the SC state reduces the oxidation rate. With increasing air-to-propane ratio (Fig. 9.22), the propane conversion increases a monotonically, whereas the oxygen conversion passes through a maximum at an air-to-propane ratio of five. It is at this ratio (one oxygen molecule consumed per hydrocarbon molecule, as in the partial oxidation of methane) that the maximum selectivity of formation of methanol, methane, and some other organic compounds is achieved. The experimental results in the presence and absence of catalysts showed complete identity (Fig. 9.22).

Of the other available works on the partial oxidation of methane homologues under SC conditions, it is worthwhile to single out a rather detailed comparative review on the oxidation of isobutane in the liquid phase and under SC conditions [232]. However, the oxidation of butanes is so complex that, at the current stage of studying the basic pattern of the partial oxidation of light alkanes under SC conditions, it is hardly advisable to analyze these data.

As in the case of the gas-phase partial oxidation, the oxidative conversion of alkanes under SC conditions is complicated by a much more rapid oxidation of methanol and other oxygenates. Therefore, in the process of oxidation, they accumulate in relatively small quantities, only as intermediate products. However, it is worth briefly discussing the stability of major products in oxygen-free conditions after reaction completion. The authors of [224] studied the pyrolysis of methanol under SC conditions. At a pressure of 600 atm and a reaction time of 1 h, the temperature rise in SCW from 400 to 500 °C led to an increase in the methanol conversion from 3.8% to 20.6%. The products were methane and, in lesser concentrations, CO_2 and H_2. However, at the same pressure of 600 atm, a temperature of 400 °C, and a reaction time of 1 h, the degree of its conversion in the SC nitrogen was significantly higher,

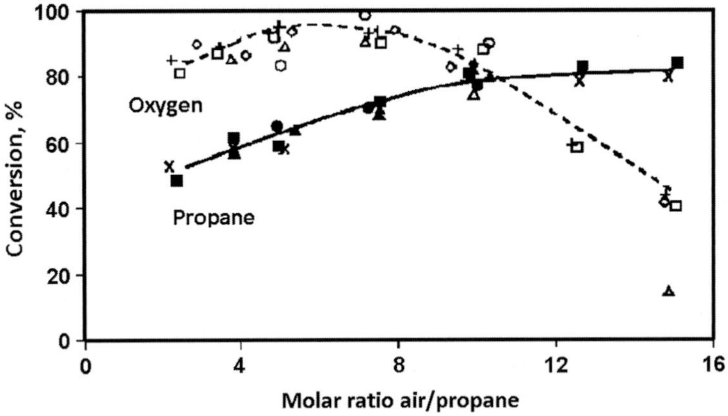

FIGURE 9.22 Noncatalyzed and heterogeneously catalyzed oxidation of propane: influence of the air-to-propane ratio on the conversion of propane and oxygen at $T = 400\,°C$, $P = 240$ atm, and $\tau = 0.4$ s. Propane conversion: (x) without a catalyst, (●) MnO_2/Al_2O_3, (▲) Co_2O_3/Al_2O_3, (◆) $MnO_2{-}Co_2O_3/Al_2O_3$, and (■) MnO_3/Al_2O_3. Oxygen conversion: (+) without a catalyst, (○) MnO_2/Al_2O_3, (Δ) Co_2O_3/Al_2O_3, (◇) $MnO_2{-}Co_2O_3/Al_2O_3$, and (□) MnO_3/Al_2O_3 [231].

34.4%, than that in SCW; the products contained small amounts of dimethyl ether, formaldehyde, and methyl formate. A high stability of methanol in SCW at 450 °C and 250 atm was also observed in [225], where under these conditions, and reaction times of tens of seconds, no any conversion products were detected.

By contrast, formaldehyde is unstable in SCW, with its decomposition rate being strongly temperature-dependent. At temperatures above the critical, formaldehyde completely decomposes to methanol, formic acid, carbon oxide, and carbon dioxide. The main decomposition product at low temperatures is methanol, which gives way to carbon monoxide at high temperatures. At 300 atm, 300–500 °C, and a residence time of 2 min, formaldehyde decomposes almost completely. The rapid decomposition of formaldehyde seems to be the main reason for its absence in the reaction products in a number of works. The Cannizzaro reaction, leading to the formation of methanol and CO, makes it possible to purify dilute formaldehyde wastewater to form easily separable methanol [233].

Since the authors of [224] concluded that most of the CO_2 formed in the partial oxidation of methane in SCW at temperatures below 400 °C is produced by the water-gas reaction rather than directly by the oxidation reaction, the equilibrium constant of this reaction was determined, the value of which passes through unity at a temperature within 500–600 °C. It was suggested that the approach to equilibrium of the water-gas shift reaction at a high water density occurs through the intermediate formation of formic acid:

$$CO + (n + 1)H_2O \leftrightarrow HCOOH + nH_2O \leftrightarrow CO_2 + H_2 + nH_2O, \quad (9.23)$$

where n is the hydration number of the polar formaldehyde molecule. This assumption is supported by studies of formic acid decomposition in SCW, which decomposes under these conditions mainly to CO_2 and H_2.

Formic acid decomposes rapidly in SCW via two parallel pathways. At temperatures below 300 °C, the main channel is dehydration with the formation of CO and water.

At temperatures above 380 °C, the main products are CO_2 and H_2, with small amounts of CO and, possibly, H_2O being also formed, as a result of parallel dehydration. The process is well described by first-order kinetics, with the kinetic parameters known for different temperatures and pressures. Water apparently catalyzes the decomposition of formic acid, since the rate of decomposition in SCW is higher than in the gas phase [220,233].

Summarizing the works discussed above, we would like to point out that, in general, the conversion of methane under SC conditions has been studied more extensively than its homologues. Nevertheless, the number of works on SC methane conversion, the results of which can be compared is very limited, whereas contradictions in the experimental data obtained by different authors are significant, exceeding what could be accounted for by differences in the parameters of the process. For this reason, it is too early to judge the applicability of various approaches to the description of the kinetics of the oxidative conversion of methane under SC conditions, let alone particular models. At the current stage of research, it is more relevant to elucidate the causes of possible differences in the results of various authors and to develop approaches to obtaining consistent experimental data.

CHAPTER

10

Partial Oxidation of Methane Homologues. General Principles of the Partial Oxidation of Light Alkanes to Oxygenates

Methane is the predominant component of most natural and industrial hydrocarbon gases and, therefore, it is not surprising that the vast majority of studies on the partial oxidation of hydrocarbon gases have dealt with the oxidation of methane, all the more that the conversion of this most stable alkane is a major technological challenge. However, almost all hydrocarbon gases contain admixtures of heavier homologues of methane, even small concentrations of which have a marked effect on the kinetics of its oxidative conversion. Furthermore, the production, transport, and processing of hydrocarbons are accompanied by the formation of significant amounts of gases with high or even predominant content of methane homologues, a circumstance that makes the recovery and recycling of such gases an important technological issue. In addition, the oxidative conversion of methane and other hydrocarbon gases produces a set of nearest lighter and heavier hydrocarbons, which influence the kinetic characteristics of the process. Therefore, studying the partial oxidation of methane homologues and mixtures of hydrocarbons similar to real natural gas is a topical problem.

PARTIAL OXIDATION OF ETHANE AND METHANE–ETHANE MIXTURES

A first special study of the partial oxidation of ethane at high pressure was apparently performed in [22]. The oxidation of mixtures containing 12–15% O_2 in ethane was investigated under static conditions at pressures from 15 to 100 atm, temperatures from 260 to 315 °C and a reaction time of a few minutes. As the oxidation of methane, the process has a long induction period, significantly exceeding the time of the conversion itself. Under these conditions, the predominant condensable product was typically ethanol, the maximum concentration of which reached 36.5%, whereas the maximum yield of the sum of oxygenates

Direct Methane to Methanol
http://dx.doi.org/10.1016/B978-0-444-63253-1.00010-6

TABLE 10.1 Pressure Dependence of the Selectivity of Formation of Ethane Oxidation Products in a Static Reactor [22]

		Selectivity, %;									
P, atm	*T*, °C	EtOH	CH_3OH	CH_3CHO	HCHO	AcOH	HCOOH	CO	CO_2	CH_4	Sum of Liquids
15	315	16.0	19.4	1.9	4.5	0	0	–	–	–	41.8
50	286	24.4	14.1	8.3	2.0	1.7	0.9	34.8	10.0	0	51.4
75	279	18.0	16.6	6.8	0.4	3.6	0.6	–	–	–	46.0
100	266	23.7	11.2	6.0	0.05	23.8	0.6	9.3	14.2	8.6	65.3

was 71.7%. Along with carbon oxides, methane, water, and ethanol, methyl alcohol, detectable amounts of formaldehyde, acetaldehyde, formic acid, and acetic acid were formed.

Raising the pressure increased the yield of ethanol, acetaldehyde, and acetic acid, i.e., C_2-products formed without breaking the C—C bond, and reduced the yield of methanol and formaldehyde (Table 10.1). Note that, in these experiments, the reaction temperature was also changed, possibly to keep the reaction time within certain limits (2.5—4.5 min) (Table 10.1); achieving the maximum yield of alcohols at each pressure was not pursued. At pressures above 50 atm, the maximum yield of alcohol and acetaldehyde was weakly dependent on pressure, but the yield of acetic acid increased with pressure up to 100 atm, at which the maximum selectivity, ~24%, was reached. Up to 100 atm, the formaldehyde yield decreased monotonically with increasing pressure.

As the initial concentration of oxygen in the mixture was increased from 11% to 15.5%, the selectivity of formation of most (except formaldehyde) oxygenates, especially ethanol and acetic acid, dropped markedly. However, the maximum temperature of heating of the mixture also increased from 283.5 to 316.8 °C, which could give a significant contribution to this reduction. Adding sequential portions of oxygen to the reaction mixture had little effect on the concentrations ethanol, aldehydes, acids, and methane — only the concentrations of methanol and carbon oxides increased monotonically [22].

Oxidation of a $C_2H_6{:}O_2{:}N_2 = 90{:}3{:}7$ mixture of in a flow reactor at $P = 50$ atm was studied in [17]. At $T = 360$ °C and a contact time of 4 s, the selectivity to ethanol was 63%. However, as the contact time was increased to 20 s, the selectivity of ethanol formation dropped to 14.5%, whereas the methanol formation selectivity increased from trace amounts to 7.8%.

The formation of ethanol, acetaldehyde, methanol, formaldehyde, and acetic acid during the oxidation of a $C_2H_6{:}O_2 = 2{:}1$ mixture in a static quartz reactor at atmospheric pressure and $T = 320$—345 °C was observed in [24]. At atmospheric pressure, the predominant products were formaldehyde and acetaldehyde (with a selectivity of 10—12%), whereas selectivities of ethanol and methanol formation were 1.5—3%, with methanol being generally predominant.

The reaction conditions greatly influence the composition and yield of the products of partial oxidation of ethane, not only the pressure but also the temperature (and hence, duration) of oxidation. This is evident when comparing the pressure dependences of the products of partial oxidation of ethane for static (Table 10.2) and flow (Table 10.3) conditions. At low

TABLE 10.2 Effect of Pressure on the Yield of the Main Products of Gas-Phase Oxidation of an $8C_2H_6 + O_2$ Mixture [18]

			Selectivity, %						
P, atm	T_{in}, °C	t_r, min	C_2H_5OH	CH_3OH	CH_3CHO	HCHO	CH_3COOH	HCOOH	Σ
15	315	3.0	16.0	19.4	1.9	4.5	0	0	41.8
50	294	3.25	17.2	14.1	5.2	1.9	0	0.7	39.1
75	279	2.5	18.0	16.6	6.8	0.4	3.6	0.6	46.0
100	270.5	4.5	23.6	14.0	9.7	0.1	12.5	0.5	50.4

TABLE 10.3 Effect of Pressure and Temperature on the Yield of the Main Products of Gas-Phase Oxidation of Ethane [93]

	Composition of the Oxidation Products, mol%			
Compound	**7.03 atm 595 °C**	**7.03 atm 460 °C**	**70.3 atm 360 °C**	**140.6 atm 340 °C**
Methanol	13	42	41	30
Ethanol	2	4	16	23
Formaldehyde	18	36	12	5
Acetaldehyde	3	5	4	4
Ethylene	46	0	0	0

temperatures of static experiments, high pressure promotes the yield of C_2-products, dominate under these conditions (Fig. 10.1). In harsher conditions of flow experiments (at higher temperatures and much shorter reaction times), C_1-products are predominant even at highest pressures (Fig. 10.2).

Note that, as in the oxidation of methane, low pressure favours the formation of formaldehyde (Table 10.3), whereas temperature rise rapidly increases the yield of ethylene, although the mechanism of its formation during oxidation of ethane is fundamentally different from the mechanism of its formation in the oxidation of methane. Although ethylene is the major primary product of the oxidation of ethane-rich mixtures at high temperatures, experiments with isotope-labelled ethylene showed that the bulk of oxygen-containing products are formed directly from ethane, bypassing the stage of ethylene formation [234].

A detailed study of the partial oxidation of ethane under flow conditions at pressures up to 90 atm was conducted in [235]. The reactor described in [42], with a slow flow of the reagents (residence time of a few minutes) and a Pyrex inner surface, consisted of two separately heated sections: the preheating section and the reactor itself. The experiments on the oxidation of methane conducted earlier in this reactor [42] gave a surprisingly high yield and selectivity of methanol, not consistent with the results of the other studies. The possible reasons for these

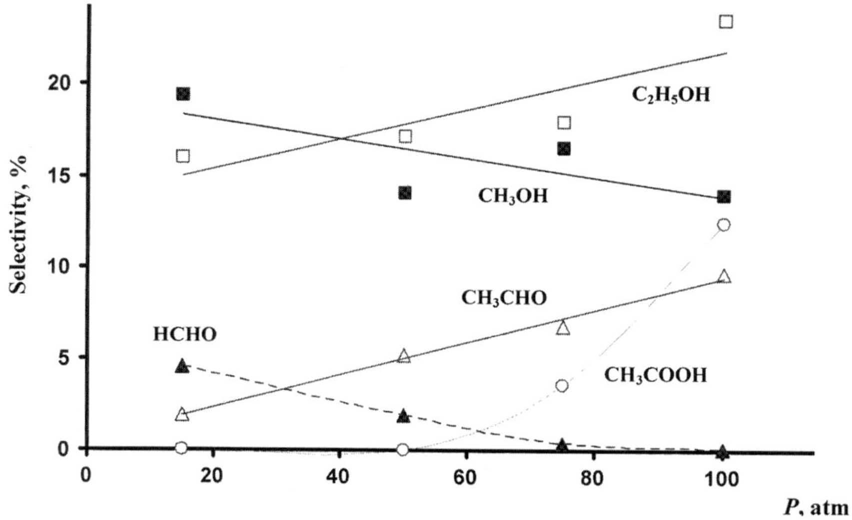

FIGURE 10.1 Pressure dependence of the selectivity of formation of the main ethane oxidation products in an $8C_2H_6 + O_2$ mixture under static conditions: (■) methanol, (□) ethanol, (▲) formaldehyde, (△) acetaldehyde, and (○) acetic acid. *According to the data from Table 10.2.*

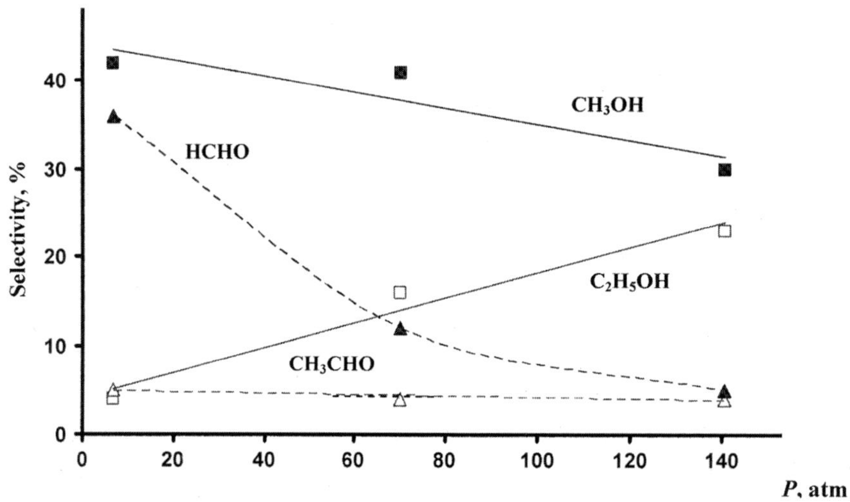

FIGURE 10.2 Pressure dependence of the selectivity of formation of the ethane oxidation products in flow conditions: (■) methanol, (□) ethanol, (▲) formaldehyde, (△) acetaldehyde. *According to the data from Table 10.3.*

results and the role of structural features and conditions of the oxidation process in such a reactor were discussed in Chapters 2 and 3.

The authors of [235] reported a high (up to 60%) selectivity of formation of alcohols. However, while at pressures below 40 atm, the ethanol-to-methanol ratio of the products was less than unity, at 90 atm, it was as high as 6 (Fig. 10.3).

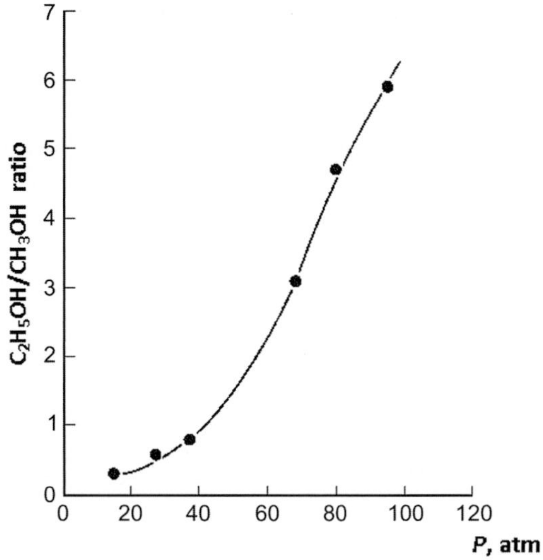

FIGURE 10.3 Ethanol-to-methanol ratio as a function of pressure. $[O_2] = 3\%$ [235].

Furthermore, this ratio increased drastically with decreasing flow rate of the reagents through the reactor (however, this was accompanied by a noticeable decrease in the temperature, from 311 to 275 °C) (Fig. 10.4). Since the flow rate of the reagents in experiments with increasing pressure was maintained constant, the residence time of the mixture in the reactor increased with the pressure: from 1 min at 15 atm (284 °C) to 7 min at 90 atm (236 °C). Given

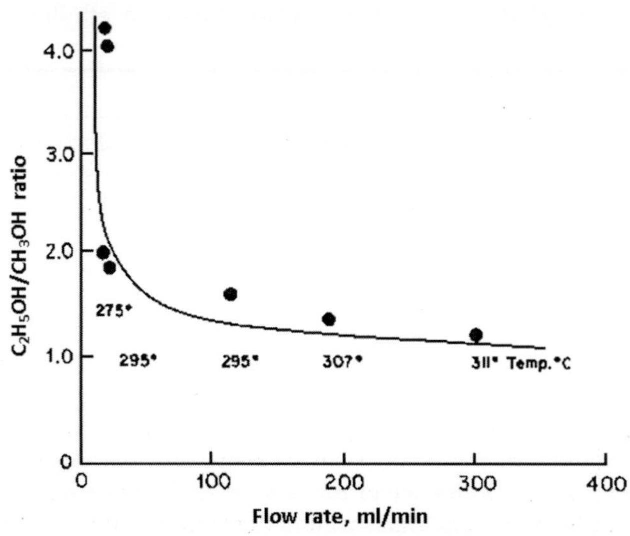

FIGURE 10.4 Ethanol-to-methanol ratio as a function of the reaction mixture flow rate. $[O_2] = 1.5\%$, $P = 84$ atm, and a reactor volume of 1.4 ml [235].

that the ethanol-to-methanol ratio increases with decreasing linear velocity of the reactants in the reactor and the temperature changes, it is difficult to separate the effect of these factors on the actual influence of pressure. It should be noted that, in [17], the residence time of the mixture in the reactor had the opposite effect: at the pressure of 50 atm, the increase in the residence time from 7 to 20 s reduced the ethanol-to-methanol ratio from 19 to 2. The authors also noticed the growth of this ratio with decreasing oxygen concentration.

Increasing the concentration of oxygen (from 2% to 12%) caused a slight decrease in the C_2H_5OH-to-CH_3OH ratio and a significant reduction in the yield of aldehydes. In contrast to ethanol, the yield of acetaldehyde decreased substantially with increasing pressure and decreasing flow rate of the reagents.

The authors of [236] performed a more thorough analysis of the products of this reaction in the same two-section stainless steel flow reactor with a Pyrex liner at a pressure of 34 atm. The main oxidation products detected in the gas and liquid phases were methanol, ethanol, carbon oxides (primarily CO), water, methane, ethylene, propane, n-butane, formaldehyde, acetaldehyde, formic acid and acetic acid, dimethoxymethane, dimethyl ether, acetone, and hydrogen. The selectivity of formation of incomplete oxidation products was high, while the selectivity to carbon oxides did not exceed 16–28% at an oxygen concentration of 3.6–12.8%. The best results were obtained at 287 °C and an oxygen concentration of 6.6%: the ethane conversion was 6.2% whereas the total selectivity of alcohols was 57% at an ethanol-to-methanol ratio of 0.47. The liquid products contained 37% methanol and 17% ethanol.

According to this work, conditions optimal from the viewpoint of the formation of alcohols should provide a low concentration of oxygen, a minimum temperature necessary for complete ethane conversion, the presence of a preheating section, and a low flow rate of the gas in the reactor. With increasing temperature and flow rate, the selectivity of formation of alcohols decreased, especially of ethanol. In addition to alcohols and carbon oxides, chromatographic analysis revealed the formation of water, methane, ethylene, propane, *n*-butane, formaldehyde, acetaldehyde, formic acid, acetic acid, dimethoxymethane, dimethyl ether, acetone, and hydrogen. The presence of *n*-butane in the reaction products was interpreted as a consequence of the recombination of ethyl radicals. With increasing temperature, the process switched to the oxidative coupling of ethane, with a selectivity to *n*-butane of 33% at 425 °C. The rates of formation of CH_4, C_2H_4, and propane also increased. At high flow rate and the maximum oxygen concentration (12.8%), temperature oscillations with an amplitude of up to 80 °C were observed.

The presence of the preheating section produced a positive effect on the yield of alcohols, because of, as the authors believe, a decrease in the temperature required for complete oxygen conversion in the main reactor. A similar effect was observed in the oxidation of methane in the works [42] and [79], wherein the authors pointed out that, at high oxygen concentration, preheating is a crucial prerequisite for obtaining a high selectivity of methanol formation. However, as discussed in Chapter 2, a long stay in the feed mixture in the preheat section may significantly change the composition of the mixture flowing into the main reactor. This, along with the inevitable influence of the reactor surface at a reaction time of a few minutes, may turn out to be one of the main factors determining the yield of the products.

Experimental and kinetic simulation studies of the gas-phase partial oxidation of methane–ethane mixtures were performed in [237,238]. Two different experimental setups were used. In experiments on the setup operating at pressures of up to 15 atm (a reactor with a quartz insert 7 mm in diameter and 100 mm in length), the oxygen concentration was varied from 2% to 8% (oxidizer was air), while that of ethane from 0% to 100%. The residence time of the mixture in the reactor and the temperature ranged within of 10–50 s and 350–500 °C, respectively. Experiments at 100 atm pressure were performed on the apparatus described in [69] at ethane concentrations of 0–22%, oxygen concentrations of 2.7–3% (oxidant was air), and a temperature of 350 °C. The kinetic scheme used for modeling the process was presented in [237].

The experiments showed that the selectivity to methanol grows with increasing ethane concentration and decreases (less dramatically compared to pure methane) with increasing oxygen concentration (Fig. 10.5). Despite the monotonic decrease of the selectivity to methanol, its yield increased up to a concentration of $[O_2] = 7\%$. The simulations demonstrated a qualitative agreement with the experimental results.

The temperature dependences of the concentrations of the main oxidation products are shown in Fig. 10.6. In this case, too, the experimental and calculation results on the yields of methanol and ethylene are in qualitative agreement, while the formation of carbon oxides is not adequately described. The model gives an overestimated yield of CO and greatly underestimated the yield of CO_2. Most likely, this is because the model ignores the heterogeneous transformation of CO to CO_2.

With increasing ethane concentration in the mixture, the $[C_2H_5OH]/[CH_3OH]$ ratio increases sharply, whereas the ratio of the concentrations of formaldehyde and methanol exhibits a more complex behaviour (Fig. 10.7). However, even at $P = 100$ atm, the $[C_2H_5OH]/[CH_3OH]$ ratio was less than unity and almost ceased to grow at ethane concentrations above 20%, observations inconsistent with the results presented in [235], where at a pressure of 90 atm, though in pure ethane, this ratio reached a value of 6.

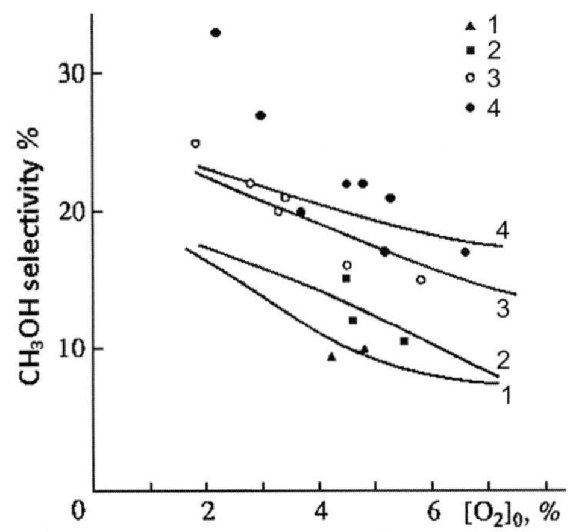

FIGURE 10.5 Dependence of the methanol formation selectivity in the oxidation of methane–ethane mixtures on the oxygen concentration at $P = 10$ atm, $t_r = 26$ s, and various ethane concentrations: (1) 0, (2) 4.5, (3) 12.5, and (4) 30%. The symbols and curves represent the experimental and simulation results, respectively [237].

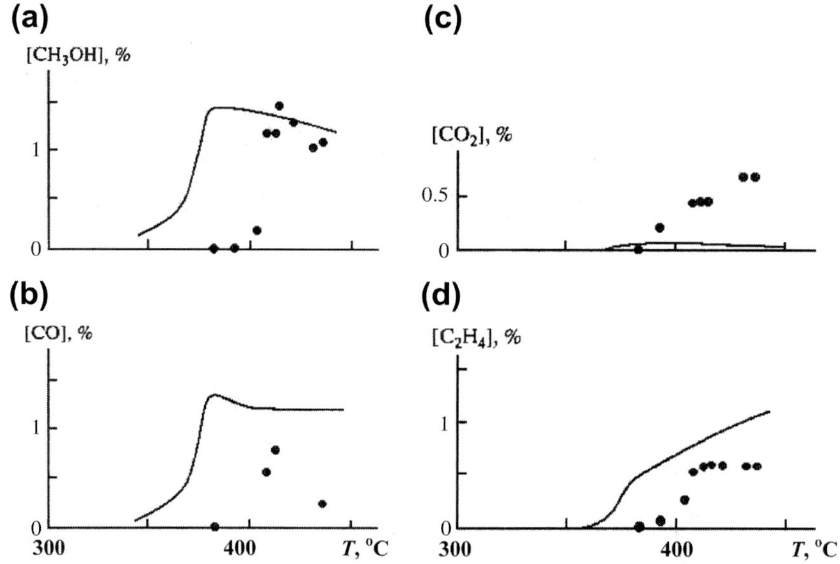

FIGURE 10.6 Temperature dependence of the concentrations of (a) CH_3OH, (b) CO, (c) CO_2, and (d) C_2H_4 for the oxidation of a methane–ethane mixture at $[C_2H_6]_0 = 30\%$, $[O_2]_0 = 3.5\%$, $P = 5.5$ atm, and $t_r = 26$ s. The symbols and curves represent the experimental data and kinetic simulation results, respectively [237].

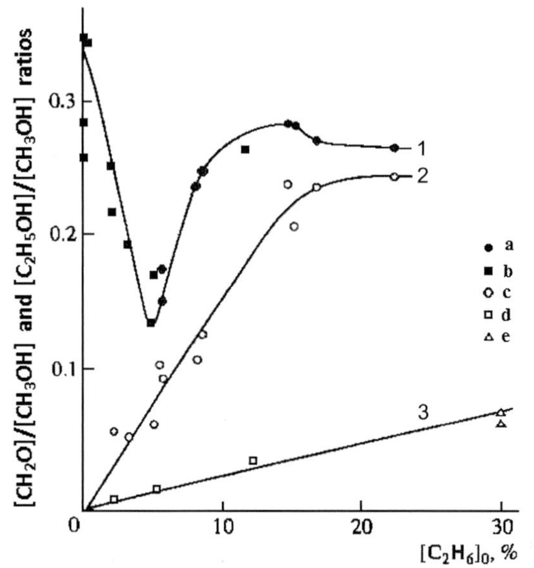

FIGURE 10.7 Dependence of (1) $[CH_2O]/[CH_3OH]$ and (2, 3) $[C_2H_5OH]/[CH_3OH]$ ratios for the oxidation of methane–ethane mixtures on the ethane concentration: (curve 1, points a) $[O_2] = 2.7–3.0\%$, $P = 100$ atm, and $T = 350\ °C$ [237]; (points b) $P = 40$ atm [235]; (curve 2, points c) $[O_2] = 2.7–3.0\%$, $P = 100$ atm, and $T = 350\ °C$ [237]; (curve 3, points d) $P = 40$ atm [235]; (points e) $P = 10$ atm [237].

Experiments under flow conditions on the oxidation of ethane [108] also yielded results somewhat different from those obtained by Newitt [18,22] and Gesser [235]. As in the oxidation of methane, the main product of the gas-phase oxidation of ethane in these conditions was carbon monoxide. Its concentration in the reaction products increased with the initial oxygen concentration, reaching ~6 vol % at the maximum oxygen concentration. Moreover, appreciable amounts of hydrogen, ethylene, and carbon dioxide were observed (Fig. 10.8). The hydrogen concentration in the product grew approximately linearly with increasing

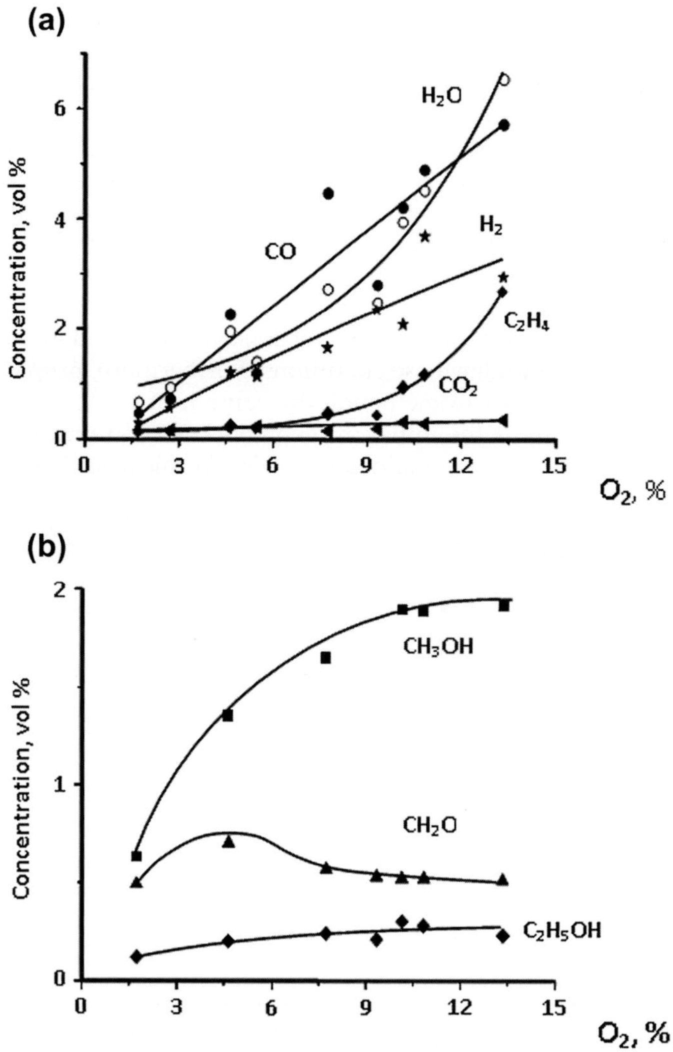

FIGURE 10.8 Dependence of the concentrations of products of the gas-phase oxidation of ethane on the initial oxygen concentration at $P=26$ atm, $Q=400$ l/h, and $T_{max}=550$ °C: (a) (○) H_2O, (●) CO, (◀) CO_2, (★) H_2, (◆) C_2H_4, (b) (■) CH_3OH, (▲) CH_2O, and (◆) C_2H_5OH [108].

initial oxygen concentration, reaching 3%. With growing initial concentrations of oxygen, the ethylene concentration exhibited an incremental increase, also reaching almost 3%. Since the maximum self-heating was virtually independent of the oxygen concentration, the increase in the concentration of ethylene is associated namely with the change of the initial concentration of oxygen. The final concentration of carbon dioxide remained almost constant as the initial oxygen concentration increased from 1.3% to 13.3%, remaining on average an order of magnitude lower than the carbon monoxide concentration (for methane oxidation, $[CO]/[CO_2] \approx 6-7$). However, according to sampling at intermediate stages of the oxidation, the carbon dioxide concentration was much higher than its final value. As in the case of methane oxidation, the reaction products contained small amounts of oxygen (a few percent of its initial concentration), which, probably, had a secondary origin.

The main liquid organic product of the oxidation of ethane, like of methane, was methanol. At low initial oxygen concentrations (up to 5%), its concentration in the liquid oxidation products was ~40%, i.e., practically identical to that for methane. The main difference between the compositions of the oxidation products of ethane and those of methane was a significantly higher concentration of formaldehyde, which, at a low oxygen concentration achieved almost 30%, as compared to ~5–8% for the oxidation of methane. However, with increasing initial oxygen concentration, the content of formaldehyde in the liquid oxidation products decreased sharply (Fig. 10.9).

As compared to methane oxidation, the ethanol concentration under these conditions was almost tenfold higher, ~10% under these conditions. Furthermore, propanol was formed in amounts of 0.2–0.5%. In contrast to methanol, the content of C_2–C_3 alcohols in the liquid remained essentially constant as the initial oxygen concentration was increased from 1.3% to 13.3% (Fig. 10.9). The liquid also contained 0.5–1.5% formic acid. The total concentration

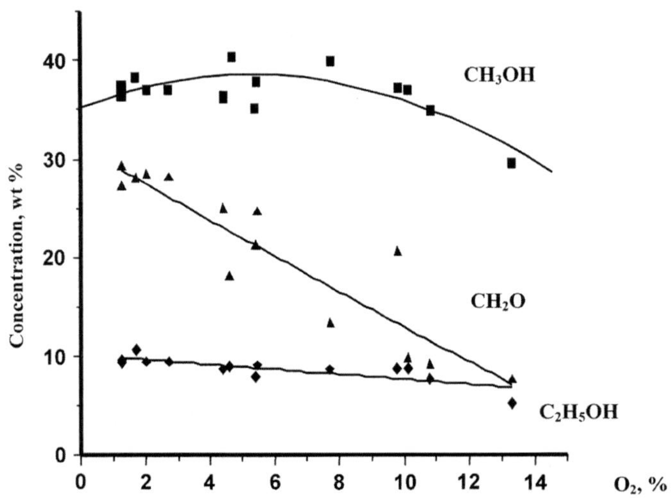

FIGURE 10.9 Dependence of the concentrations of the main organic compounds in the liquid oxidation products on the initial oxygen concentration at $P = 26$ atm, $Q = 400$ l/h, and $T_{max} = 550$ °C: (■) CH_3OH, (▲) CH_2O, (◆) C_2H_5OH [108].

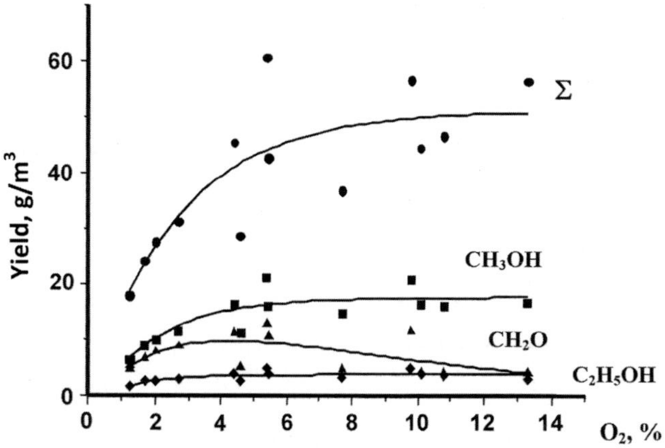

FIGURE 10.10 Dependence of the yield (g/1000 l passed gas) of (●) the sum Σ of liquid products, (■) methanol, (▲) formaldehyde, and (◆) ethanol on the initial concentration of oxygen at $P = 26$ atm, $Q = 400$ l/h, and $T_{max} = 550$ °C [108].

of organic products in the liquid oxidation products, represented mainly by methanol, ethanol, and formaldehyde, reached 75%, i.e., was significantly higher than in the oxidation of methane (<50%); however, it decreased with increasing initial oxygen concentration.

The yield of the liquid products in the oxidation of ethane grew as the initial oxygen concentration was increased up to ~10% (Fig. 10.10). The methanol yield reached its maximum at an oxygen concentration above 6%, being, as in methane oxidation, ~20 g/1000 l of passed gas. With a further increase in the initial concentration of oxygen, it decreased slightly.

The ethanol yield reached almost steady level at an initial oxygen concentration above 4%. The yield of formaldehyde, in contrast to methane oxidation, passed through a maximum at an initial oxygen concentration of ~4–5%. The ratio of the yields of methanol and ethanol was ~0.2–0.3, in agreement with the results obtained at the same pressure in [236].

An important advantage of the direct oxidation of ethane and ethane-containing mixtures in comparison with methane is the possibility of conducting the oxidation process at much lower pressures. Since the pressure of ethane in gas-phase oxidation experiments is limited by its saturation vapour pressure, studies in a wider range of pressures are usually performed with ethane–nitrogen mixtures. Dilution with nitrogen did not reveal any significant effect on the kinetics of the process and the formation of the products. Figure 10.11 shows the pressure dependences of the concentrations of the basic organic oxidation products in the liquid for an initial mixture containing 28.5% ethane in nitrogen. Similar dependences were obtained for the yields of these products. As can be seen, significant changes occurred only at pressures below 30 atm, whereas, for methane oxidation, a pressure reduction from 70 to 30 atm results in a twofold decrease in the yield of methanol and its concentration in the liquid. Ethanol behaved similarly to methanol. The yield and percentage of formaldehyde slightly increases as the pressure decreased to 30 atm and then increases more sharply as the pressure was reduced still further, which made it the main liquid organic oxidation product at pressures below 15 atm. As in the case of oxygen concentration variation, the $[C_2H_5OH]/[CH_3OH]$ ratio remained essentially constant and significantly less than unity

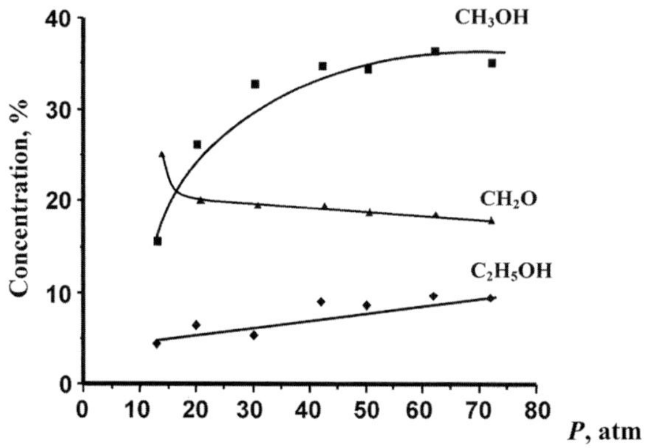

FIGURE 10.11 Pressure dependence of the concentrations of (■) methanol, (▲) formaldehyde, and (◆) ethanol in the liquid oxidate for the oxidation of an ethane–nitrogen mixture at $T_{max} = 460\,°C$, $[C_2H_6]_0 = 28.5\%$, and $[O_2]_0 = 11.5\%$ (the rest is nitrogen) [108].

when the pressure was varied, in contrast to the data from [235], where this ratio increased rapidly with the pressure, reaching a value of 6 at a pressure of ~100 atm.

When the gas flow rate Q was decreased almost fourfold, from 650 to 180 l/h, no changes in the concentrations of C_1–C_2 alcohols in the liquid was observed, whereas the formaldehyde concentration decreased almost twofold. That the formaldehyde concentration decreases with increasing residence time of the mixture in the reactor may be explained by an intensification in its homogeneous–heterogeneous conversion into CO, the concentration of which increases with decreasing mixture flow rate.

The oxidation of methane–ethane mixtures was investigated at pressures of 36–45 atm. An extremely interesting fact is that the concentration of methanol in the liquid (Fig. 10.12), its yield, and the total yield of liquid grow as the methane content in the mixture

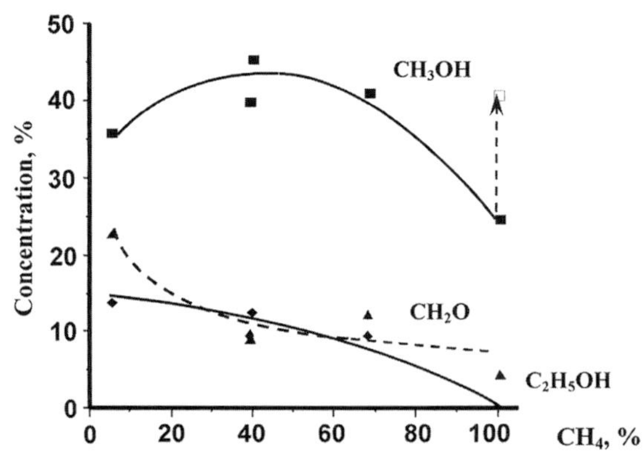

FIGURE 10.12 Dependence of the concentrations of (■) methanol, (▲) formaldehyde, and (◆) ethanol in the liquid oxidate on the methane–ethane mixture composition at $P = 35$–45 atm, $T_{max} = 450$–$500\,°C$, and $[O_2]_0 = 3.6$–4%. The arrow indicates how the methanol concentration would increase if the pressure was raised to $P = 70$ atm [108].

is increased to 40%. This may be indicative of a certain promoting effect of methane on the oxidation of ethane. Unfortunately, the presence of a sufficiently large admixture of methane (~8%) in used ethane did not allow the authors to completely clarify this issue. A rapid decline in the yields of methanol and other oxygenates with a further increase of the concentration was mainly due to the need to heighten the pressure to optimize the oxidation of such mixtures. The arrow in Fig. 10.12 shows the increase in the methanol concentration caused by a pressure rise to 70 atm. Thus, upon pressure optimization for mixtures with high methane content, the curves in Fig. 10.12 are expected to become flatter.

Information on the partial oxidation of ethane at high pressures is almost exhausted by that presented in the aforementioned works, which are at odds on the question of whether the main products of the process can be C_2-oxygenates formed without rupture of the C–C bond. At least at short reaction times and flow conditions, the main products, as in the oxidation of methane, were methanol and formaldehyde. Since these works are quite different in many experimental parameters, the causes of differences in the formation of products remain unclear.

PROPANE

Propane occupies an intermediate position between methane and ethane, which have no internal CH_2 groups, and the other members of the paraffin series. The oxidation of propane at high pressures was studied in static [18,23] and flow [26] conditions. The data reported by Newitt (Tables 10.4 and 10.5) show that the products are acetaldehyde, propionic aldehyde, methanol, ethanol, propanol, and acetic acid, as well as small amounts of propionic and formic acids.

TABLE 10.4 Effect of Pressure and Temperature on the Yield of the Main Products of the Gas-Phase Oxidation of Propane [18,23]

	Composition of Oxidation Products, mol%			
Products	**1.055 atm 373 °C**	**21.09 atm 281 °C**	**63.28 atm 252 °C**	**105.5 atm 250 °C**
Sum of aldehydes	20.5	21.8	13.5	13.7
Normal alcohols	19.7	21.0	17.5	15.2
Isopropanol	1.3	2.8	6.2	16.0
Acetone	0.5	4.3	12.5	7.9
Acids	4.3	17.0	19.0	18.9
Carbon dioxide	7.3	17.1	21.4	20.6
Carbon monoxide	21.3	16.0	9.9	7.7
Propylene	25.1	0	0	0

TABLE 10.5 Relative Yields of Alcohols in the Oxidation of a Propane: Air = 1:3.6 mixture [18,23]

	Yield, %			
Alcohol	**5 atm**		**30 atm**	**65 atm**
CH_3OH	76.3	58.0	55.3	50.1
C_2H_5OH	19.1	17.6	13.2	9.8
n-C_3H_7OH	4.6	4.3	7.3	12.2
iso-C_3H_7OH	–	20.0	24.2	27.9

It should be noted that, in flow conditions at atmospheric pressure, a propane-to-oxygen ratio of 9, and $T = 465\,°C$, the yield of peroxides was high enough: 15 mol per 100 mol of propane consumed, in particular, 9 mol of hydrogen peroxide. The rest is accounted for mainly by propylene [239]. An appreciable concentration of peroxides in the products (up to 1.7 mol %), represented almost exclusively by hydrogen peroxide, was obtained under similar conditions ($P = 1$ atm, $T = 350–475\,°C$, $[C_3H_8]/[O_2] = 5.6–12.3$, $t_r = 3.2–11.8$ s) [240]. Under these conditions, at a large excess of propane ($[C_3H_8]/[O_2] = 5.6$), the ratio of the number of moles of reacted propane to the number of moles of reacted oxygen was close to unity, increasing with temperature from ~0.8 at 375 °C to ~1.4 at 475 °C. In this case, the main saturated hydrocarbon product was methane. Also detected was a small amount of ethane ($[CH_4]/[C_2H_6] \approx 10$), as well as traces of normal and iso-butanes. Olefins were represented by ethylene (up to 8.9%) and propylene (up to 11.1%), with no detectable amounts of acetylene. Along with hydrogen peroxide, the fraction of oxygenates included methanol (3%) and aldehydes (3.8%), mainly formaldehyde.

The growth of pressure was accompanied by a significant increase in the yields of isopropyl alcohol, acetone, and organic acids, a decrease in the yields of aldehydes and normal alcohols with a redistribution their abundance in favour propanol, an increase in the yield of CO_2, and a decrease in the CO yield. At 20 atm and above, propylene disappeared completely from the products (Table 10.4).

The relative fractions of methanol and ethanol decrease, whereas that of propanol increases, with increasing pressure (Table 10.5). The presented data show that the most significant changes occur as the pressure increases to 20–30 atm. A further increase in the pressure, to 100 atm [23] and 200 atm [26], does not provide any appreciable advantages, except for increasing the yield of isopropanol. In addition, increasing the pressure significantly reduces the temperature of the process (Fig. 10.13).

Note that the data on the pressure dependence of the yield of propane oxidation products listed in Tables 10.4 and 10.5 were obtained at temperatures providing an approximately constant rate of the process rather than the maximum yield of alcohols. Since the same data show that a temperature rise increases the overall yield of normal alcohols (from 21% at 260 °C to 40% at 286 °C, both at $P = 30$ atm), optimum conditions are expected to provide a significantly higher yield of alcohols.

As in the oxidation of methane and ethane, a decrease of the oxygen concentration in the mixture results in an increase in the selectivity of alcohols [23]. However, a sufficiently high

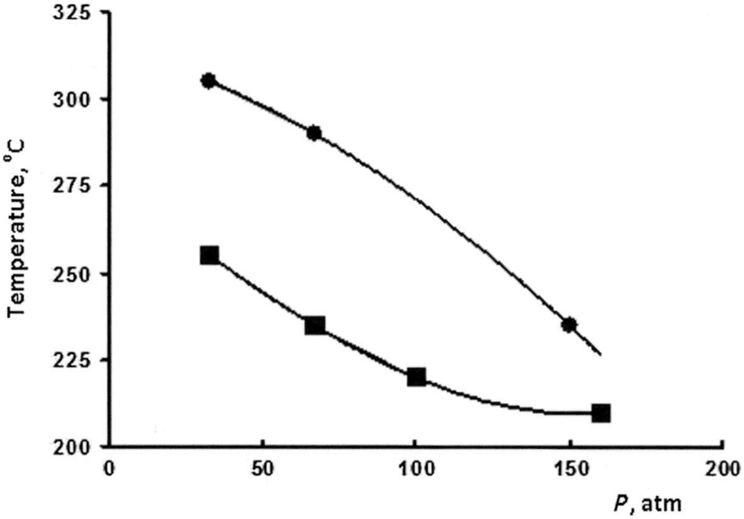

FIGURE 10.13 Pressure dependence of the temperature of onset of the oxidation of (●) propane and (■) butane at $[O_2] = 6\%$ and a long reaction time. *Based on data [26].* (For colour version of this figure, the reader is referred to the online version of this book.)

selectivity of formation of alcohols (30–40%) during the oxidation of propane could be achieved even at a propane-to-oxygen ratio of ~1 (Table 10.6). It was even stated that the optimal conditions for producing alcohols are realized at an oxygen concentration providing the highest possible safety of the process [26].

TABLE 10.6 Influence of the Propane-to-Air Ratio on the Yield of the Main Products of the Gas-Phase Oxidation of Propane at $P = 30$ atm and $T = 275$ °C [18]

	Composition of Oxidation Products, mol %				
Products	**1:20**	**1:5**	**1:3.6**	**1:1.25**	**1:0.5**
Sum aldehydes	12.5	8.8	12.0	16.1	16.7
Normal alcohols	17.3	25.5	23.0	33.1	31.5
Isopropanol	2.7	6.9	5.2	5.2	14.4
Acetone	1.2	1.4	1.3	0.3	7.4
Acids	13.9	13.4	15.2	8.9	12.5
Carbon dioxide	31.5	25.0	22.1	10.5	7.0
Carbon monoxide	20.9	19.0	21.3	25.9	8.0
Sum of condensable products	47.6	56.0	56.7	63.6	85.5

TABLE 10.7 Oxygen Selectivity of Formation of Propane Oxidation Products at $P = 170$ atm, $T = 350\,°C$, and $[O_2] = 7.9\%$ [26]

Product	Selectivity, %
Acetaldehyde (and formaldehyde)	9.4
Acetone	2.3
Methanol, Ethanol, Propanol in a ratio of ~13.4: 8.0:5.2	26.6
Butanols	0.7
Acetic (and formic) acid	4.1
A total of target products	43.1
CO_2	10.1
CO	9.7
Water	30.7
Unreacted oxygen	7.0

The authors of [26] determined the oxygen selectivity of formation of the products in a hollow steel tubular reactor (Table 10.7) at a sufficiently low concentration of oxygen under conditions of incomplete oxygen conversion.

A vast body of experimental studies on the direct oxidation of methane and its homologues, propane and butane, aimed at exploring the possibility of developing an industrial process was performed in the 1950s at the State Institute of Nitrogen Industry (Moscow). The authors of [241,242] analyzed the yield of alcohols, aldehydes, and acids in the oxidation of propane with air under isothermal conditions. It was shown that, as in the oxidation of methane and ethane, the main product is methanol, the yield of which is several times greater than that of ethanol (Fig. 10.14). The authors attribute this to a lower energy of hydrogen abstraction from the central carbon atom as compared the energy of abstraction of the terminal atoms. The interaction with oxygen gives isopropylperoxy radicals, leading to the formation of acetaldehyde and methoxy radicals and then methanol. However, rising temperature increases the probability of reaction of the hydrogen atom with the terminal group and, accordingly, that of formation of ethanol, as was observed in these experiments.

The yield of alcohols increased markedly with the pressure, but the maximum pressure of 15 atm used in the experiments was apparently close to the optimal. In the temperature range covered, 325–375 °C, the yield of alcohols only slightly changed with rising temperature: decreased for methanol and even more slightly increased for ethanol. As can be seen from Fig. 10.14, under these conditions, the reaction is practically completed in just a few seconds.

The same researchers used labelled atoms, more specifically, methanol and aldehydes containing ^{14}C radiocarbon, to demonstrate that the main product formed from methanol during the oxidation of propane was CO_2, with a pressure decrease promoting this process. Aldehydes yield mainly carbon monoxide. In addition, acetaldehyde is partially converted to formaldehyde.

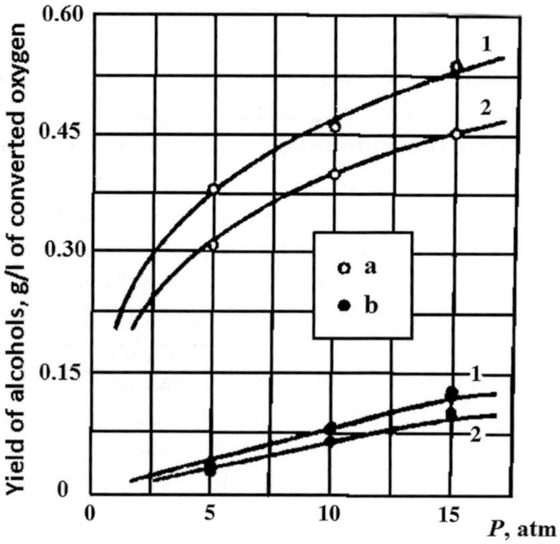

FIGURE 10.14 Pressure dependence of the yield of (a) CH_3OH and (b) C_2H_5OH in the oxidation of propane at $T = 350\,°C$ and various reaction times τ: (1) 16 s and (2) 4 s [241].

Another series of studies on the oxidation of propane under pressure performed at the State Institute of Nitrogen Industry [243,244] showed that, at concentrations below 27 vol %, carbon monoxide formed during the oxidation process produces no effect on the yield and composition of the reaction products. This enables to carry out the oxidation process in the recycling mode. At 350 °C, $P = 10$ atm, oxygen concentration of 3%, and residence time of 2.7 s, the carbon selectivity of formation of target products was as high as 78%. Introduction of catalysts and various packing materials into the reaction zone provided no advantages over homogeneous oxidation.

As noted, even at elevated pressures, a temperature rise causes a rapid changeover from oxygenates to unsaturated compounds. A similar transition was observed for the oxidation of lean propane-air mixtures, studied to elucidate the mechanisms of ignition and knocking in engines, phenomena in which the processes of low- and medium-temperature oxidation play a key role. The experimental results [245] at pressures of 3–10 atm, temperatures of 850–900 K, and a propane to oxygen equivalence ratio of $\varphi = 0.3$, consistent with the results of their kinetic simulations, show that, under these conditions, the main products are C_3H_6, C_2H_4, and CO with trace amounts of CH_4, CO_2, H_2, CH_2O, CH_3CHO, C_3H_6O, and C_2H_5CHO. Moreover, with rising pressure, lowering temperature, and increasing oxygen concentration, the yield of C_3H_6 increases substantially with respect to that of C_2H_4 as a result of a change in the relative importance of channels (10.1)–(10.4)

$$\text{n-}C_3H_7^{\bullet} \rightarrow C_2H_4 + CH_3^{\bullet} \tag{10.1}$$

$$\text{n-}C_3H_7^{\bullet} + O_2 \rightarrow C_3H_6 + HO_2^{\bullet} \tag{10.2}$$

$$\text{i-}C_3H_7^{\bullet} \rightarrow C_3H_6 + H^{\bullet} \tag{10.3}$$

$$\text{i-}C_3H_7^{\bullet} + O_2 \rightarrow C_3H_6 + HO_2^{\bullet} \tag{10.4}$$

(Channels (10.2) and (10.4) are not elementary reactions, representing the results of occurrence of a set of elementary processes).

In experiments [246] at a propane-to-oxygen equivalence ratio of $\varphi = 0.3$, $P = 15$ atm, and $T < 685$ K, the major products were CO_2 and formaldehyde; at higher temperatures, CO and propylene, with propylene being most abundant at $T > 745$ K. The changes in the mechanism of low- and medium-temperature oxidation of propane caused by variation of temperature and pressure, as well as the main pathways of formation of the products, including oxygenates, have been examined in detail.

In the absence of works devoted to the kinetic modeling of the gas-phase oxidation of rich propane-oxygen mixtures at moderate temperatures ($T < 800$ K) and high pressures ($P > 10$ atm), it is worth mentioning, apart from the above works, the simulation results for similar ranges of parameters: $1 \leq P \leq 10$ atm, $900 \leq T \leq 1200$ K, and $0.15 \leq \varphi \leq 4$ [247].

BUTANE

As regards the gas-phase oxidation of butane, pentane, and heavier hydrocarbons at high pressures, there are only disembodied data, which enable to suggest only general ideas concerning the expected yields of alcohols and the ratio thereof. In [26], on the same equipment and approximately under the same flow conditions as in the study of propane oxidation, the oxidation of normal butane, pentane, and heptane was investigated. As in the propane oxidation, increasing the pressure from 33 to 160 atm reduces the temperature of oxidation of *n*-butane from 255 to 210 °C (Fig. 10.13). Among the products of the reaction of *n*-butane with oxygen at pressures from 33 to 160 atm, along with the products found in the oxidation of propane, butyl alcohols were detected, their fraction increasing with the pressure. The selectivity of formation of the oxidation products with respect to oxygen consumed measured on a semi-industrial flow facility at an oxygen content of 5–7 mol % is given in Table 10.8.

Based on these results, the authors concluded that the optimum conditions for the production of higher alcohols from methane series hydrocarbons are as follows: pressure, from 130 to 200 atm; the lowest possible temperature; the residence time in the heated reactor, less than 10 s; and as high as possible (within safe operation limits) oxygen concentration. It was concluded that, in the oxidation of alkanes, low pressure favours the formation of lower alcohols and organic acids, whereas high pressure promotes the formation of higher alcohols and aldehydes [26].

The effect of temperature on the yield of the main groups of butane oxidation products is shown in Fig. 10.15. The yield of oxygenates is promoted by a temperature reduction, whereas temperature rise increases the yield of olefins, first primary and then lower ones [248]. It is interesting that the composition of the gas-phase butane oxidation products is strikingly different from that of the liquid-phase oxidation products obtained at high pressure, when mainly acetic acid and methyl ethyl ketone are produced, with alcohols being present in small amounts [249].

In [250], the oxidation of *n*-butane at 385 °C, pressures of 3.5–12.5 atm, and process times of 0.8–4.5 s was carried out under turbulent jet conditions, to minimize the radial gradients of concentration and temperature, and then quickly quenched by cooling with water. The

TABLE 10.8 Selectivity of Formation of the Butane Oxidation Products with Respect to Oxygen [26]

	Selectivity, %	
Product	**P = 33 atm**	**P = 133 atm**
Acetone	1.4	0.7
Methanol	15.0	8.4
Ethanol	2.5	6.3
Propanol	3.2	8.9
Butanols	0.6	2.5
Aldehydes	6.9	15.9
Acids	16.7	4.6
Sum of target products	46.3	47.3
CO_2	9.4	6.1
CO	3.8	9.4
Unreacted oxygen	15.0	9.0

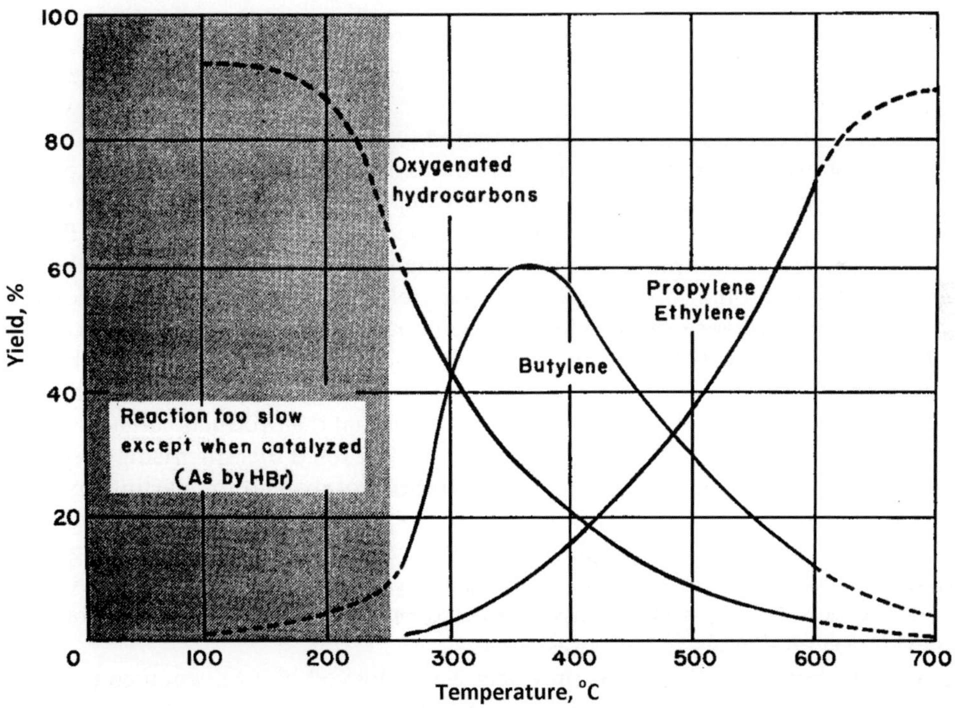

FIGURE 10.15 Effect of temperature on the distribution of products for butane oxidation at atmospheric pressure and constant reactant ratio [248].

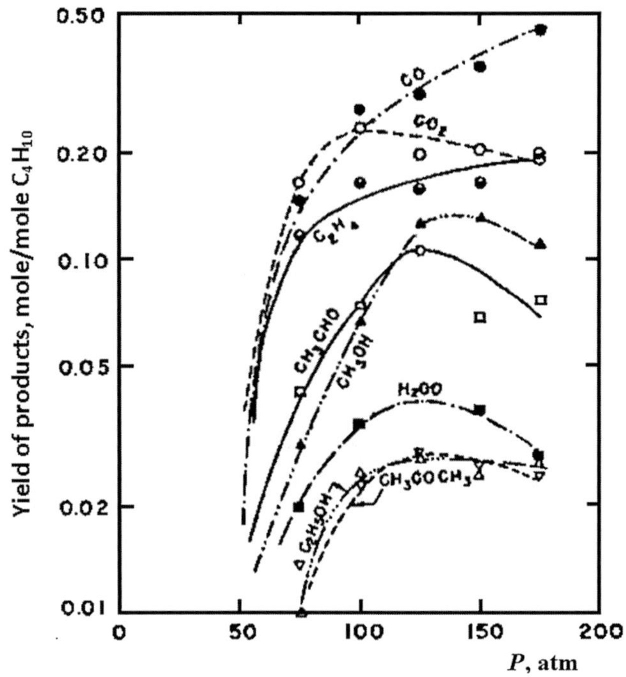

FIGURE 10.16 Effect of pressure on the yield of the butane oxidation products at $T = 386\,°C$, reaction time of 3.0 s, and C_4H_{10}:O_2:$N_2 = 3$:3:94 [250].

concentration of each of the reagents, butane and oxygen, was varied within 1.5–6 mol %, The rest of the mixture was nitrogen. The pressure dependence of the yield of the target products (alcohols, aldehydes, acids, and acetone) passed through a maximum at $P \approx 8.8$ atm (Fig. 10.16). Pressure rise caused no significant changes in the relative content of the different products with the exception of an increase in the yield of carbon monoxide.

Studying the dependence of the yield of the products on the reaction time (Fig. 10.17) clearly shows the existence of an induction period, ~1 s. The concentrations of the main oxygenates pass through a maximum at a reaction time of ~2 s. Further, the composition of products does not change significantly. Note, however, that the products contained an appreciable amount of oxygen, up 1.0–1.5%. The authors interpreted this surprising observation as an indication of the existence of limiting concentrations of oxygen for starting and stopping the process. In their experiments, the reaction did not occur when the initial oxygen concentration was less than 1.5%, whereas the ongoing reaction stopped when the oxygen concentration decreased below this value. Although an incomplete oxygen conversion was observed in other studies, none of them reported on the existence of concentration limits for the oxidation of alkanes at high pressures.

The maximum yield of the target products was achieved at an oxygen concentration of 4.5%. Under optimal conditions ($P = 8.4$–9.8 atm, reaction time of 1.2–1.8 s, and concentrations of *n*-butane and oxygen in a mixture with nitrogen of 4–5% for each), the conversion

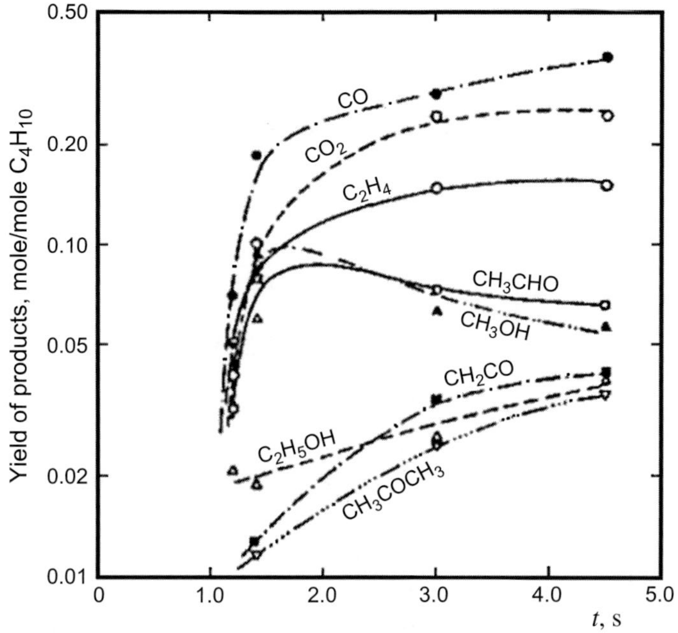

FIGURE 10.17 Effect of the reaction time on the yield of butane oxidation products at $T = 386\,°C$, $P = 7$ atm [250].

of the fed hydrocarbon into the target products was as high as 20% (Fig. 10.18), whereas the selectivity of formation of these products was about 50%.

How the pressure and temperature influence the composition of the products of the gas-phase oxidation of isobutane [93] is illustrated by the data presented in Table 10.9.

The gas-phase oxidation of *n*-butane in butane–air mixtures composed of 18 vol % C_4H_{10}, 3 vol % O_2, 79 vol % N_2 at pressures of 5, 10, and 15 atm and temperatures of 325, 350, and 375 °C was studied in [251,252]. The residence time of the gas mixture in the reaction zone was varied from 2 to 12 s. The heated laboratory reactor was made of a massive stainless steel cylinder with a diameter of 100 mm and a wall thickness of 37.5 mm. The gases were fed into the reactor from a heated rapid-mixture-injection vessel. The quantitatively analysed liquid products were methanol, ethanol, formaldehyde, sum of higher aldehydes, and sum of acids, whereas acetone and ethers were determined qualitatively.

The yield of the liquid oxidation products increases with pressure, but differently for different components: for example, the yields of the aldehydes are virtually identical at 10 and 15 atm, whereas the methanol yield continues to increase within this pressure interval (Fig. 10.19, Table 10.10). The absolute yield of the liquid oxidation products only slightly depended on the temperature within the range covered, but the carbon selectivity of formation of the products decreased with increasing temperature. The maximum yield of the products was achieved within 4–8 s and then, up to 12 s, decreased insignificantly, probably because of their deeper oxidation to CO and CO_2.

The oxidation of butane was also studied in a flow reactor at a pressure of 15 atm and temperatures of 370–390 °C in the mixture circulation mode at a flow rate of 25 m^3/h. Oxygen

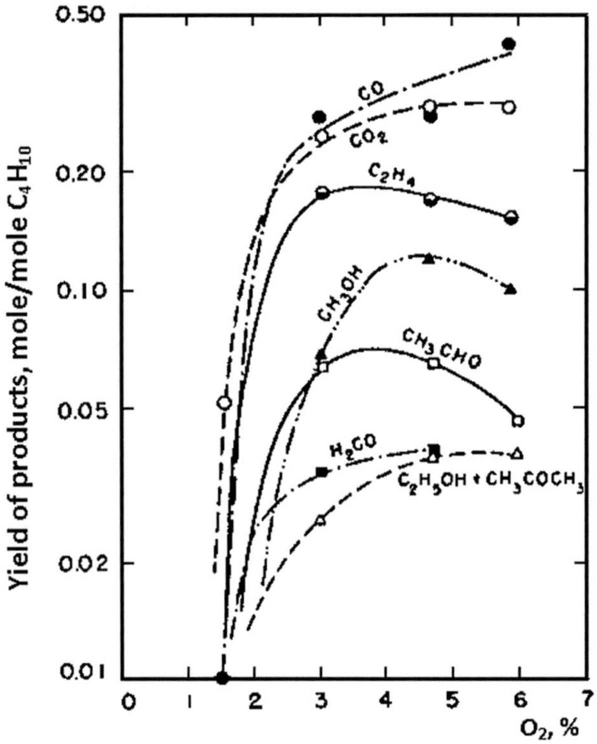

FIGURE 10.18 Effect of the oxygen concentration on the yield of the butane oxidation products at $T = 386\,°C$, $P = 7$ atm [250].

TABLE 10.9 Effects of Pressure and Temperature on the Composition of the Gas-Phase Isobutane Oxidation Products [93]

	Composition of Oxidation Products, mol % with Respect to Introduced Isobutane			
Products	**7.03 atm 452 °C**	**35.15 atm 350 °C**	**140.6 atm 265 °C**	**140.6 atm 215 °C**
Methanol	44	44	33	17
Formaldehyde	44	19	1	1
Acetaldehyde	30	20	9	2
Acetone	31	47	53	37
tert-Butanol	2	6	30	43
tert-Butyl hydroperoxide	0	0	2	9
Water	155	70	42	8

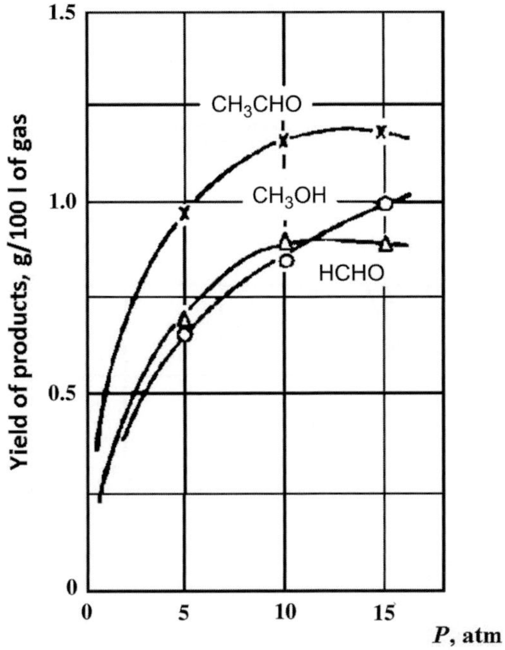

FIGURE 10.19 Effect of pressure on the oxidation of *n*-butane at $T = 350\,°C$ and $\tau = 8$ s [251].

TABLE 10.10 Experimental Results on the Oxidation of *n*-Butane in a Static Reactor Gas Composition: C_4H_{10}, 18 vol %; O_2, 3 vol %; and N_2, 79 vol %. The Reaction Time was 8 s [251]

Temperature, °C		325			350			375	
Pressure, atm	5	10	15	5	10	15	5	10	15
Weight of condensate, g/100 l of gas	3.09	3.64	3.90	2.68	3.62	4.31	2.14	4.19	4.45
Yield of liquid oxidation products, g/100 l of gas									
Methanol	0.64	0.81	0.96	0.66	0.85	1.00	0.51	0.90	1.12
Formaldehyde	0.82	0.86	0.92	0.69	0.90	0.89	0.59	0.94	0.92
Higher aldehydes*	0.99	1.11	1.08	0.97	1.17	1.19	0.89	1.27	1.23
Sum of acids**	0.08	0.07	0.09	0.04	0.05	0.06	0.03	0.07	0.05

* *In terms of CH_3CHO.*
** *In terms of CH_3COOH.*

was mixed with the circulating gas to a concentration of 3–5 vol %. The circulation of the gas continued until the butane content in the mixture dropped to 12%, after which fresh butane was fed into the gasholder. When the CO concentration at the reactor inlet increased to 10%, typically after four to five cycles, the remaining gas mixture was discharged. The reactor was a stainless steel tube with a diameter of 102 mm and a length of 1.4 m.

These data (Fig. 10.20, Table 10.11) are in good agreement with the results obtained in the static laboratory reactor (see above). Under these conditions, the average yield of dewatered

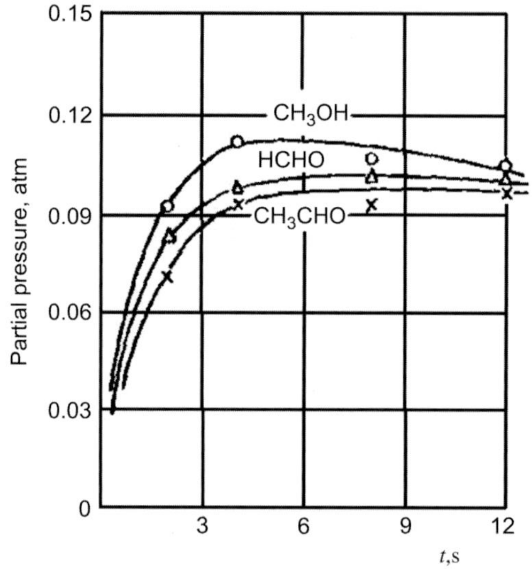

FIGURE 10.20 Kinetics of accumulation of *n*-butane oxidation products at $T = 350\,°C$ and $P = 15$ atm [251].

TABLE 10.11 Experimental Results in the Oxidation of *n*-Butane in the Circulation Mode [251]

Temperature, °C	340–350		370–380		390–400	
Reaction time, h	4	3	2	4	3	3
Average amount of circulating gas, m^3	15.5	17.8	18.5	22.25	21.0	20.0
Butane content, vol %:						
At the beginning of the experiment	20.1	18.2	18.0	21.0	21.8	22.7
At the end of the experiment	11.63	11.7	12.0	12.6	11.7	13.2
Amount of butane reacted, $m^3/100\,m^3$ of gas	1.31	1.55	2.2	1.68	2.82	2.53
Composition of the liquid products, $g/100\,m^3$ of gas						
CH_3OH	536	655	842	753	834	785
C_2H_5OH	28	48	20	–	88	77
CH_3CHO	869	1000	976	910	1104	2100
HCHO	736	816	1020	821	952	
CH_3COOH	70	79	34	41	44	70
Sum	2239	2598	2892	2525	3022	3032

liquid products was 2.6–2.9 kg per 100 m^3 of feed gas mixture, including 0.75–0.8 kg of methanol, 0.9–1.1 kg of acetaldehyde, 0.9–1.0 kg of formaldehyde, and 0.03–0.04 kg of organic acids. Higher alcohols (propyl alcohols and, possibly, butyl) and ketones have not been analyzed, which somewhat reduces the yield of target products. It was demonstrated that reducing the temperature to 340 to 350 °C improves the efficiency of conversion of the hydrocarbon, whereas increasing the pressure enhances the yield of the target products, especially methanol. A significant accumulation of carbon monoxide in the circulating gas (up to 10–12 vol %) did not influenced the oxidation process. The yield of the target products in experiments with different concentrations of CO in the gas remained approximately at the same level.

Because of the complexity of the mechanism of the low- and medium-temperature gas-phase oxidation of rich butane–oxygen mixtures at elevated pressures, there are no works on a detailed kinetic modeling of the process. The mechanism of the low- and medium-temperature oxidation of butane and heavier alkanes at atmospheric pressure has been studied more extensively in connection with the modeling of the self-ignition processes in engines [253–255], to mention only a few. Processes at elevated pressures $9 \leq P \leq 11$ atm ($700 \leq T \leq 900$ K), but for almost stoichiometric mixtures, $0.8 \leq \varphi \leq 1.2$, are considered, for example, in [256]. The authors of this work experimentally identified 22 molecular intermediate products, which gives a clear idea about the level of complexity of the problem.

Little is known about the gas-phase oxidation of heavier alkanes to oxygenates in rich mixtures. It was demonstrated that the products of oxidation of a mixture of 60% *n*-pentane and 40% iso-pentane premixed with 5–6% oxygen consisted essentially of alcohols, aldehydes, acetone, and acids with two to three carbon atoms. In contrast, the oxidation products of heptane (a fraction with boiling point from 70 to 97 °C, containing nearly all of the heptane isomers present in the gasoline fraction head) were mainly alcohols belonging to the hexyl–heptyl group, with a boiling point from 140 to 180 °C. That these data are inconsistent with the results for pentane oxidation was tentatively explained by the occurrence of this reaction in the liquid phase [26].

GENERAL FEATURES OF THE OXIDATION OF ALKANES AT HIGH PRESSURES

First of all, it should be pointed out that the rate of oxidation in the series of normal paraffin hydrocarbons increases rapidly with the length of the hydrocarbon chain. Note also that branched-chain paraffinic hydrocarbons are oxidized more slowly than normal paraffins with the same number of hydrocarbon atoms. This may seem surprising, since the hydrogen atom is more easily separated from a tertiary carbon atom than from a secondary, let alone a primary carbon atom. In this case, the cleavage of the C–H bond is apparently not the limiting step, with the rate of the process being determined by the stability of intermediate oxidation products. The oxidation of normal paraffinic hydrocarbons produces aldehydes, more reactive compounds, whereas the oxidation of isoparaffinic hydrocarbons yields ketones, more stable species. This dependence of the relative reactivity of paraffins on their structure is directly related to their motor properties (octane number) and explains why branched paraffins exhibit higher antiknock properties [93].

The mechanism of the oxidation of hydrocarbons depends on the temperature range in which it occurs. Below about 300 °C, the oxidation of C_4+ hydrocarbons proceeds by the low-temperature mechanism with predominant formation of oxygen-containing compounds: aldehydes, ketones, alcohols, CO, and CO_2. Within 300–450 °C, in the so-called transition region, the formation of oxygenates declines in favour of olefins with the same number of atoms as in the original hydrocarbon. Above 450 °C, cracking to light olefins and partially paraffins takes place (Fig. 10.15). In the transition region, the reaction rate decreases with increasing temperature (NTC phenomenon). For propane, ethane, and especially methane, the boundaries of these regions are shifted to higher temperatures, but the general pattern of the changeover of the oxidation mechanism holds.

The type and structure of the oxidized hydrocarbon are the most important parameters affecting the yield of the products. For example, under identical conditions, acetone is formed much more efficiently from isobutane than from normal butane, whereas the yield of methanol and formaldehyde from it are considerably lower than from the normal hydrocarbon (Table 10.12).

The oxidation of olefins is characterized by a substantially higher yield of valuable products, such olefin oxides, as compared to the oxidation of saturated hydrocarbons (Table 10.13).

Another important parameter of the oxidation process is the ratio of the reagents, i.e., of oxidized hydrocarbon and oxygen. Changing the propane-to-air ratio from 1:20 to 1:0.5 results in a nearly twofold increase (from 47.6% to 85.5%) in the selectivity of formation of liquid oxidation products in a static isothermal reactor at $P = 30$ atm and $T = 275$ °C, mainly due to an increase in the yield of alcohols and acetone (Table 10.14).

A similar effect was observed for the oxidation of propane under adiabatic conditions in a flow reactor (Table 10.15). With increasing hydrocarbon-to-oxygen ratio, the target products have a higher molecular weight, i.e., the degree of destruction of the initial hydrocarbon is smaller.

TABLE 10.12 Influence of the Type and Structure of the Hydrocarbon on the Yield of the Product [248]

	Yield of Liquid Products, wt%		
Product	**Propane**	***n*-Butane**	***i*-Butane**
Acetaldehyde	28.0	30.6	17.0
Formaldehyde	26.4	22.9	19.8
Methanol	23.8	19.4	13.8
Acetone	1.7	4.9	26.0
C_3-aldehydes	2.9	4.5	3.8
Propylene oxide	2.5	–	3.0
Butylene oxide	–	2.8	1.1

TABLE 10.13 Role of Olefins in the Formation of Oxides [248]

	Yield, mol/100 mol of Oxidized Hydrocarbon				
Product	**Propane**	**Propylene**	***n*-Butane**	**Butene-1**	**Butene-2**
1,2-Propylene oxide	2	11.5	–	–	–
1,2-Butylene oxide	–	–	1	8	–
2,3-Butylene oxide	–	–	1.5	–	6.8

TABLE 10.14 Effect of Ratio of the Reagents on the Yield of the Products of Oxidation of Propane in Air at $P = 30$ atm and $T = 275$ °C [23, 248]

Propane-to-Air Ratio	**1:20**	**1:15**	**1:3.6**	**1:1.25**	**1:0.5**
Products	**Selectivity with Respect to Carbon, %**				
Sum of aldehydes	12.5	8.8	12.0	16.1	16.7
Normal alcohols	17.3	25.5	23.0	33.1	34.5
Isopropanol	2.7	6.9	5.2	5.2	14.4
Acetone	1.2	1.4	1.3	0.3	7.4
Acids	13.9	13.4	15.2	8.9	12.5
Total: condensable products	**47.6**	**56.0**	**56.7**	**63.6**	**85.5**
Carbon dioxide	31.5	25.0	22.1	10.5	7.0
Carbon monoxide	20.9	19.0	21.3	25.9	8.0

TABLE 10.15 Effect of the Ratio of the Reactants on the Yield of the Products of Propane Oxidation with Oxygen in a Flow Reactor under Adiabatic Conditions [248]

Propane-to-Oxygen Ratio	**4.5**	**9.4**	**18.0**
Products	**Products Yield, wt%**		
Carbonyl compounds	33.0	45.7	51.0
Alcohols	15.1	21.1	24.7
Carbon oxides	37.2	20.8	13.3

Pressure produces a significant effect on the rate of oxidation of hydrocarbons by increasing the oxidation rate (or lowering the reaction onset temperature) (Table 10.16, Fig. 3.1, Figs 10.13 and 10.21) and changing the composition of the products (Tables 10.3, 10.5, 10.6, and 10.9). Pressure rise, as well as temperature lowering, increases the yield of oxygen-containing products with the same number of atoms in the molecule in the initial hydrocarbon.

TABLE 10.16 Influence of Pressure on the Oxidation Onset Temperature [248]

	Oxidation Onset Temperature, °C			
Hydrocarbon	**1 atm**	**50 atm**	**100 atm**	**150 atm**
Methane	420	350	330	320
Ethane	285	276	260	—
Propane	270	255	245	232
n-Butane	—	248	220	210

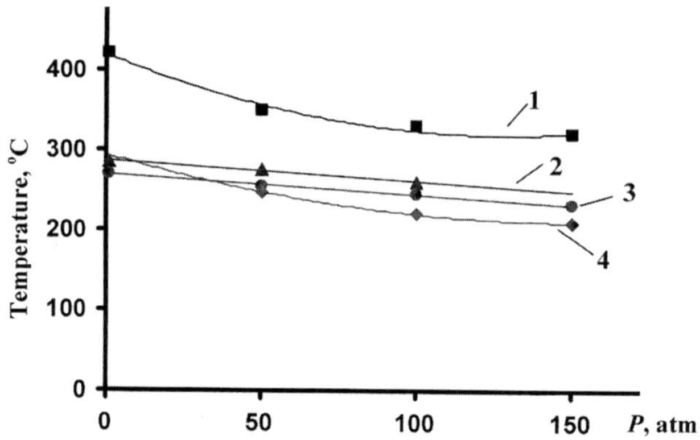

FIGURE 10.21 Temperature of onset of the oxidation of C_1–C_4 alkanes as a function of the pressure: (1) methane, (2) ethane, (3) propane, and (4) butane. *Based on data [248].*

The general pattern of influence of temperature on the yield of the products of the partial oxidation of alkanes is shown in Fig. 10.15. There are three main groups of oxidation products: oxygen-containing compounds, which are predominantly formed at low temperatures, olefins corresponding to the initial hydrocarbon, and olefins formed via the decomposition of the hydrocarbon skeleton of the molecule. Increasing pressure shifts the temperature limits of the regions optimal for the formation of these groups of products towards higher temperatures. For example, at a pressure of 7 atm, the peak yield of butenes in the oxidation of butane occurs at ~600 °C as compared to 375 °C at atmospheric pressure (Fig. 10.15).

A specific feature of the oxidation of lower alkanes is the aforementioned mutual promoting effect of methane and heavier homologues, observed experimentally and predicted by kinetic simulations. It is interesting that not only a more reactive additive increases the methane oxidation rate, but also, under certain conditions, addition of methane, a reactant providing a more effective chain branching, increases the rate of oxidation of its gaseous homologues. The promotion of methane oxidation by its heavy homologues is well known, being mainly associated with a higher rate of initiation of radicals

$$C_nH_{2n+2} + O_2 \rightarrow C_nH_{2n+1}^{\cdot} + HO_2^{\cdot} \quad (10.5)$$

compared to the rate of the reaction

$$CH_4 + O_2 \rightarrow CH_3^{\cdot} + HO_2^{\cdot} \quad (10.6)$$

which largely ensures a higher reactivity of heavy alkanes in the low-temperature oxidation. An increase in the rate of oxidation of less stable hydrocarbon by adding methane, a stable hydrocarbon, is apparently due to a more efficient chain branching in the methane oxidation mechanism, which manifests itself at higher temperatures. In fact, methylperoxy radicals are most likely to provide the branching through the following sequence of reactions:

$$CH_3^{\cdot} \rightarrow CH_3OO^{\cdot} \rightarrow CH_3OOH \rightarrow CH_3O^{\cdot} + OH^{\cdot} \quad (10.7)$$

In the case of heavier alkanes, along with a similar branching mechanism, there are additional possibilities, such as the rapid decomposition of the peroxy radical to the respective olefin and the hydroperoxy radical $HO_2^{\cdot}$, virtually nonreactive under these conditions:

$$C_nH_{2n+1}^{\cdot} \rightarrow C_nH_{2n+1}OO^{\cdot} \rightarrow C_nH_{2n} + HO_2^{\cdot} \quad (10.8)$$

This channel is equivalent to switching-off the branching step and switching-on an additional chain-termination step. Therefore, the addition of methane, for which the channel leading to olefin formation is nonexistent, increases the overall rate of the process. This is confirmed by the kinetic calculations presented in Fig. 10.22, which show that, at relatively low temperatures, the complete conversion of oxygen in its reaction with ethane takes much

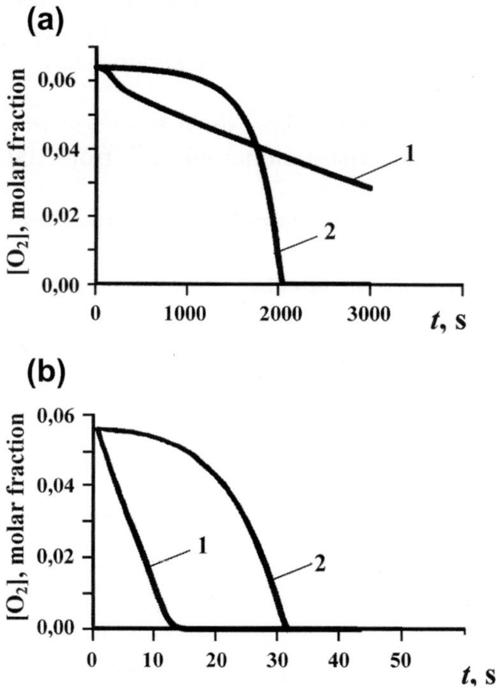

FIGURE 10.22 Calculated kinetic curves of oxygen consumption during the oxidation of (1) methane and (2) ethane at a pressure of 30 atm and temperatures of 573 K (a) and 673 K (b).

shorter time than in its reaction with methane (Fig. 10.22(a)). However, at a higher temperature, at which methane oxidation occurs in the branched-chain mode, the conversion of oxygen in the oxidation of methane proceeds significantly faster (Fig. 10.22(b)).

This conclusion is supported by calculations of the electronic structure, according to which the formation of $HO_2^{\bullet}$ and C_2H_4 by the concerted mechanism,

$$CH_3CH_2OO^{\bullet} \leftrightarrow C_2H_4 + HO_2^{\bullet} \tag{10.9}$$

dominates over the isomerization of this radical

$$CH_3CH_2OO^{\bullet} \leftrightarrow C^{\bullet}H_2CH_2OOH \tag{10.10}$$

since, at similar values of the preexponential factor, the energy barrier for reaction (10.9) is several kcal/mol lower than that for reaction (10.10) [257]. A similar concerted mechanism is likely to operate for heavier alkyl peroxy radicals [258].

Based on the above data, one can draw some general conclusions on the optimal conditions (Table 10.17) and basic patterns of the oxidation of methane and its homologues to oxygenates at high pressures:

1. The oxidation of C_1-C_4 alkanes in optimum conditions can provide a selectivity to liquid organic products of 50% or higher.
2. Increasing the oxygen concentration reduces the selectivity to liquid organic products, but the changeover from methane through to butane sharply decreases the hydrocarbon-to-oxygen ratio (~30 for methane, ~10 for ethane, and ~1 for propane and butane) at which it is still possible to achieve a high selectivity. At the same time, the degree of conversion of the hydrocarbon to useful products increases (~2% for methane, ~5% for ethane, and ~20% for propane and butane).
3. In the methane-to-butane series, the temperature and pressure optimal for producing liquid hydrocarbon products decrease monotonically (from 100 atm and 450 °C for methane, 30 atm and 400 °C for ethane, 20 atm and 350 °C for propane, and 10 atm and 350 °C for butane). These data were obtained in flow conditions at reaction times from a fraction of a second to several seconds.
4. In the methane-to-butane series, the effect of pressure on the composition of the products diminishes.
5. In the oxidation of mixtures of alkanes, even small amounts of higher alkanes substantially lower the temperature of the process, almost to the temperature characteristic of their oxidation.

TABLE 10.17 Optimal Parameters for the Formation of Liquid Organic Products in the Oxidation of Alkanes at High Pressures

Hydrocarbon	*P*, atm	Initial Reaction Temperature, °C	$[C_nH_{2n+2}]/O_2$	Conversion to Useful Products per Passage, %
Methane	70–100	450	30	2
Ethane	25–30	400	10	5
Propane	15–20	350	1	20
Butane	7–10	350	1	20

RELATIVE CONVERSION OF ALKANES IN THEIR JOINT OXIDATION

The process of oxidation of complex hydrocarbon mixtures, corresponding to real natural gases, is not equivalent to the sum of oxidation processes of the individual components. The interaction of the initial reagents and numerous intermediates with each other can qualitatively change the mechanism and behaviour of the process, leading in some cases to phenomena untypical of the oxidation of the individual components. The following describes the results of experimental and theoretical studies of the oxidation of methane–ethane and methane–ethane–propane–butane mixtures that mimic real natural gases [259].

Since methane and ethane are chemically quite different from the rest of the hydrocarbons of the methane series, the process of their co-oxidation is considered separately. As a measure of the relative conversion of methane and ethane, it is convenient to use the ratio of the concentrations of these gases at the inlet (index 0) and outlet (index f) of the reactor:

$$\alpha = ([CH_4]_f/[C_2H_6]_f)/([CH_4]_0/[C_2H_6]_0) \tag{10.11}$$

A more universal parameter for mixtures of arbitrary composition is the relative change in the concentration of each of the hydrocarbons during the oxidation:

$$\beta\ (\%) = ([C]_0 - [C]_f)/([C]_0 \tag{10.12}$$

A negative value of β corresponds to an increase of the concentration of the component in the mixture relative to its initial concentration.

Figure 10.23 shows the experimental and kinetic modeling results on the degree of conversion α on the initial oxygen concentration. For mixtures with a high (50–70%) concentration of ethane, the value of α increases sharply with the initial the concentration of oxygen. Up to the initial concentration of $[O_2]_0 \approx 5$–6%, when the heating of the mixture is relatively low (the adiabatic heating in the partial oxidation of methane to methanol is ~40 °C per

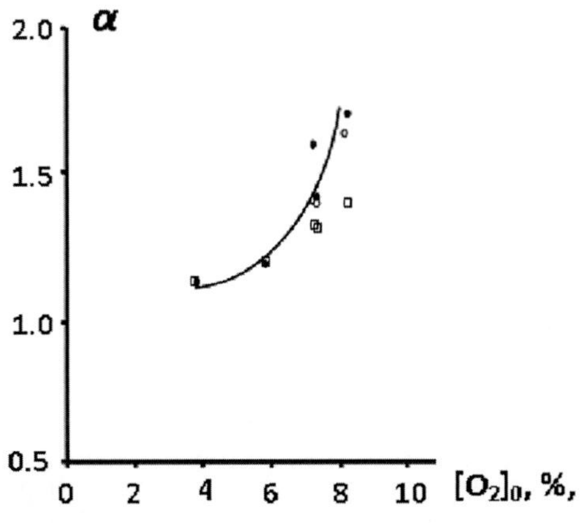

FIGURE 10.23 Dependence of the relative conversion α on the initial oxygen concentration. The solid line and symbols (●) represent the experimental results for mixtures with $[C_2H_6]_0 = 50$–70%. The symbols (○) and (□) represent simulations under adiabatic and isothermal conditions, respectively, all other things being equal [259].

percentage of oxygen in the mixture; for the oxidation of ethane this heating is much lower), the results of isothermal calculations are close to the experimental data. At higher oxygen concentrations, the experimental values of α grow much faster, which is virtually adequately reproduced by calculations for adiabatic conditions, which are more appropriate for these experiments. Simulations for adiabatic conditions almost exactly describe the behaviour of the experimental curves for all tested values of the initial oxygen concentration.

An interesting result of the kinetic modeling is a striking contrast between the dependences of the relative conversion on the initial oxygen concentration for mixtures of with low and high initial concentrations of ethane. While at high ethane concentrations (50–70%) and initial oxygen concentrations less than 5%, simulations under isothermal and adiabatic calculations give nearly identical results (Fig. 10.23), at low ethane concentrations (3%), these values not only differ significantly, but also have different signs, depending on the oxygen concentration. There are presumably two main reasons for this difference. First, the heat capacity of mixtures with low ethane concentration is significantly lower, which results in a considerably higher heating under adiabatic conditions. Second, at very low concentrations of ethane, which is one of the major gas-phase products of the oxidative conversion of methane, it is actually not consumed, because its concentration is maintained at a certain quasi-steady-state level. At higher oxygen concentrations, providing the adiabatic heating of the mixture above 200 °C, the process shifts to the temperature region of oxidative dimerization of methane, with the preferential formation of ethane and ethylene. Under these conditions, the ethane concentration remains practically constant during the adiabatic heating of the mixture, a behaviour quite natural from the standpoint of the kinetics of the process.

The results of kinetic simulations show that the variation of the pressure within 25–70 atm has little effect on the relative conversion at all values of the initial concentration of ethane. There is only a slight decrease in the value of α with increasing ethane concentrations, with the simulation results being in good agreement with experimental data.

The dependence of α on the ethane concentration in the mixture is shown in Fig. 10.24. Determination of the parameter α at zero concentration of methane or ethane is meaningless,

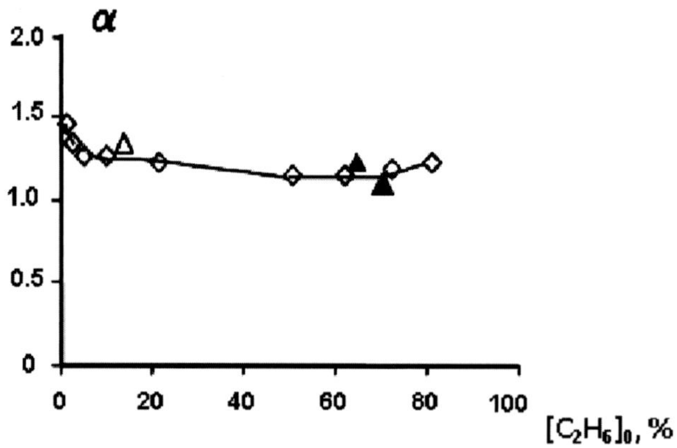

FIGURE 10.24 Dependence of the relative conversion α on the initial concentration of ethane in the mixture: (▲, △) experimental results; the curve and points (◇) represent the results of isothermal calculations at $T = 673$ K, $P = 70$ atm, and $[O_2]_0 = 5\%$ [259].

so calculations were performed at ethane concentration from 1% to 80%. In isothermal conditions, the parameter α very weakly depends on the ethane concentration, but it somewhat increases at the boundaries of the specified interval. Taking into account the weak effect of pressure on the parameter α, the calculation results can be compared with experimental data obtained at a similar initial concentration of oxygen: the agreement is good. In general, a close agreement between the available experimental and theoretical results enables to theoretically predict of the relative conversion α for conditions for which there are no experimental data.

Experiments and calculations show that, in the co-oxidation of methane and ethane at an ethane concentration above several percent, not only a preferential conversion of ethane takes place, but also an increase in the concentration of methane in the mixture is observed due to the oxidative destruction of ethane. However, at an initial concentration of ethane close to 1%, it remains almost unconsumed, since its concentration is maintained by the processes of its formation as a product of the oxidative conversion of methane and by the oxidative conversion of ethane formed. Thus, for the partial oxidation of methane, ethane, and mixtures thereof, it is fundamentally impossible to achieve a complete conversion of only one of these hydrocarbons.

Figure 10.25 demonstrates how the relative changes in the concentrations of C_1–C_4 hydrocarbons during their joint oxidation depend of the most important factor, the initial concentration of oxygen in the mixture. While the relative conversions of propane and butane increase rapidly with the initial oxygen concentration, that of methane at oxygen concentrations above 5% has a negative value, i.e., its concentration in the mixture increases. The formation of methane, as in the oxidation of ethane, is due to the oxidative destruction of higher alkanes. Note, however, that, at oxygen concentrations below 5%, the relative conversion of C_3–C_4 hydrocarbons decreases drastically, as does the methane concentration. This is apparently due to the role of methane in the branched-chain reaction at the initial stage of the oxidation process. For example, the oxidation of propane–butane mixtures under these conditions in the absence of methane is substantially slower. The same effect is predicted by kinetic simulations of the oxidation of methane–ethane mixtures.

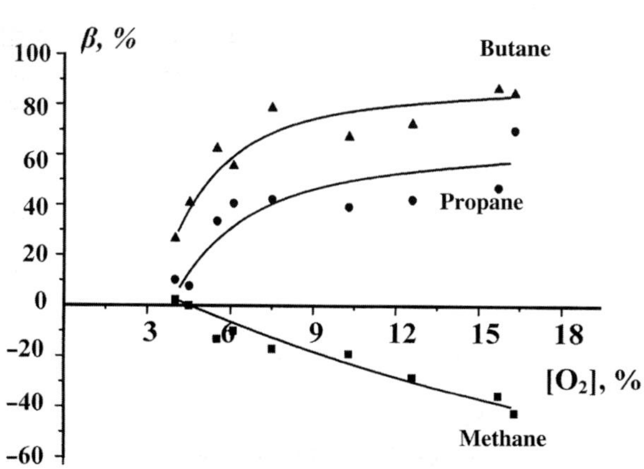

FIGURE 10.25 Relative changes of the concentrations of methane, propane, and butane as functions of the initial concentration of oxygen [259].

Thus, the partial oxidation of complex mixtures of C_1–C_4 hydrocarbons at initial oxygen concentrations above 5% is characterized by a drastic decrease in the concentrations of C_3–C_4 hydrocarbons, an increase in the methane concentration, and the formation of significant amounts of hydrogen and carbon oxides. However, in the oxidation of methane–ethane mixtures, the hydrocarbons are converted into each other, so that their concentrations are coupled more strongly and, therefore, it is difficult to achieve a preferential conversion of ethane.

CHAPTER

11

Promising Technologies Based on the DMTM Process

INDUSTRIAL EXPERIENCE OF THE PARTIAL OXIDATION OF HYDROCARBON GASES TO OXYGENATES

At the time, the partial oxidation of hydrocarbon gases to oxygenates was one of those technological areas in which practical advances were far ahead of the theoretical understanding of the underlying chemical processes. Industrial processes based on the gas-phase partial oxidation of hydrocarbons appeared almost simultaneously with the first data on the possibility of obtaining a high yield of oxygenates and other products in this reaction. The first pilot plant of the Cities Service Company (Tallant, Oklahoma) was built as early as 1926 [11,12], just a few years after the birth of the petrochemical industry, but it turned out to be ineffective because produced too many low-value by-products (Fig. 11.1).

FIGURE 11.1 Works of the Cities Service Company (Tallant, Oklahoma).

Direct Methane to Methanol
http://dx.doi.org/10.1016/B978-0-444-63253-1.00011-8

Then, the Celanese Corporation built three plants in Bishop (Texas), Pampa (Texas), and Edmonton (Canada). At that time, it was the cheapest method for producing three basic oxygen-containing petrochemical products: methanol, formaldehyde, and acetaldehyde [260]. According to estimates, by 1941, these processes accounted for 1/4 to 1/3 of the total production of methanol and formaldehyde in the United States [261].

Thus, by the middle of the last century, the processes of partial oxidation of propane, butane, and mixtures thereof, extracted from associated gas and stabilization and refining gases, became widely spread in the U.S. and grew into a full-fledged branch of the traditional petrochemical industry [93]. Initially, the oxidation was performed with air, with oxygen being first applied in 1949. The main products of the oxidation of associated gas were formaldehyde, methanol, acetaldehyde, acetic acid, acetone, and a mixture of liquid oxygenates, which was used as a solvent or a raw material for producing other products. Unfortunately, the available literature contains little information that would give a clear idea of the operation of the major facilities of the Celanese, McCarthy, Warren, and Cities Service Oil Co companies in the U.S., while the methods developed in the research laboratories of these firms are still kept secret [262].

In general, the applied technological schemes for the gas-phase oxidation of associated gas were similar to the process of methane oxidation. Associated gas mixed with oxidant gas and recycle gas was heated to 355–370 °C at a pressure of 7.0–10.5 atm and fed into a mild-steel reactor. Due to the reaction self-heating, the temperature rose to 425–455 °C, considered optimal for the production of oxygenates. Then the gas mixture flowed from the reactor into cooling, separation, and purification units. The conversion of gas per pass was approximately 20–25%. The optimal mode of the gas-phase oxidation was achieved at a minimum contact time, 0.25–2.0 s, and a maximum oxygen conversion. Although a number of processes involved heterogeneous catalysts, they have not found widespread use in the gas-phase oxidation of petroleum gases [93]. Despite the fact that a wide variety of catalysts has been patented, they provided no significant advantages [262]. An approximate composition of the liquid products of oxidation of C_2–C_4 hydrocarbons are given in Table 11.1.

TABLE 11.1 Approximate Composition of the Products of Gas-Phase Oxidation of C_2–C_4 Hydrocarbons (mol % of Initial Hydrocarbon Converted) [174]

	Ethane	Propane	*n*-Butane	Isobutane
Methanol	15–60	25–50	8–25	5–25
Formaldehyde	5–20	10–15	5–12	5–10
Acetaldehyde	1–6	10–20	2–10	2–4
Ethanol	1–8	2–4	–	–
Acetone	–	3–7	1–6	15–35
Glycols	–	–	5–15	1–10
Other oxygen-containing products	1–4	3–6	7–25	3–27

These results suggest that a moderate pressure of ~7 atm suppresses the formation of olefins and promotes the output of oxygenated products. Oxidation at elevated pressures and low temperatures enables to obtain the maximum yield of the products with the same number of carbon atoms as in the starting hydrocarbon, i.e., without the destruction of the hydrocarbon molecule. Raising the pressure increases the yield of alcohols at the expense of aldehydes. The oxidation of hydrocarbons heavier than ethane is characterized by a heightened yield of secondary and tertiary alcohols. The process was carried out at a low oxygen-to-hydrocarbon ratio: 1 kg of associated gas per ~1 kg of oxygen. According to economic assessments of that time, an efficient production process should consume at least 85–100 tons of O_2 per day; i.e., the plant must process not less than 132,500 m^3 of gas a day. The advantages of using oxygen are the possibility of decreasing the sizes of the equipment for recirculation of oxidized gas, lesser loss of hydrocarbons through exhaust gas discharge, and a high yield of oxidation products (Table 11.2).

Walker process. The first successful industrial facility for the oxidation of hydrocarbon gases is apparently that based on the process developed by the Cities Service Oil Co Company in Tallant (Oklahoma), also known as the Walker process. The process was discovered by accident, as a result of studying the oxidation of natural gas as a probable cause of pipeline corrosion because of the formation of carboxylic acids. Partially stripped natural gas was used, into which air compressed to ~21 atm was injected. The process was conducted over an aluminium phosphate catalyst without recycling at low conversion, pressure of 21 atm, and temperature of 425 °C. One of the subsequent patents describes the oxidation of natural gas to a mixture of methanol, formaldehyde, and acetaldehyde over a mixed catalyst composed of aluminium phosphate and metal oxides. The exhaust gas was discharged into the gas supply system for sale. The main products were methanol and formaldehyde [93,260,264].

Later, a few small plants were built, the process parameters for which were, apparently, quite different. According to [262], the process occurred without catalyst. In this case, natural gas containing 25% ethane yielded a liquid product containing 35% methanol, 20% formaldehyde, 5% acetaldehyde, and some amounts of acetone and dimethyl acetal. According to [263], 1 m^3 of natural gas containing 60% methane (the rest, propane and butane) was

TABLE 11.2 Yield of the Products of Oxidation of Hydrocarbons with Air or Oxygen by Means of the Technology Developed by Celanese Company (kg/l of Liquid Raw Material) [263]

	Air			Oxygen		
Oxidation Products	**Propane**	***n*-Butane**	**Isobutane**	**Propane**	***n*-Butane**	**Isobutane**
Acetaldehyde	0.165	0.180	0.091	0.217	0.191	0.090
Formaldehyde	0.180	0.191	0.120	0.191	0.240	0.144
Methanol	0.144	0.118	0.074	0.168	0.120	0.024
Solvent mixture	0.043	0.072	0.073	0.120	0.180	0.191
Acetone	0.012	0.024	0.144	0.018	0.024	0.168
Sum	0.544	0.585	0.502	0.714	0.755	0.618

TABLE 11.3 Yield of the Products of Butane Oxidation in the Walker Process at the Plant in Oklahoma City [260]

	Approximate Yield of Oxidation Products per Feed Liquid Butane		Annual Capacity	
Product	**lb/gal**	**kg/l**	**lb/year**	**ton/year**
Methanol	0.95	0.114	5,000,000	2268
Formaldehyde (100%)	0.85	0.102	4,300,000	1950
Acetaldehyde	0.39	0.047	2,000,000	907
Other solvents	1.30	0.153	6,500,000	2948
Total	3.49	0.416	17,800,000	8073

converted to 0.333 l of liquid products, including 15% methanol, 22% formaldehyde, 3% acetaldehyde 3%, and 60% mixture composed of solvents and water.

Due to the abundance and cheapness of butane in the U.S. at the time, the Cities Service switched its process in 1953–1954s from natural gas to butane. According to the data from [260], the flow entering the separation unit usually contained 15% methanol, 22% formaldehyde, 3% acetaldehyde, and 60% mixture of various solvents and water. The total yield of the product, as is evident from Table 11.3, was low.

The Celanese firm process (Bludworth process) was launched into operation in 1945 (Bishop, Texas). The initial mixture consisting of ~7 volumes of recycle gas, 1 volume of fresh propane–butane fraction, and 2 volumes of air under a pressure of 7 atm was heated in a furnace to 370 °C and fed into an empty tubular steel reactor (Fig. 11.2), where the ongoing

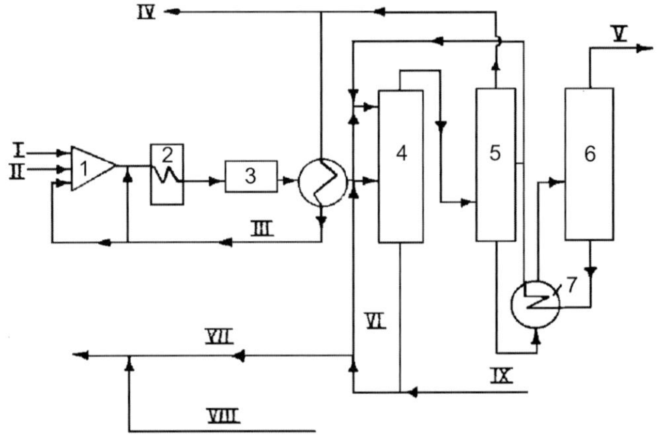

FIGURE 11.2 Scheme of the installation for the gas-phase oxidation of hydrocarbons used by the Celanese firm [263]: (1) mixer, (2) heater, (3) reactor, (4) formaldehyde scrubber, (5) water scrubber, (6) distillation column, (7) heat exchanger. (I) Hydrocarbon gas; (II) air; (III) cycle gas; (IV) gas for the hydrocarbon recovery unit; (V) gas for light products purification unit; (VI) recycled formalin solution; (VII) crude formalin for purification; (VIII) formaldehyde from chemical products purification unit; (IX) caustic soda to adjust the pH value.

reaction raised the temperature to 455 °C. The hot products then flowed into the extended part at the end of the coiled reactor filled with ceramic packing for decomposition of peroxides. Next, the products passed into the bottom part of a quench absorber, where they were cooled to 93 °C with a cold aqueous formaldehyde solution (12–14%). The same solution removed most of the formaldehyde formed in the reaction mixture. At this point, the pH value was strictly controlled by adding an alkali to avoid loss of product due to the formation of acetals.

After the quench absorber, the gas mixture was fed into a water scrubber, where the rest of the condensable products were absorbed. The gas mixture leaving the scrubber through the top outlet contained unreacted hydrocarbons, carbon oxides, and nitrogen. Three-quarters of this gas went for recycling. The remaining part was subjected to oil absorption or cooling to collect hydrocarbons, which were also recycled.

The aqueous solution obtained in the scrubber, containing 15–25% formaldehyde, was fed into stripping tower to remove volatiles; aqueous formaldehyde was collected as a side-cut distillate. High-boiling components were pumped off from the bottom of the column. The resulting 35–40% aqueous solution of formaldehyde was purified from acids with ion exchange resins, whereas other impurities were purified by solvent extraction to match industrial standards.

The aqueous solution of oxygenates from the scrubber bottom was fed into the distillation column from the top of which acetaldehyde was withdrawn and converted to acetic acid by oxidation with air. The remaining mixture was hydrogenated to convert all olefinic hydrocarbons and aldehydes to the corresponding alcohols and paraffins and directed to azeotropic distillation, where methanol in the form of azeotropic mixture with water and heptane was distilled off and then purified from acetone and acetals by rectification. In addition to acetaldehyde and methanol, in some cases, acetone was isolated from the reaction products, with the mixture of remaining compounds (mainly higher alcohols) being discharged, after drying, as a mixed solvent. On average, 1 ton of propane yielded 0.384 tons of formaldehyde, 0.433 tons of acetaldehyde, 0.333 tons of methanol, and 0.270 tons of other oxidation products [93]. An approximate yield of the products in the Bludworth process is given in Table 11.4.

According to other data, presented in [262], a similar process of oxidation of propane or butane with air in a large excess of hydrocarbon and steam was performed at pressures of

TABLE 11.4 Approximate Yield of Oxidation Products in the Bludworth Process [260]

	Yield					
	Propane		*n*-Butane		*i*-Butane	
Product	**lb/gal**	**kg/l**	**lb/gal**	**kg/l**	**lb/gal**	**kg/l**
Acetaldehyde	1.38	0.165	1.5	0.180	0.76	0.091
Formaldehyde	1.5	0.180	1.6	0.192	1.0	0.120
Methanol	1.2	0.144	0.98	0.117	0.62	0.074
Mixture of solvents	0.36	0.043	0.60	0.072	0.61	0.073
Acetone	0.10	0.012	0.20	0.024	1.20	0.144
Sum	4.54	0.544	4.88	0.585	4.19	0.502

up to 60 atm. Passage of 100 kg of a mixture of 1 wt part of butane, 5 wt parts of air, and 34 wt parts of steam through the reactor at 400 °C, 20–30 atm, and a residence time of the mixture in the reactor of 1.15 s, produced (in kg):

Acetaldehyde	19.6
Acetone	7.0
Methanol	19.0
Propanol	1.0
Butanol	0.5
Carboxylic acids	11.4
Formaldehyde	15.2
Total	73.7

The facility was able to process 350 tons of propane and butane per day. The reaction gases were scrubbed with pressurized water, which dissolved oxygen-containing compounds. The residual gas was passed, also under pressure, through an oil scrubber in which nitrogen was separated from the hydrocarbons. In a number of modifications to the process, the nitrogen pressure was used to drive gas turbines. In the course of processing the liquid products, acetaldehyde and formaldehyde were first separated. The remaining compounds were hydrogenated. As a result of various treatments, including additional oxidation of acetaldehyde, the daily production was 120 tons of acetic acid, 150 tons of 37% formalin, 30 tons of acetone, 30 tons of methanol, 15 tons of higher alcohols (ethanol, *n*-propanol, isopropanol, etc.), and ~5 tons of propylene glycol. Isolation of these products poses a significant challenge. A variety of processing methods was used: distillation, rectification, azeotropic distillation, adsorption, selective extraction, and separation by means of ion exchangers. The procedures of isolation of the indicated compounds and of their purification according to the specifications for commercial products are described in more detail in [263].

In the process of oxidation of butane to produce mainly acetaldehyde, converted then to acetic anhydride, one vol part of butane (entering the apparatus in the liquid state) was heated to 150 °C and introduced into the mixer, along with 10 vol parts of air heated to 320 °C and 70 vol parts of steam (400 °C). After the reactor (residence time, 0.3 s; pressure, 4–14 atm), the gas mixture was immediately cooled to 150 °C with injected water. After a sophisticated sequence of treatments, the following set of products was obtained (wt%):

Acetaldehyde	23
Formaldehyde	15
Acetone	10
Organic acids	7
Alcohols	3

Thus, the production of 1 ton of acetaldehyde from *n*-butane required 3.9 tons of butane, ~20 tons of air, and 85 tons of steam.

The oxygen-based process of the Celanese Company was used at the plant in Bishop (Texas) (Fig. 11.3). Oxygen, 95% pure, was mixed with hydrocarbon using a special mixing device. The size of the holes in the mixer through which hydrocarbon was supplied was chosen such as to maintain a certain pressure drop between the mixing chamber and the hydrocarbon pipeline. When the pressure decreased below a permissible safe level, the oxygen flow was automatically stopped and the discharge and purge valves were opened.

The use of oxygen instead of air made is possible to increase the output without using additional equipment and to oxidize a wider variety of raw materials. The increase in the upper limit of the hydrocarbon-to-oxygen ratio in the feed mixture enhanced the efficiency of use of the raw materials.

The technological schemes of the processes employed by the Celanese Company with the use of air (Pampa plant) and oxygen (Chemcel plant) as the oxidant and the description thereof are given in [265].

The process of the McCarthy Company was designed for processing partially stripped natural gas supplied under a pressure of 56 atm at the gasoline plant in Winnie, Texas. The

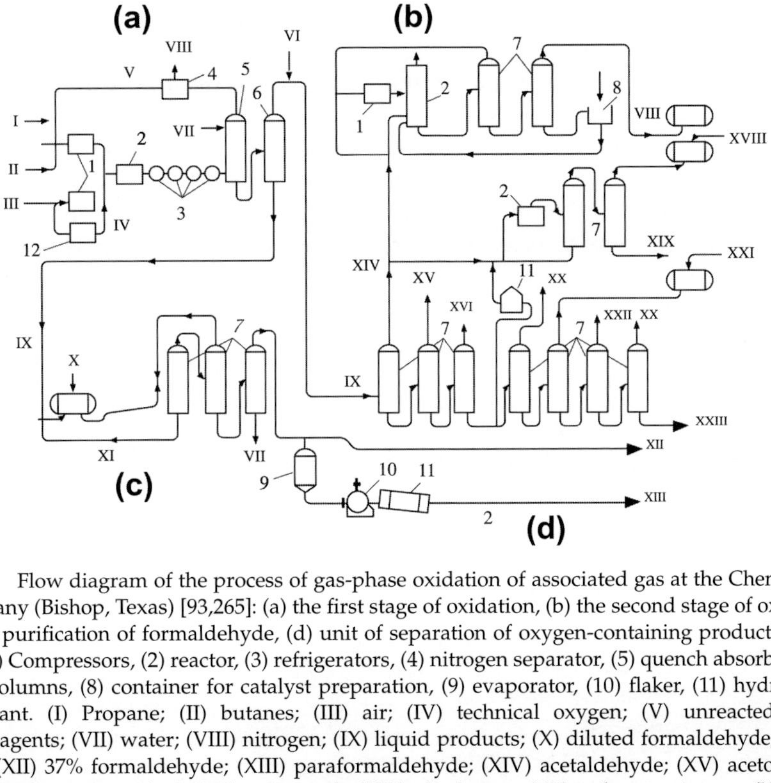

FIGURE 11.3 Flow diagram of the process of gas-phase oxidation of associated gas at the Chemcel plant of the Celanese Company (Bishop, Texas) [93,265]: (a) the first stage of oxidation, (b) the second stage of oxidation, (c) unit of isolation and purification of formaldehyde, (d) unit of separation of oxygen-containing products (acetaldehyde, acetone, etc.). (1) Compressors, (2) reactor, (3) refrigerators, (4) nitrogen separator, (5) quench absorber, (6) separator, (7) distillation columns, (8) container for catalyst preparation, (9) evaporator, (10) flaker, (11) hydrogen generator, (12) oxygen plant. (I) Propane; (II) butanes; (III) air; (IV) technical oxygen; (V) unreacted hydrocarbons; (VI) auxiliary reagents; (VII) water; (VIII) nitrogen; (IX) liquid products; (X) diluted formaldehyde; (XI) circulating formaldehyde; (XII) 37% formaldehyde; (XIII) paraformaldehyde; (XIV) acetaldehyde; (XV) acetone; (XVI) methanol; (XVII) glacial acetic acid; (XVIII) *n*-butanol; (XIX) discharge; (XX) solvent mixture; (XXI) *n*-propanol; (XXII) isobutanol; and (XXIII) glycols.

gas was mixed with 95% oxygen and heated to 370 °C, so that ethane and heavier hydrocarbons were oxidized to valuable oxygenates. The reaction products were washed with water, whereas the exhaust gas leaving the apparatus at a pressure of ~50 atm, virtually containing no nitrogen, was used as a fuel. Like the Walker process, this process is a single-pass one, as opposed to the recycling processes of the Celanese and Warren companies. The process was designed so as to oxidize only ethane, propane, and butane present in natural gas in a total amount of 5 mol %. However, its operation has shown that, along with heavy hydrocarbons, a significant amount of methane is oxidized. Therefore, the process was dismissed as uneconomical, and in 1950, in less than a year of its exploitation, it was abandoned. The main liquid product (70 wt%) was methanol, with some amounts of formaldehyde, acetaldehyde, and other products, a total of 72.7 tons/day of liquid products at a feed gas flow rate of ~68 thousand m^3/day [260]. The process of the McCarthy Company is also known for the fact that it was the first in the petrochemical industry that mastered the large-scale production and use of 95% pure technical oxygen. The daily production of oxygen, supplied for the oxidation process, was 175 tons at a pressure of 56 atm [12].

Process of the Warren Company (Meyer process) was launched into commercial operation in 1954 (Fig. 11.4). In this process, oxygen was mixed with propane or butane in a ratio of 1:2 and added to the recycle flow of hydrocarbons, which was preliminary stripped of oxygen-containing oxidation products, so that the resultant oxygen concentration was 3–6 vol %. The mixture was heated to 315–370 °C and fed into an unpacked tubular reactor made of mild steel. The length of the reactor was sufficient for completing the reaction and oxygen conversion. In the course of the process, the reaction mixture was heated to 427–455 °C, an optimum temperature. The gas mixture leaving the reactor gave up its heat in a heat exchanger to the incoming gas flow, cooling itself to 150 °C, and then entered a formaldehyde scrubber where it was further cooled to 60–70 °C and washed off from nonvolatile compounds, including formaldehyde and nonvolatile acids. After water washing, part of the gas was directed to recirculation, whereas the rest was stripped of propane and butane. The aqueous solution, containing mainly acetaldehyde, methanol, and higher alcohols, was fed into a stripping column, from top of which organic products were

FIGURE 11.4 Plant for the oxidation of propane with oxygen (Meyer process) in Conroe (Texas).

TABLE 11.5 Product Yields in the Meyer Process [260,266]

	Propane		*n*-Butane		Isobutane	
Product	**lb/gal**	**kg/l**	**lb/gal**	**kg/l**	**lb/gal**	**kg/l**
Acetaldehyde	1.80	0.216	1.6	0.192	0.75	0.090
Formaldehyde	1.6	0.192	2.0	0.240	1.2	0.144
Methanol	1.4	0.168	1.0	0.120	0.20	0.024
Acetone	0.15	0.018	0.2	0.024	1.40	0.168
Solvent mixture	1.0	0.120	1.5	0.180	1.6	0.192
Total	5.95	0.714	6.3	0.756	5.15	0.618

collected, whereas water was returned for adsorption. Purification of the final products was completed in several distillation columns. The yields of the products in the process are listed in Table 11.5 [93,260,266].

These processes played a prominent role in the petrochemical industry in the U.S. in the 1950s. Suffice it to note that, of the total of 770,000 tons of acetaldehyde annually produced in the United States during that period, over 120,000 tons were produced by means of the gas-phase oxidation of propane and butane [93,174]. The rapid progress of technology for the direct oxidation of petroleum gases (propane and butane) in the United States in the 1930s—1950s was due to several factors. First, in connection with the burgeoning oil industry, the volume of production of these gases increased dramatically, but there were no sufficiently large markets for these gases and no technologies in which they could be used as raw materials. Therefore, the problem of their utilization as an alternative to flaring was acute. Second, before the war, and especially in the wartime, the industry was in dire need of relevant chemical products, the modern production methods for which have not yet been developed or implemented. The subsequent decline of the technology in the late 1950s was primarily associated with the rapid development of the market of propane and butane, which are now widely used in household and as raw materials in many industrial processes. The second factor that contributed to this was a serious difficulty with extracting individual components from mixtures of a wide variety of products formed by the nonselective gas-phase oxidation of hydrocarbons. Although the developed method provided the necessary purity of the products and met the needs of the wartime, they were too complex and cumbersome to become the basis of a competitive commercial production. Finally, these partial oxidation technologies were based solely on empirical data because of the absence of fundamental studies on the mechanism of the process, a factor that seriously impeded their development.

The current revival of direct oxidation technologies must be based on the following principles: use of less deficient sources of hydrocarbons (natural and associated gas of remote fields, deethanization gases); synthesis of less-difficult-to-isolate products (crude methanol for inhibition of hydrate formation or production of olefins and motor fuels, fuel alcohol mixtures), development of new and effective methods for production of valuable products, such as olefin oxides. Certain economic incentives to the revival of this technology may arise from its potential role in solving environmental and resource problems of the gas and energy industries associated with the development of small-yield sources of hydrocarbons and the transportation of their resources, as well as addressing the problem of utilization of associated gas. However, in all cases, the focus should be on a better understanding of the kinetics of the oxidation processes and methods of their optimization.

Production of ethylene oxide. The National Research Corporation patented a method for the direct noncatalytic oxidation of ethane to ethylene oxide [263]. According to this method, oxygen is mixed with fresh ethane and recycle gas so as to obtain a molar proportion of $C_2H_6:O_2 = 10:15$. The reaction is conducted at 500–700 °C under a pressure of 1–7 atm. The gas mixture leaving the reactor passes through an alumina layer to decompose hydrogen peroxide, and then cooled and fed into a scrubber for the condensation of ethylene oxide, formaldehyde, and acetaldehyde. Part of the noncondensable gases is discharged while the remainder is fed into the reactor for recycling. The specific yields per kg of ethane consumed are 0.38 kg of ethylene oxide, 0.40 kg of formaldehyde, and 0.09 kg acetaldehyde.

Preparation of *tert*-butyl hydroperoxide. Of considerable interest is the possibility of synthesizing *tert*-butyl hydroperoxide by the vapour-phase oxidation of hydrocarbons with a tertiary carbon atom in the presence of hydrogen bromide. For example, at 160 °C, a mixture of isobutane, oxygen, and hydrogen bromide in a proportion of 10:10:1 gives *t*-butyl hydroperoxide with a yield of ~75%:

$$(CH_3)_3CH\ (+O_2) \rightarrow (CH_3)_3COOH \tag{11.1}$$

Isobutanol and di-*tert*-butyl hydroperoxide are formed as by-products. *Tert*-butyl hydroperoxide and di-*tert*-butyl hydroperoxide are used as initiators in radical polymerization and vulcanization of polyolefin rubbers.

There are reports on the gas-phase oxidation of C_5–C_8 saturated hydrocarbons. For example, the gas-phase oxidation of hexane gives α-oxides and derivatives of oxacyclobutane and tetrahydrofuran. There is evidence of the possibility using the gas-phase oxidation of hydrocarbons to increase the octane number of gasoline without changing their boiling limits [174].

Liquid-phase oxidation of associated gas. For comparison, it is worthwhile to briefly mention the liquid-phase oxidation of associated gas. In contrast to the gas-phase oxidation, wherein the bulk of the products are oxygenated compounds with a number of carbon atoms smaller than in the original hydrocarbon, the liquid-phase oxidation makes it possible to introduce oxygen into the hydrocarbon molecule without changing the structure of the latter. For example, the oxidation of n-butane gives methyl ethyl ketone with a high yield. The main product of the liquid-phase oxidation of n-butane is acetic acid, which enables to organize its production by this method. The oxidation is carried out at a temperature of 175–200 °C and a pressure of 45–60 atm. In addition to acetic acid, methyl ethyl ketone, ethyl acetate, methyl acetate, acetone, isobutanol, and other compounds are formed.

The Celanese Company constructed an installation in Pampa that performed the liquid-phase oxidation of butane at a pressure of ~60 atm; the solvent was acetic acid, and the catalyst was cobalt, manganese, or nickel acetate [260].

A great contribution to understanding the mechanism of the liquid-phase oxidation of hydrocarbons has been made by Emanuel and his co-workers (Institute of Chemical Physics RAS). It has been shown that this process is a degenerate branched-chain autocatalytic reaction with an induction period of 10–12 h. To reduce the induction period, it was suggested to initiate the reaction by a short-term blowing of air containing 1% NO_2 through the liquid butane for 15 min, after which the reaction proceeded in the autocatalytic mode.

It was also proposed to carry out oxidation in a solution the critical temperature of which is higher than the critical temperature of *n*-butane (152 °C), because the oxidation near the critical temperature is inefficient. The most suitable solvent appeared to be acetic acid produced in the process, or the unseparated mixture of *n*-butane oxidation products.

The preparation of low-molecular oxygen-containing oxidation products from heavier alkanes is apparently unpromising because of the sharply increasing number of different primary and secondary products and the related difficulties of isolation of individual compounds. For example, while the oxidation of propane can produce up to 16 primary oxygen-containing products, in the oxidation of n-hexane, their number increases to approximately 60, including unsaturated oxygen-containing products, as well as secondary products, such as ethers, formals, and acetals. It is clear that the separation of such mixtures is too complex for the process to be cost-effective. However, as a method for increasing the octane number, the liquid-phase partial oxidation of hydrocarbons with the formation of oxygenates can prove effective.

Oxidation of methane to formaldehyde. One of the first studies in this area was reportedly an experimental factory production of formalin in the United States from natural gas (Empire Refining Co., 1930) with a capacity of 70 million gallons (265 million litres) of a mixture of formaldehyde, methanol, and acetaldehyde. The description of the installation and the method, as well as the yields, has not been published. However, in contrast to the oxidation of propane and butane (associated gas), the processes of direct oxidation of methane have not received widespread in the United States. Two industrial processes for production of formaldehyde from methane were developed in Germany. To produce formaldehyde, methane was oxidized with molecular oxygen in the presence of 1–2% of nitrogen oxides or a heterogeneous catalyst (94% Cu, and 6% Sn). The oxidation of methane in the presence of platinum or palladium yielded mainly formic acid. In this case, the reaction proceeds at a very high rate, so it is impossible to isolate oxidation intermediates, formaldehyde, and methanol [174].

The process of the Gutehoffnungshutte Company involves the oxidation of methane at atmospheric pressure and high temperature in the presence of small amounts of NO_2 (Fig. 11.5). A mixture of methane and air in a volumetric ratio of 1:3.7 is added to the recycle gas (1 volume of mixture to 9 volumes of circulating gas). After being heated in the heat exchanger by the gases from the reactor to 400 °C and being seeded with 0.08% NO_2 (with respect to the volume of the fresh methanol–air mixture), the mixture was fed into the reactor. Nitrogen dioxide was produced just before the mixture entered the reactor, by burning ammonia in air over a platinum catalyst.

The reaction is carried out at 600 °C. The products are cooled in the heat exchanger by the incoming gas to 205 °C and directed into the refrigerator and then into the water scrubber, where a 5–10% solution of formaldehyde is prepared. The solution is neutralized to remove formic acid and then distilled under pressure to obtain a 34% solution of formaldehyde with 3% methanol. Consuming 203.3 m^3 of methane per hour, the installation produced 26.4 kg/h of formaldehyde, i.e., selectivity of its formation was 9.7% [93]. Thus, although the patent description claimed a yield of 35%, the actual yield was only ~10%. The rest of the methane

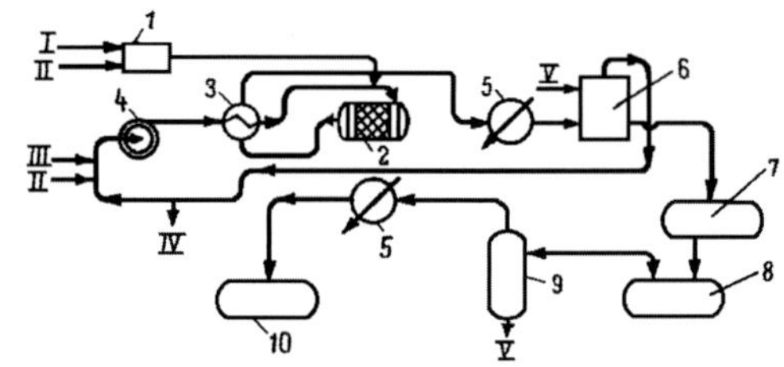

FIGURE 11.5 Flow diagram of the oxidation of methane to formaldehyde (process of the Gutehoffnungshutte Company) [93]: (1) reactor for the synthesis of nitrogen dioxide, (2) reactor for formaldehyde production, (3) heat exchanger, (4) air blower, (5) refrigerator, (6) wash column, (7) receptacle for the crude product, (8) converter, (9) distillation column, (10) receptacle for finished product. (I) ammonia; (II) air; (III) methane; (IV) discharge gas; and (V) water.

oxidized was converted into 3% methanol and 97% of carbon oxides and formic acid, with the latter being formed in trace amounts [267].

Based on this process, the Gutehoffnungshutte Company built the first industrial-scale plant for production of formaldehyde by direct oxidation of natural gas in Romania (Copsa Mica) in 1942 (Fig. 11.6). The process was conducted at 400–600 °C with a 10-fold gas circulation and use of nitrogen oxides as a homogeneous catalyst. The optimum concentration of nitrogen oxides was 0.08%. Its further increase caused a decrease in the selectivity to formaldehyde because of the formation of carbon oxides. The efficiency of methane use (in terms of fresh gas) was ~8%. About 30% of the methane was oxidized to CO, CO_2, and H_2O, with the rest being the unconverted methane. As a result, the formation of one formaldehyde molecule is accompanied by the formation of no less than 7–8 water molecules. The concentration of formaldehyde in the reaction gas was 0.16–24%. The reaction gases, cooled to

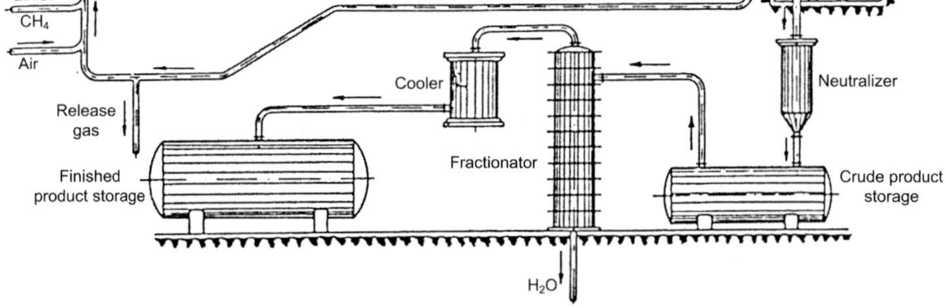

FIGURE 11.6 Flow diagram of the methane oxidation plant operated in Copsa Mica, Romania [35].

200 °C in a heat exchanger, entered a scrubber. Absorption with water produced solutions containing 5–7% formaldehyde. The steel reactor was lined inside with ceramics. The production capacity was 18 tons of formaldehyde per month [35,174,267].

The process of the Hibernia Company [93]. In this process, methane is oxidized by oxygen with ozone addition in the presence of a barium peroxide at 120 °C at atmospheric pressure, and an oxygen-to-methane ratio of 2:3. The temperature must be maintained very precisely, and the content of hydrogen in methane should not exceed 3% vol. The yield was 26.4% of the theoretical one. Industrial plants were not built because of a high cost of ozone.

Process of the Institute of Chemical Physics RAS. The gas-phase oxidation of natural gas to formaldehyde promoted by NO_2 was developed in the 1950s at the Laboratory of Oxidation of Hydrocarbons of the institute of Chemical Physics RAS under guidance of Nalbandyan. The process was based on the ability of potassium tetraborate to accelerate the reaction of methane oxidation and on the ability of packing materials to inhibit the oxidation of formaldehyde to CO_2 [174,268]. A flow diagram of an installation with a capacity of 150–180 m^3 of gas–air mixture per hour, which was built at an experimental plant of the Gazprom VNIIGAZ, is shown in Fig. 11.7.

A mixture of natural gas and circulating gas is taken by compressor 5 and, after the receiver, is mixed with air and nitrogen oxides. Nitrogen oxides are prepared in contact apparatus 6 by the catalytic oxidation of ammonia on a platinum catalyst. The ratio of fresh natural gas to air is 1:2. The methane content in the gas mixture is about 30–33%. The air–gas mixture passes through tubular heater 7 heated by flue gases from furnace 2 and enters reactor 8.

The reactor (Fig. 11.8) was a vertical cylindrical vessel with a conical bottom, made of sheet stainless steel. The diameter of the cylindrical part was 480 mm, whereas the total height, 600 mm. The reactor had an outer insulation jacket. The reaction zone was packed with ceramic Raschig rings pretreated with a solution of potassium tetraborate. The gas–air mixture entered the tapered bottom of the reactor from below at a temperature of

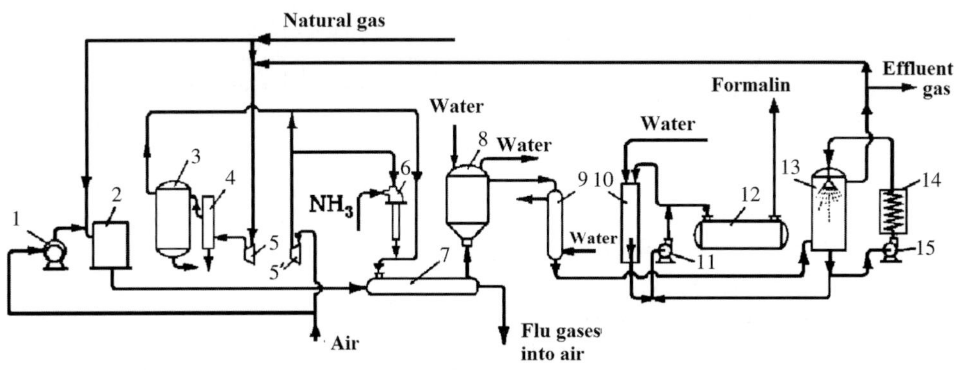

FIGURE 11.7 Flowsheet of formaldehyde production by the gas-phase oxidation of methane in the presence of nitrogen oxides (process of ICP RAS) [174]. (1) Air blower, (2) furnace, (3) receiver, (4) separator, (5, 5′) compressors, (6) contact apparatus for ammonia oxidation, (7) tubular heater, (8) reactor, (9) pipe cooler, (10) absorber, (11, 15) pumps, (12) receptacle for formalin, (13) scrubber, and (14) cooling coil.

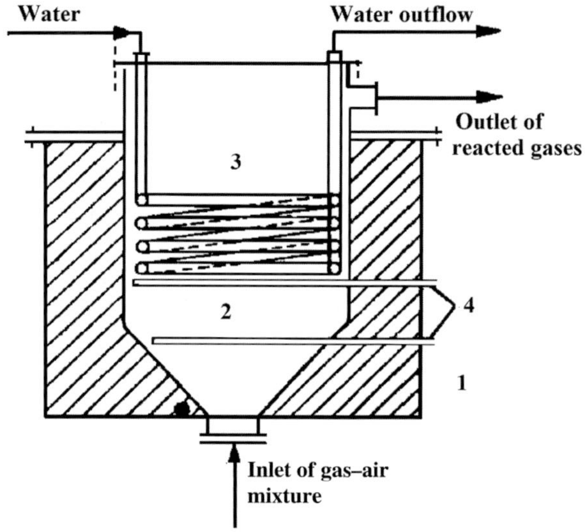

FIGURE 11.8 Scheme of the reactor for the gas-phase oxidation of methane to formaldehyde [268]. (1) Reactor casing, (2) reaction zone, (3) quenching refrigerator, and (4) pockets for thermocouples.

630–650 °C and was heated to 690–700 °C by the reaction. The residence time in the reactor was 0.08–0.1 s. At the outlet of the reaction zone, the reaction gases were quenched. For this purpose, the upper part of the reactor had a cooling coil with circulating cold water. As a result of cooling, the temperature of the gas at the reactor outlet lowered to 400 °C.

The gas outflowing from the reactor was cooled in tubular refrigerator 9–70 °C and fed into scrubber 13, irrigated with a cold (3–5 °C) formaldehyde solution circulating through cooling coils 14 cooled with a brine. As a result, the temperature of the gas–air mixture lowered to 5–7 °C, with provided a 96% capture of formaldehyde. To maintain a constant concentration of formalin, part of the formaldehyde solution from the bottom of scrubber 13 was driven by centrifugal pump 11 into adsorber 10, and irrigated with water. Excess formaldehyde was collected in receptacle 12. The entire hardware of the unit was made of stainless steel to prevent corrosion. The gases unabsorbed in scrubber 13 were directed for recycling, whereas part of the circulating gas was discharged into the atmosphere.

The experiments were conducted in the flowthrough and circulation modes. In flowthrough mode, the optimal residence time of the gas–air mixture in the reaction zone was 0.14–0.16 s. The maximum yield of formaldehyde, achieved at a content of nitrogen oxides of 0.14–0.16%, was 2.4–2.8% with respect to methane, which corresponds to a formaldehyde content in the gas mixture of ~0.80–0.93 vol %. The supplied air oxygen was oxidized only partially, its content in the collected gas was 11.5–12.5%. The discharge gas also contained 30–31% unreacted methane, 56–57% nitrogen, 1.0–1.5% carbon oxides, ~1% water vapour, 0.1% nitrogen oxides, and trace amounts of formaldehyde. Its calorific value was 2800–3000 kcal/m^3.

In the sixfold circulation mode, 150 m^3 of gas mixture per hour circulated in the system. The rate of gas discharge into the atmosphere was 25 m^3/h, which was compensated for by adding fresh mixture with a natural gas-to-air ratio of 1:2 and containing 0.2% NOx.

The average yield of formaldehyde in many-day experiments was 10.6% with respect to methane consumed, or 12.5–13.0 g/m^3. The concentration of formaldehyde in the gas mixture flowing into the scrubber-condenser was ~0.6 vol %. In this case, the formation of one formaldehyde molecule was accompanied by the formation of about three water molecules. The oxygen was also consumed incompletely: the discharge gas contained 5–7% residual oxygen, its calorific value being 2500–2700 kcal/m^3. When air was replaced by oxygen, the concentration of formaldehyde in the reaction gas was 1.5- to 2-fold higher, whereas the optimum temperature of the reaction was 30–40 °C lower. The process was stabile and easy to control during a long period of operation [174,268]. The results were deemed acceptable for the organization of industrial production. Based on the data obtained in 1956–1957, a pilot plant with capacity of up to 2000 tons of formaldehyde per year was designed and built at the Shchekino chemical enterprise. However, later, for reasons not related to technological issues, the works on the process were stopped.

This installation was also used for the oxidation of methane–propane mixtures, since it was demonstrated in laboratory experiments that the yield of formaldehyde in the oxidation of propane is 2.5–2.7 times higher than in the oxidation of methane under similar conditions. Experiments were carried out with an 80% CH_4–20% C_3H_8 hydrocarbon mixture. The hydrocarbon-air working mixture contained 67 vol % air and was fed at a flow rate of 150 m^3/h The yield of formaldehyde was 4.4 vol % with respect to the sum of the hydrocarbons, i.e., 1.8 times greater than for the oxidation of CH_4 under the same conditions. The number of moles of water formed per mole of formaldehyde was higher than for the oxidation of methane alone, so the concentration of the solution obtained by formaldehyde condensation was below 30–35 wt% as compared to 40–45 wt% for the oxidation of methane [196].

As of 1977, the share of the oxidation of hydrocarbon gases in the overall world's production of formaldehyde did not exceed 3–5% [269]. At present, there is no information on the existing facilities of formaldehyde production by the method. Since the mid-1950s, for a number of reasons, the industrial use of the direct oxidation of hydrocarbon gases to oxygenates virtually ceased. First of all, the propane–butane fraction, extracted as a byproduct of natural gas production, has been an important household fuel and a raw material for the petrochemical industry, circumstances that deprived the direct oxidation process of low-cost and convenient-to-handle raw materials. On the one hand, the partial oxidation of dry natural gas, which requires much more stringent conditions and gives a lower yield of the target products, has never been mastered by the industry. Moreover, the low selectivity to certain products and their wide variety, especially in the oxidation of C_3–C_4 hydrocarbons, required a large effort to separate the individual components. But most importantly, there was no clear understanding of the mechanism of the process, knowledge of which was based on disembodied and rather contradictory empirical data.

For these reasons, DMTM technology was replaced by the competing process for producing methanol and other chemical products by converting natural gas to syngas, which, despite being more complex, provides a highly selective large-scale production of many staple chemicals under consideration of easy access to enormous gas reserves. This technology remains the foundation of modern gas chemistry. However, the rapid change in the raw material base of the gas industry and the inevitable increase in it of marginal and dispersed sources (shale gas, gas hydrates, low-yield fields, associated gas, etc.) requires a new, more simple, low-capacity, and flexible technology. The use of most of these resources on the basis of the existing technologies for the transportation and processing of natural gas is practically impossible, which opens new perspectives for alternative technologies, including DMTM.

CONCEPTUAL SCHEMES OF THE PARTIAL OXIDATION OF NATURAL GAS TO OXYGENATES

With the development of the theory of the branched-chain oxidation of hydrocarbons and the progress in understanding the DMTM mechanism, new versions of technological processes based on it have been suggests with account of these basic concepts. The 'niches', most promising from the economic and technological points of view, where its application can provide tangible benefits, have been identified in [6,270]. These processes are essentially based on the concepts of the circulation and flowthrough modes of operation.

The flowthrough mode of the process, with the use of air as the oxidant (Fig. 11.9), is the most simple and attractive, particularly for the development of small-sized installations that meet the own needs of gas companies in methanol hydrate inhibitors. For the conditions of Russian gas fields, the consumption of methanol for these purposes is up to 1–2 kg per 1000 m^3 of gas produced [271]. On average, a gas processing plant (GPP) consumes about 4500 tons of methanol per year. Given that the delivery of methanol in remote areas on average doubles its price, providing the GPP with locally produced crude methanol would give a significant economic effect. Due to the high heat of the reaction, the recovery of the heat of the exhaust gas is not only sufficient for heating the fresh reaction gas, but can also provide the formation of steam for heating the liquid products to be distilled and, moreover, can provide heat for the gas production complex.

The optimum conditions of the process are as follows: the temperature of the gas mixture at the reactor inlet ~450 °C, pressure 7–10 MPa, and initial oxygen concentration 3–4%. At this concentration of oxygen, without special measures taken, the temperature of the reaction mixture will rise by 120–160 °C. The yield of methanol per pass through the reactor is 17–20 kg per 1000 m^3 of the gas passed. The approximate composition of the resulting liquid (wt%): methanol 38–42, formaldehyde 4–8, ethanol, 0.5–1.0, acids 0.2–0.5, and the rest is

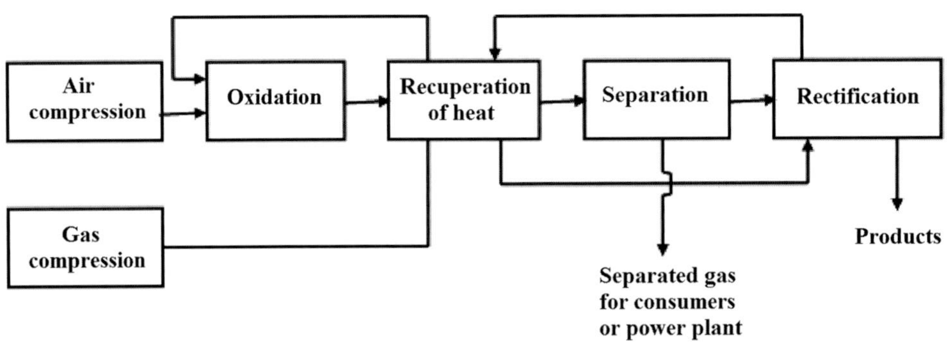

FIGURE 11.9 Flow diagram of the process of partial oxidation of natural gas.

water. The methane conversion is almost equal to the initial concentration of oxygen, whereas the exhaust gas, in addition to ~1% CO, ~0.2% CO_2, and small amounts of hydrogen, contains only nitrogen, in accordance with its volume in the supplied air. An optimal solution is to use this gas as fuel for local needs. It can also be returned back into the pipeline if its calorific value and content of nitrogen permit. In the latter case, since the purpose is only to obtain a certain amount of methanol, with the unreacted gas being returned, there is no sense in intensifying the process of oxidation: a small degree of conversion, 3–5%, attainable in a single pass through the reactor, would do. If the inlet gas pressure is not less than 5 MPa, there is no need in additional compression. The use of only one distillation column produces crude methanol with a concentration of the main component no less than 91%, whereas the water concentration is below 9%, which is quite suitable for using this product as a hydrate inhibitor [271]. As a byproduct, the distillation residue contains a weak (8–12%) formalin solution, which has numerous applications, including gas fields. For example, it can be used as a curing agent for phenol-formaldehyde resin in treatment of wells against flooding and as a microbicide against sulphate-reducing bacteria, which deteriorate the collector properties of the reservoir.

If the discharge gas cannot be returned to the gas pipeline and its energy content is much higher than the existing energy needs, its conversion can be increased. The most obvious way is to increase the number of sequentially connected reactors (cascade connection). However, this involves multiple cooling, separation, and reheating of the reaction mixture, i.e., excessive complication of the technological scheme, so that the process is unlikely to be economically justified. A more interesting approach may be based on an incomplete cooling of the reactants with simultaneous extraction of methanol by means of absorption or adsorption. Such a scheme has been proposed to eliminate the recirculation of the technological gas during methanol production from syngas. Tetraethylene glycol dimethyl ether (tetraglyme, TEGDME), boiling at 275 °C and having a good thermal stability, was selected as the best-suited absorbent [272,273]. However, in any cascade scheme, due to the gradual increase of the volume of the reaction gas because of air nitrogen addition (the volume which is about four times that of the converted methane), the size the reactors of succeeding cascades stages necessarily increases, while the productivity with respect to methanol decreases.

A more promising option is a spread-oxygen introduction at several points along the reactor. This makes it possible to not only provide a high rate of the process and, hence to reduce the size of the reactor, but also to increase the yield of methanol at the same flow rate of oxygen. Another important prerequisite for obtaining a high selectivity to methanol at a relatively high conversion of methane is to limit the heating of the reaction mixture to ~550–570 °C. At $T > 600$ °C, in addition to a reduction in the yield of oxygenates, soot is formed, which contaminates the liquid oxidation products. This factor limits the initial oxygen concentration and, accordingly, the conversion of natural gas per pass through the reactor to 4–5%. One solution for this problem is to carry out the oxidation in a sectioned reactor with a spread oxidant supply and excess heat removal between the sections, for example by generating steam (Fig. 11.10). Two options are possible: to carry out the reaction in the void volume followed by driving the gas mixture into the steam generator section or conducting the process directly in the steam generator with continuous heat removal [274]. In the latter case, by changing the parameters of steam supply, it is possible to smoothly adjust the temperature of the oxidation process.

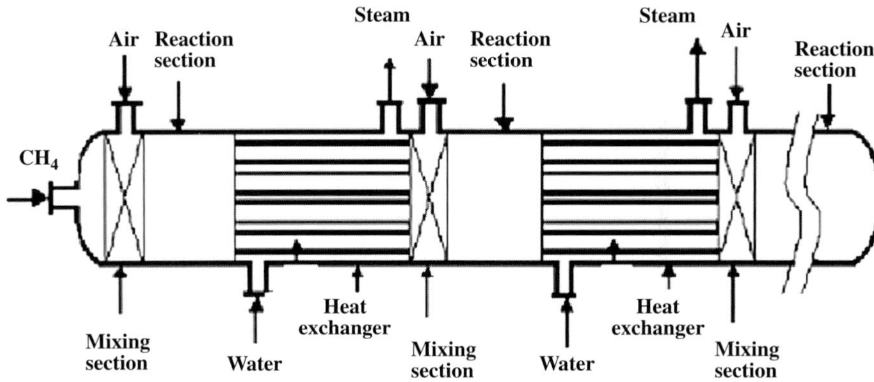

FIGURE 11.10 Reactor-generator for the partial oxidation of natural gas with stepped heat removal [274].

To increase the overall conversion of natural gas, especially if energy expenditures for compression are high, it is advantageous in some cases, even for oxidation with air, to resort to partial recycling of the exhaust gas. Such a scheme, implemented, for example, in a three-section reactor (Fig. 11.10), with supply of 3–4% oxygen in each section and excess heat removal between the sections, enables to bring the total conversion up to 9–10% per pass almost without reducing the selectivity.

Another factor conducive to high yields of methanol is the homogeneity of the reaction mixture; therefore, it is necessary to ensure a complete and prompt mixing of the gaseous reagents for no longer than 1/10 of the time of the process. Figure 11.11 shows a flowsheet of the cascade process of partial oxidation of natural gas in reactors with controlled heat removal and without separation of liquid products between the stages [274].

Natural gas from pipeline 1 is mixed with the circulation flow, compressed with compressor 2 to the reaction pressure, heated in regenerative heat exchanger 3 by the outgoing gas, and directed into the mixing section of the first oxidation stage 5. Air, compressed by compressor 4 to a pressure slightly above the reaction pressure to prevent hydrocarbon penetration into the air pipeline, enters the same section. Natural gas passes through several successive stages of oxidation, usually two or three, which may be constructively implemented in the form of one or more sequential reactor units. Each stage of oxidation is independently supplied with the appropriate amount of oxidizer from the air compressor. The reaction temperature is regulated by the parameters of produced steam with controls 6. After regenerative heat exchanger 3, the gas–liquid mixture is cooled in heat exchanger (or several heat exchangers) 7 to ambient temperature and fed into separator 8. If the discharge gas is then directed to the power plant, it makes sense to subject the gas–liquid mixture (or at least that part of the exhaust gas which is not returned into the process) to throttling to reduce the temperature of separation or to obtain additional cold, since even at normal temperature, methanol entrainment by separated gas can reach 10% or more. The separated liquid products flow into receptacle 10 and thence into the separation column. Before entering the column, the products are treated with alkali to neutralize acids (not shown in the scheme). Methanol (and ethanol), corresponding to the standard specifications for raw methanol, is removed through an outlet at the top of

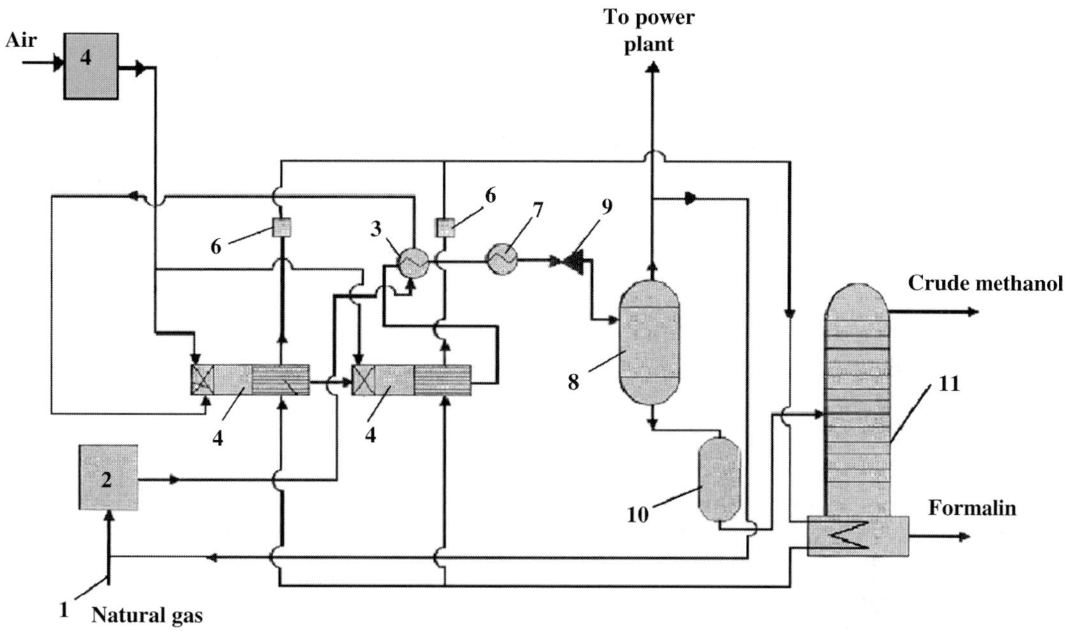

FIGURE 11.11 Flowsheet of the cascade process of the partial oxidation of natural gas [274]. (1) Gas pipeline, (2) compressor (3) recuperative heat exchanger, (4) air compressor, (5) reactor (oxidation stage), (6) controls of steam parameters, (7) heat exchanger, (8) separator, (9) throttle, (10) receptacle of liquid products, and (11) rectifying column.

the column, whereas a dilute aqueous solution of formaldehyde is removed through the bottom outlet.

A three-stage reactor with a spread-oxygen introduction (Fig. 11.11) operating at a natural gas flow through the reactor of ~13,000 m^3/h (110 million m^3/year) actually consumes only about 10 million m^3 of gas per year, whereas the remainder may be used for energy production or returned into the main pipeline. In addition to the approximately 5 million tons/year of methanol, about 0.5 million tons/year of formaldehyde can be produced, which in terms of 37% commercial formalin constitutes approximately 1400 tons/year. However, the production of commercial formalin and rectified methanol on small-scale installations is hardly justified, since it requires additional distillation steps and associated equipment. The ethanol yield will be less than 100 tons/year, so it is separated from methanol as inappropriate. The discharge gas with a calorific value of ~7400 kcal/m^3 is suitable for use at power plants of any type.

The use of a steam generator to remove the heat of reaction is effective but not the only way of controlling the temperature of the oxidation process. If methanol is produced from a small part of the pipeline gas, the oxidation is carried out substantially at the pressure of the incoming gas, with only small additional compression, whereas the discharge gas is returned back into the gas pipeline, with the volume of the gas passing through the reactor being unimportant. In this case, a simpler scheme of cooling the reaction mixture, based on additional supply of cold gas into the reactor (Fig. 11.12), is more advantageous.

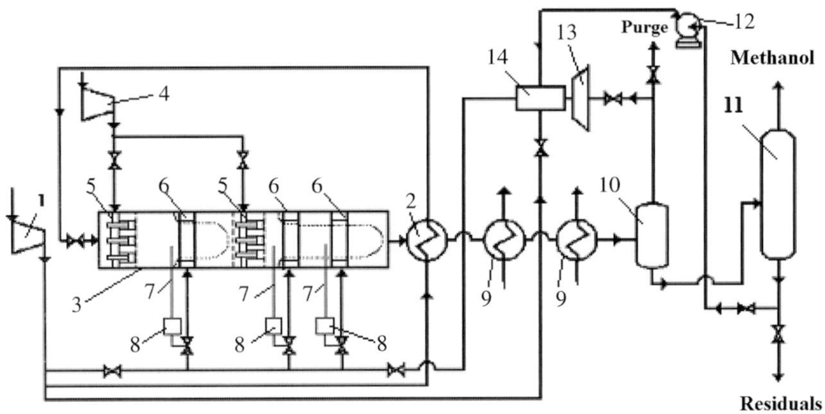

FIGURE 11.12 Flow diagram of the flowthrough process of methanol production by the partial oxidation of natural gas in a sectioned reactor with temperature control by means of introducing cold gas into the reactor. (1) Compressor, (2) recuperative heat exchanger, (3) reactor, (4) air compressor, (5) mixers, (6) cold gas supply collectors, (7) thermocouples, (8) adjusting devices of cold gas supply, (9) heat exchangers, (10) separator, (11) distillation column, (12) pump for pumping liquid by-products to additional oxidation, (13) booster compressor, and (14) mixer.

Natural gas from the supply line is boosted by compressor 1 to the reaction pressure, mixed with the circulation gas, heated to the reaction temperature in regenerative heat exchanger 2, and fed into the first mixer of sectioned reactor 3. The required amount of air is also supplied here by compressor 4. In addition to mixer 5, each section of the reactor has a reaction zone and a cooling zone. Each cooling zone 6 has an O-collector for creating a near-wall cylindrical-shape flow of cold gas (shown by the dotted line in the diagram). The gradual mixing of the cold gas with the hot reaction mixture limits the heating of the latter. Furthermore, the displacement of the reaction mixture from the reactor wall by the cold gas reduces the effect of the metal reactor surface on the oxidation process – in particular, the rate of heterogeneous processes of deep oxidation leading to the formation of mainly carbon oxides and water, thereby increasing the selectivity of the process. This also slows down the corrosion of the reactor walls and their temperature, which makes the requirements for the material of the reactor less stringent.

To ensure a more reliable temperature control and the formation of a cylindrical layer of cold gas in each section of the reactor, several collectors for its introduction can be mounted, with thermocouple 7 and control device 8 ahead of each to adjust the volume of supplied cold gas. As the cold gas, the discharge gas or initial natural gas or any combination thereof can be used. Furthermore, the oxidation by-products can be pumped from the vat of column 11 by pump 12 into mixer 14, from which, along with the circulation gas (boosted by compressor 13) or initial gas, they move in the cooling gas flow to final oxidation. Making use of the method for temperature control of the process by supplying additional amount of cold gas, the 'AMTEK-engineering' company has developed a technical project of a plant with a capacity of 5000 tons of methanol per year.

The circulation process of the oxidation of natural gas is an implementation of the standard technological method used at a low conversion of the reactants. However, the application of gas recirculation to the partial oxidation of hydrocarbons to oxygenates is associated with certain difficulties. First, recycling involves the use of oxygen, a factor that raises by

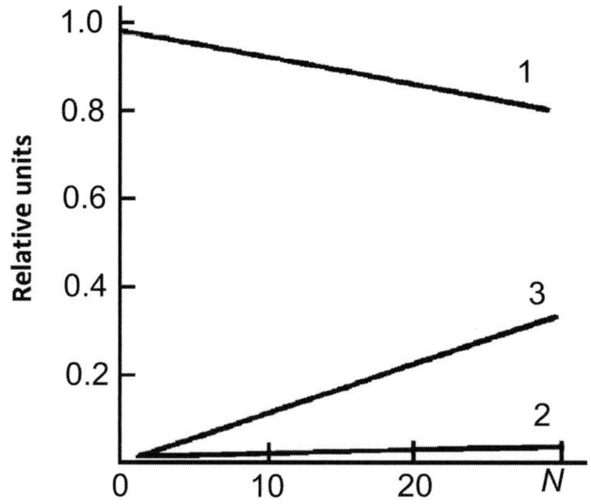

FIGURE 11.13 Calculated dependences of the (1) methane concentration in the reaction mixture, (2) methanol concentration at the outlet of the reactor, and (3) total yield of methanol on the number of cycles N at $P = 100$ atm, $T = 713$ K, and $[O_2] = 2.2\%$ [275].

30–40% the investment expenditures, increases the complexity of the entire process of production, and tightens the requirements for its maintenance and staff training. Apparently, for DMTM installations with a methanol output of up to 20,000 tons/year, the most suitable technology of producing oxygen is swing adsorption. It enables to produce technical oxygen with O_2 content of over 90% in volumes from a few 100 to several 1000 cubic metres per hour. However, its cost is still high enough.

Besides the need to use oxygen at least of commercial purity, the problem of purification of the circulation gas from carbon oxides, the main gas-phase reaction products, remains essentially unresolved. Estimates [275] show that, although the circulation mode provides a much higher conversion of methane and a higher yield of methanol at a high selectivity of its formation (Fig. 11.13), the concentration of gaseous reaction products in the circulation gas increases rapidly.

Note that, while the concentrations of the reactive products, CO and H_2, eventually reach a steady-state level, the inert product, CO_2, continues to accumulate monotonically with increasing number of cycles (Fig. 11.14).

Although it was experimentally shown [5] that the presence of up to 5% of CO and H_2 in the reaction mixture virtually does not affect the yield of methanol (there is evidence that even higher CO concentrations of up to 12% do not reduce the yield of methanol [106]), the implementation of the cyclic process, even when part of the recirculation mixture is discharged, needs at least a partial removal of carbon oxides. This requirement substantially complicates the process. While the methods of purification from gaseous carbon dioxide, such as its absorption by ethanolamine, are well established, there are no effective methods of purification from carbon monoxide. The known methods (copper-ammonia purification, methanation, selective oxidation, etc.) are too complex and make the circulation process unsuitable for small-capacity plants. Given that, in the partial oxidation to oxygenates, carbon oxides, mainly CO, account for almost 50% of the selectivity in terms of carbon, the development of more effective ways to remove, or better to recover, CO, as one of the products of the process, would be of great importance to the technology.

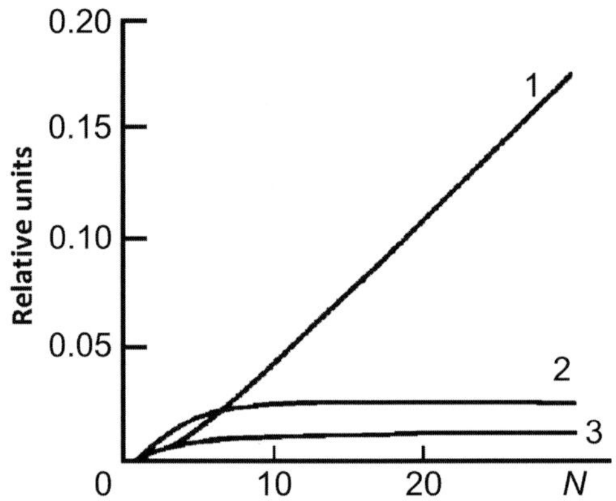

FIGURE 11.14 Calculated dependences of the concentrations of (1) CO_2, (2) CO, and (3) H_2 at the outlet of the reactor on the number of cycles N at $P = 100$ atm, $T = 713$ K, and $[O_2] = 2.2\%$ [275].

Figure 11.15 shows a flowsheet of the circulation process of the oxidation of natural gas in a three-section reactor with a heat removal between the oxidation stages by means of a steam formation. The resultant steam is used to heat the distillation columns, whereas its excess can be offered to an external consumer. The discharge gases can also be used for energy production.

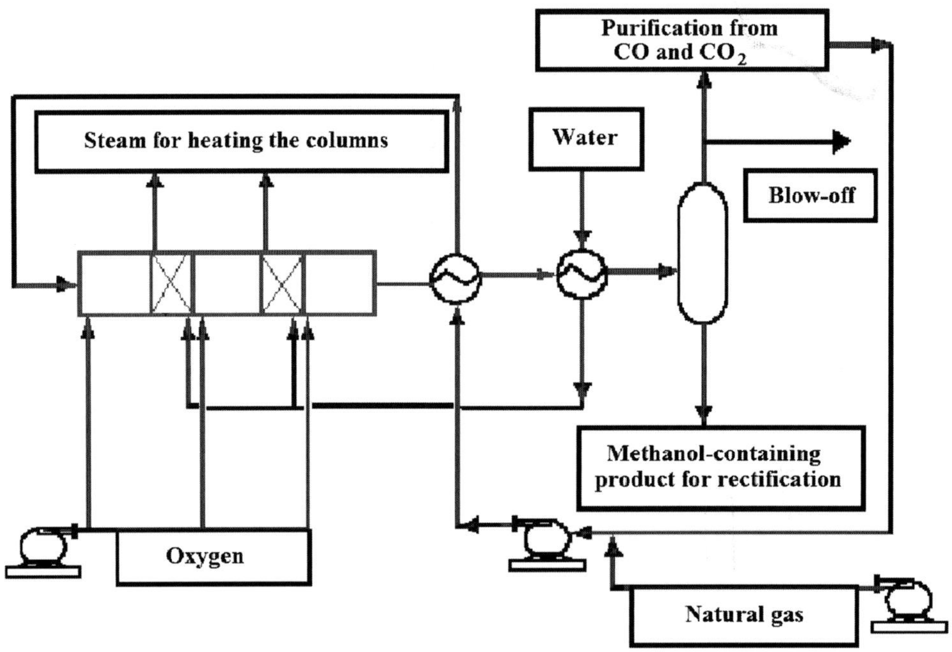

FIGURE 11.15 Flowsheet of the circulation process of methanol production. (For colour version of this figure, the reader is referred to the online version of this book.)

To take advantages of the circulation process, at least partially and, at the same time, avoid the need for purification from carbon oxides, it makes sense to use various combinations of the cyclic and flowthrough methods (hemicycle). In this case, only part of the discharge gas is subjected to recirculation, while the problem of accumulation of products, including carbon oxides and nitrogen, is solved by increasing the volume of blow-off gas. Figure 11.16 displays the calculated dependence of the methanol and the concentration of nitrogen in the exhaust gas on the volume of blow-off gas for the cyclic process of partial oxidation of natural gas with atmospheric air in a two-section reactor at a methane conversion of 4% in each section on the fraction of blow-off gas.

When the volume of blow-off gas decreases, the methanol yield increases, but the calorific value of exhaust gas declines; at the same time, the volume of gases passing through the reactor increases, which makes it necessary to increase the size and productivity of the equipment. In the case of oxidation with air, the optimal fraction of blow-off gas is ~50%, with the calculated yield of methanol reaching 60 kg/1000 m^3 of converted methane; the CO concentration at the reactor inlet is less than 3%. The air consumption is ~1400 m^3 per 1000 m^3 of process gas. Diluted by ~30% with nitrogen, blow-off gas has a high calorific value, 20 MJ/m^3 (the volume of blowing-off is ~1600 m^3 NTP at 1000 m^3 NTP of introduced gas) and can be used as a high-energy fuel with reduced production of nitrogen oxides.

The methanol yield can be substantially increased by using oxygen-enriched air, a factor that makes it possible to increase the number of gas circulations (decrease the volume of blow-off gas). For air with an oxygen content of ~50%, which can be prepared, for example, by membrane separation, the volume of blow-off gas can be reduced to 10% (Fig. 11.17). In this case, the calculated yield of methanol increases to ~200 kg/1000 m^3 of converted methane. The calorific value of blow-off gas is ~16 MJ/m^3, whereas its volume is ~1560 m^3 NTP per 1000 m^3 NTP of introduced gas. The use of technical oxygen (90% O_2) further reduces the amount of blow-off gas. However, this is technically more complex and capital-intensive solution.

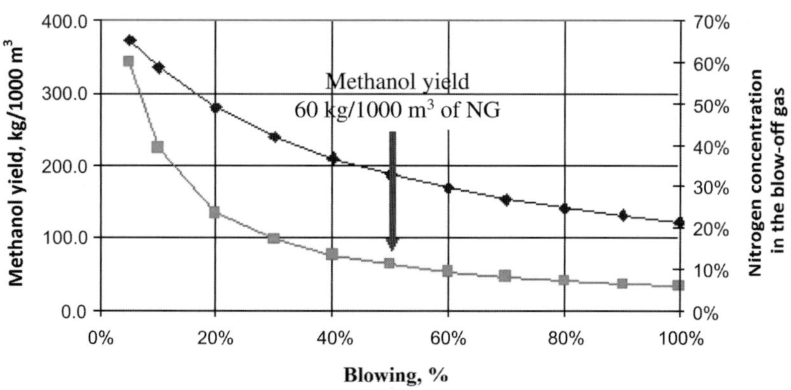

FIGURE 11.16 Calculated dependences of the methanol yield (per 1000 m^3 NTP/h of supplied natural gas at a methane content of 98%) and nitrogen concentration in blow-off gas on the fraction of blow-off gas for the cyclic partial oxidation of natural gas with atmospheric air. (For colour version of this figure, the reader is referred to the online version of this book.)

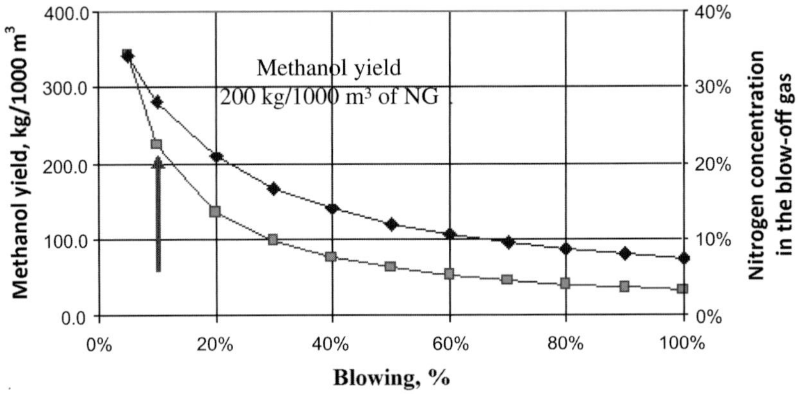

FIGURE 11.17 Calculated dependences of the methanol yield (per 1000 m^3 NTP/h of introduced gas at a methane content of 98%) and nitrogen concentration in the blow-off gas on the fraction of blow-off gas for the cyclic partial oxidation of natural gas with oxygen-enriched air ($[O_2] = 50\%$). (For colour version of this figure, the reader is referred to the online version of this book.)

INNOVATIVE TECHNOLOGIES BASED ON THE GAS-PHASE OXIDATION OF HYDROCARBON GASES TO OXYGENATES

The current state of elaboration of the technological processes of partial oxidation of hydrocarbon gases to oxygenates is far from perfect, mainly due to insufficient volume of research development. However, it is possible to clearly identify the areas (Fig. 11.18) where this technology is promising, and in some cases, the only one suitable for solving the outstanding problems [270]. Some of these applications are discussed below.

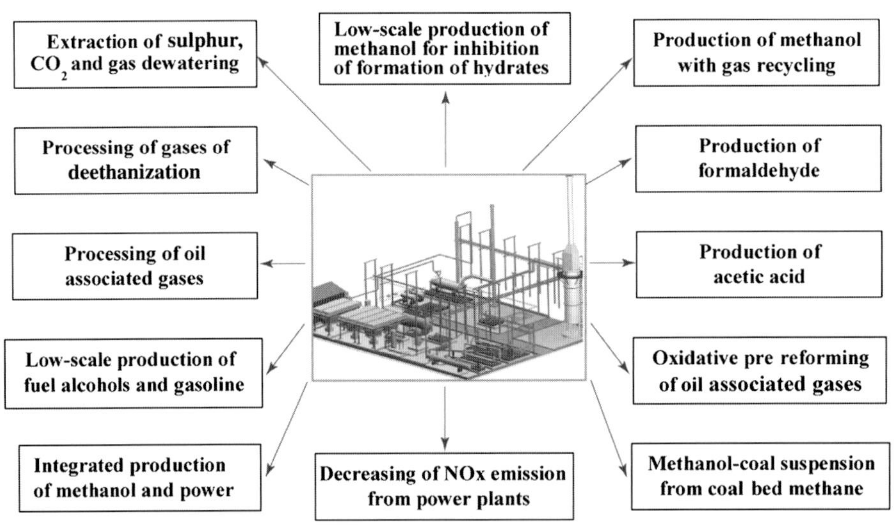

FIGURE 11.18 Promising technologies based on the direct oxidation of natural and associated gas. (For colour version of this figure, the reader is referred to the online version of this book.)

The prevention of hydrate formation during the production and transportation of natural gas remains one of the most promising areas for application of the DMTM. Hydrate formation is a serious problem for the gas industry. By some estimates, preventive measures account for 20% of the costs of the extraction and transportation of natural gas, the total consumption of methanol for this purpose in Russia already exceeding 400,000 tons/year. The main consumers are numerous remote gas-production enterprises at large distances apart, consuming annually only a few thousand ton of methanol each. Thus, the delivery of methanol from sites of its production thousands of miles away is associated with high costs. In this case, a low-tonnage on-site production of methanol can provide tangible economic benefits. No less important is the factor of reliability of supply. In addition, raw methanol, the production of which, as described in the previous section, is significantly less complicated than that of commercial-grade methanol, is quite suitable as an inhibitor.

However, the fight against hydrate formation includes not only the production of methanol, but also its introduction into the gas well or transportation system. A method for preventing hydrate formation in the gas pipeline by the partial oxidation of part of the gas and subsequent introduction of methanol-containing partially converted gas into the gas pipeline (without extracting methanol from it) is described in [276]. This eliminates the stages of isolating and distillation of the liquid products. However, this excludes possibility of methanol production for processing the upstream parts of the transport system and gas wells. Along with methanol, water and other oxidation products, including carboxylic acids, occur in the pipeline. Therefore, a more interesting method (Fig. 11.19) is such in which the treatment of transported gas is performed simultaneously with the extraction of liquid methanol, an approach that enables to optimize the performance of the systems of separation and distillation of methanol.

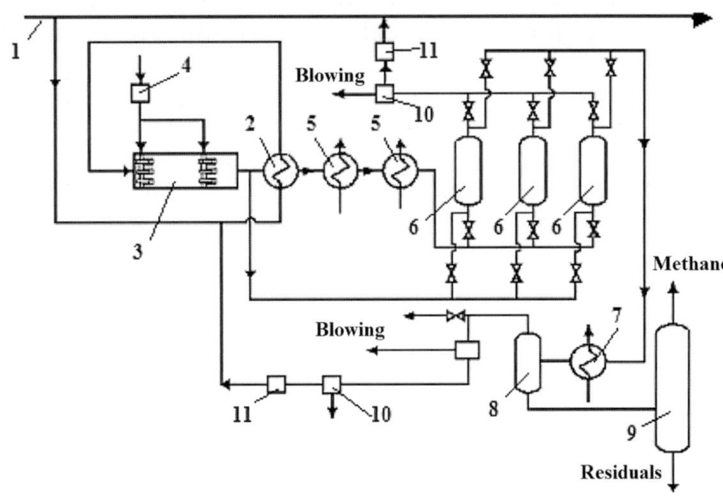

FIGURE 11.19 Flow diagram of the production of methanol and its simultaneous introduction into the pipeline. (1) Gas pipeline, (2) recuperative heat exchanger, (3) reactor, (4) air compressor, (5) heat exchangers, (6) adsorption columns, (7) heat exchanger, (8) separator, (9) rectification column, (10) separators of nitrogen-containing gas, and (11) units for separation of carbon oxides.

Part of the transported gas from line 1 through recuperative heat exchanger 2 is introduced into sectioned reactor 3, wherein air or oxygen is also supplied by compressor 4, and the gas is partially oxidized with a conversion of 5–6%. The outgoing partially oxidized gas is divided into two flows. One flow, cooled in heat exchangers 2 and 5, enters periodically operating adsorption column 6, where it is stripped of water, organic by-products (formaldehyde and acid), and methanol excess, after which this flow, containing the desired concentration of methanol vapour, is returned into the pipeline. The second hot gas flow is used for desorbing the products adsorbed from the first flow, followed by its cooling in heat exchanger 7 and recovering methanol by standard techniques with separator 8 and distillation column 9. The resulting liquid methanol can be introduced into the wells and/or gas pipeline. The separated gas can be recirculated for oxidation, returned into line 1, or used to generate energy. To reduce the concentration of nitrogen in the gas returned into the pipeline and in the recycle gas, they can be directed to separators 10 to remove part of the nitrogen and obtain a low-calorific nitrogen-containing gas usable as a fuel. Furthermore, carbon oxides can be extracted on separators 11.

Purification of natural gas from sulphur compounds can also become one of the most attractive areas of commercial applications of the DMTM process. Dewatering of produced gas and purification from sulphur compounds and carbon dioxide are the most important stages of the gas production. For this purpose, a large number of industrial processes, based on various absorbents, including methanol, have been developed. Among the well-known is the Rectisol process (the licensors are the Linde AG and Lurgi GmbH) for extracting acidic gases by organic solvents at temperatures below 0 °C. Hydrogen sulphide, carbonyl sulphide, and carbon dioxide are normally extracted with methanol, in some cases with other solvents. After purification, the fraction of sulphur is below 10^{-6}, and that of carbon dioxide is even lower. The main advantages of the process are the use of cheap and readily available solvent, ease of technological design, and low power consumption [277]. The Rectisol process is economical, with a significantly lower specific energy consumption compared to other purification methods, such as absorption by ethanolamines. That the solubility of H_2S in methanol is substantially higher than that of CO_2 allows, on the one hand, their selective separation and, on the other, reaching a high degree of purification from sulphur compounds in the presence of CO_2 [278]. The process is best suited for the purification of gases containing a large number of different contaminants, including organosulphur compounds, the solubility of which in methanol is high [279].

Another example of using methanol for extraction of acidic gases is the Ifpexol process, developed and patented by the French Institute of Petroleum (Institut français du pétrole – IFP). Cold methanol wash is well integrated in schemes with dehydration and cold extraction of liquefied gases [280].

In field conditions, the extraction of undesirable impurities with methanol becomes attractive if methanol is produced directly from the extracted gas and on the equipment integrated into the gas processing plant, i.e., using the opportunities provided by the DMTM process. A flowsheet of the combined process is shown in Fig. 11.20.

The first part of the scheme realizes the conventional process of methanol production by means of the DMTM. Purified natural gas is fed into compression and heating unit 1 and then into reactor unit 2, where air or oxygen is also pumped by oxidant compression unit 3. In separation unit 4, liquid products are extracted and then directed into distillation

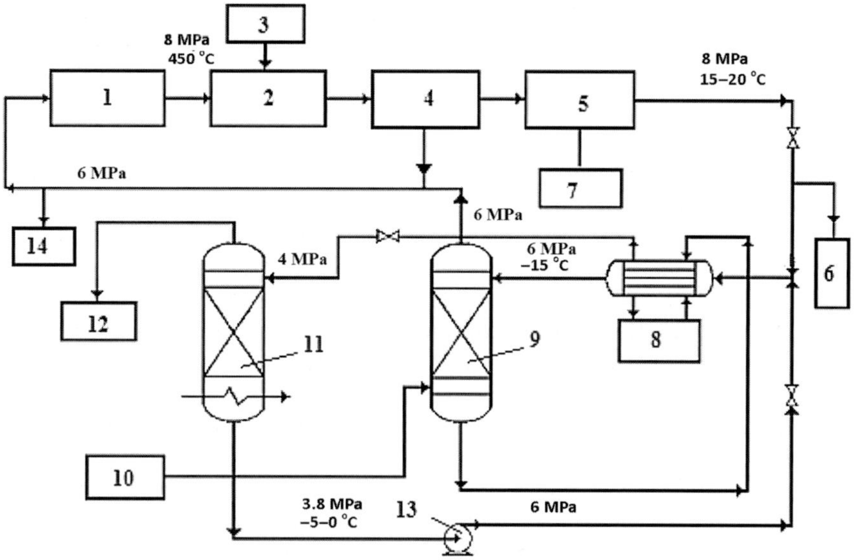

FIGURE 11.20 Technological scheme for the purification of natural gas from sulphur components with concurrent production of methanol by the DMTM: (1) gas compression and heating unit, (2) reactor unit, (3) air compression unit, (4) separation unit, (5) rectification unit, (6) receptacle for methanol, (7) receptacle for vat residue, (8) ammonia refrigerator, (9) absorption column, (10) raw gas source, (11) regeneration column, (12) hydrogen sulphide utilization unit, (13) pump, (14) sweet gas consumer.

unit 5. Methanol and vat residue come into receptacles 6 and 7, respectively. The separated gas is returned into the gas line. From receptacle 6, the resulting methanol, under a pressure of 6 MPa, enters ammonia or propane refrigerator 8, where it is cooled to −15 °C. The cooled methanol is supplied for refluxing into absorption column 9, where it encounters raw natural gas that counterflows from the bottom, being supplied by source 10. Hydrogen sulphide-containing methanol after heat exchange with lean methanol is regenerated in column 11. The hydrogen sulphide fraction is sent for utilization into Claus apparatus 12, while purified methanol is returned by pump 13 into the process. Purified hydrocarbon gas is send to consumer 14 and to conversion into methanol. The residual methanol content in the purified gas may be adjusted so as to prevent hydrate formation during its further transportation. Thus, application of such a technology, based on a combination of the DMTM and Rectisol processes enables to carry out three important operations: to effectively and economically purify natural gas from acidic components, introduce methanol into it in the amount required for its further transportation, and obtain liquid methanol for inhibition of hydrate formation in the gas transporting system upstream of the plant.

Processing of deethanization gases from gas fractionation plants can also become an area of practical applications of the DMTM process. Virtually all of the natural gas produced is processed to extract higher hydrocarbons, valuable raw materials, as well as to prevent their condensation in the cold units of the gas transporting system. Therefore, according to Russian Standard OST 51.40-93 for natural gas supplied to the gas-main pipelines, its hydrocarbon dew point must not exceed 0 °C, or even −10 °C in winter conditions. This is achieved by extracting ethane, at least partially, propane, butane, and higher hydrocarbons, valuable chemical raw materials and high-calorie ecological household and motor fuel. This treatment also provides a homogeneous composition of natural gas transported through pipelines irrespective of its origin.

Despite a wide variety of technologies used for the fractionation of hydrocarbon gases, all derived products fall into four main groups: dry natural gas purified in accordance with the requirements of transportation, liquefied gases (propane—butane fraction), liquid products (gas condensate, wide fraction of light hydrocarbons), and deethanization gases. The first three fractions are highly marketable and relatively easily transportable commercial products. However, the last fraction, which contained most of the ethane removed from natural gas, as well as some amounts of methane and propane, in many cases cannot be used otherwise than as fuel for local needs or, in some cases, is simply flared. At the same time, this fraction is an ideal raw material for the DMTM process, readily convertible at pressures below 30 atm with a high yield of products. Therefore, the integration of a partial oxidation unit with a gas fractionation plant, would allow producing valuable oxygen-containing products from otherwise waste material.

Formaldehyde production from ethane-containing gases, primarily separated on gas fractionation plants may become one of the promising areas of processing the ethane fraction. At relatively low pressures, <30 atm, and a high total yield of oxygenates, the concentration of formaldehyde in the liquid products reaches more than ~30%. This high figure makes it possible to consider the gas-phase oxidation of ethane-containing gas as a potential one-step method of formaldehyde production. The flowchart of this process is described in [281]. Nowadays, formaldehyde is produced by the oxidative dehydrogenation of methanol, which consumes more than 20% of its world's production.

Production of acetic acid and other carbonylation products is one of the most feasible ways to improve the selectivity of the partial oxidation of methane by using, along with oxygenates, carbon monoxide formed for the carbonylation of methanol and other liquid-phase reaction products. The KANMET Company (Canada) proposed a method for the production of acetic acid on the basis of the direct oxidation of methane, bypassing the step of synthesis gas production [282]. Methane is oxidized in the gas phase to methanol, which is then separated and carbonylated on a supported catalyst with carbon monoxide formed during oxidation:

$$2CH_4 + 2O_2 \rightarrow CH_3OH + CO + 2H_2O \tag{11.2}$$

$$CH_3OH + CO \rightarrow CH_3COOH \tag{11.3}$$

The technology (Fig. 11.21) was based on the DMTM process described in [2], with a conversion of methane per pass through the reactor of 6—12%, a complete oxygen consumption at a temperature above 400 °C, and working pressure of ~7 MPa. The resultant steam was used for technological purposes. At this pressure and $T = 200–250$ °C, methanol was carbonylated to CH_3COOH and CH_3COOCH_3 on Pd/C a catalyst in the presence of CH_3I. The catalytic activity persisted for 600 h, indicative that the carbonylation takes place in the gas phase. According to authors' estimates, the cost of production of 1 kg of CH_3COOH by this method must be 30% lower than that for the conventional method.

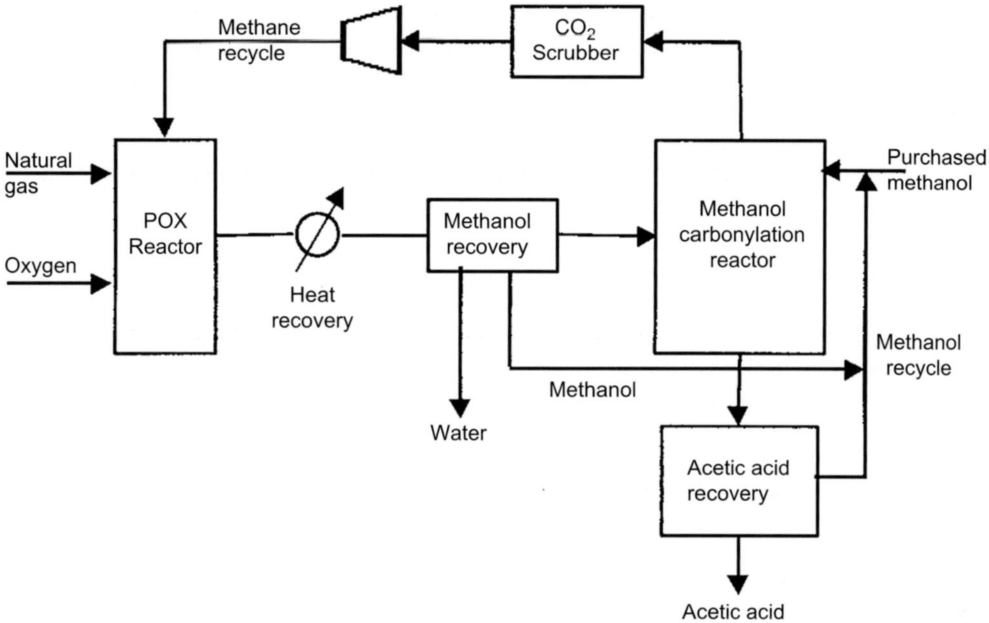

FIGURE 11.21 CANMET Integrated process for converting methane to acetic acid [282,283].

The production of acetic acid on the basis of the DMTM process with the use of carbon monoxide formed in the oxidation process for methanol carbonylation is promising primarily because it permits to reach almost a 100% selectivity of formation of liquid products in terms of carbon, i.e., eliminates one of the major shortcomings of the process. In the case of development of industrial methods for extracting CO from the recirculation gases, acetic acid can be produced directly using the known Monsanto technology of methanol carbonylation.

Research of the carbonylation of methanol to other products, such as methyl formate [284] and dimethyl carbonate [285], is also underway. Should an additional source of hydrogen be available, the possibility of using CO to increase the yield of methanol through the intermediate formation of methyl formate, proposed by Christiansen as long ago as 1919 [286], can also be considered:

$$\begin{aligned} CH_3OH + CO &\rightarrow HCOOCH_3 \\ HCOOCH_3 + 2H_2 &\rightarrow 2CH_3OH \\ \hline CO + 2H_2 &\rightarrow CH_3OH \end{aligned} \tag{11.4}$$

Methanol is first carbonylated to methyl formate, which is then hydrogenated to form twice the amount of methanol. The carbonylation reaction proceeds in the liquid phase in the presence of sodium or potassium methoxide ($NaOCH_3$ or $KOCH_3$) as a homogeneous catalyst. This fine-tuned industrial technology is used in the production of formic acid. Recently, the possibility of efficient use of heterogeneous catalysts was demonstrated. The subsequent process of hydrogenation of methyl formate can take place in the liquid or gas phase over copper-containing catalysts. With a suitable choice of the catalyst, the process can occur in a single reactor. According to the available patents of a number of firms, the process of production of methanol via methyl formate occurs at 80−120 °C and pressures from 10 to 50 atm [286], which makes it well combinable with the process of direct oxidation. Since the methyl formate can also be obtained by dimerization of formaldehyde, it opens one more possibility for increasing the yield of methanol in the direct oxidation of methane.

There are other possibilities of processing the mixture of methanol, formaldehyde, and formic acid obtained by direct oxidation of methane, without prior separation, which enables to expect a significant increase in the yield of methanol and, accordingly, the commercial attractiveness of the process. Formaldehyde may be subjected to the dimerization into methyl formate on TiO_2 and ZnO_2 catalysts.

$$2HCHO \xrightarrow{TiO_2 \text{ and } ZnO_2} HCOOCH_3 \qquad (11.5)$$

Another possibility is its conversion on CaO and MgO to methanol and formic acid, which rapidly react with each other to form methyl formate:

$$2HCHO \xrightarrow{H_2O} CH_3OH + HCO_2H$$

$$CH_3OH + HCO_2H \xrightarrow{-H_2O} HCOOCH_3 \qquad (11.6)$$

The resulting methyl formate can be catalytically hydrogenated or electrochemically reduced to form two molecules of methanol without any by-products:

$$HCOOCH_3 + 2H_2 \rightarrow 2CH_3OH \qquad (11.7)$$

Formic acid formed by the oxidation can serve as a source of hydrogen for the hydrogenation of formaldehyde in aqueous solution at 250 °C in the presence of a suitable catalyst:

$$HCHO + HCO_2H \rightarrow CH_3OH + CO_2 \qquad (11.8)$$

A suitable combination of the above reactions can significantly improve the selectivity and yield of methanol formation in the direct oxidation [286].

At an excess of methanol in the presence of acidic catalysts and zeolites, it is possible to synthesize methyl acetate [287], which, when subjected to carbonylation, yields ethylidene diacetate [288] and, further, vinyl acetate monomer, one of the most important petrochemical products [289]:

$$2CH_3COOCH_3 + 2CO + H_2 \rightarrow CH_3CH(OCOCH_3)_2 + CH_3COOH$$

$$CH_3CH(OCOCH_3)_2 \rightarrow CH_2{=}CHOCOCH_3 + CH_3COOH \qquad (11.9)$$

These syntheses were first proposed by Halcon Company in 1980 and turned out to be 30−40% more economical than the traditional methods of producing these products. Numerous modifications of carbonylation processes aimed at producing both methyl acetate and ethylidene diacetate were patented by different companies in many countries around the world.

Another valuable chemical product that can be obtained from CO and methanol (or methyl acetate), i.e., by oxidative carbonylation, is dimethyl carbonate [290,291]

$$2CH_3OH + 1/2O_2 + CO \xrightarrow{\text{catalyst}} CH_3OCO_2CH_3 + H_2O \quad (11.10)$$

Dimethyl carbonate can be a promising replacement of methyl tertiary-butyl ether as an environmentally safe antiknock additive for fuels [292]. It is a monomer used for the production of polycarbonates, a methylating agent, and a 'green' solvent [290,293].

Recent reports have claimed the possibility of selectively producing methyl formate by methanol carbonylation using nanosized copper catalysts [294].

The above reaction equations suggest that the most favourable molar ratio for the reaction between methanol and CO lies within 1:1 to 2:1, which can actually be obtained in the DMTM process.

Thus, the integration of the DMTM process with the extensive and well-studied chemistry of carbonylation processes allows the creation of entirely new group of alternative light-duty GTL-processes for the processing of natural and associated gas (Fig. 11.22) [295].

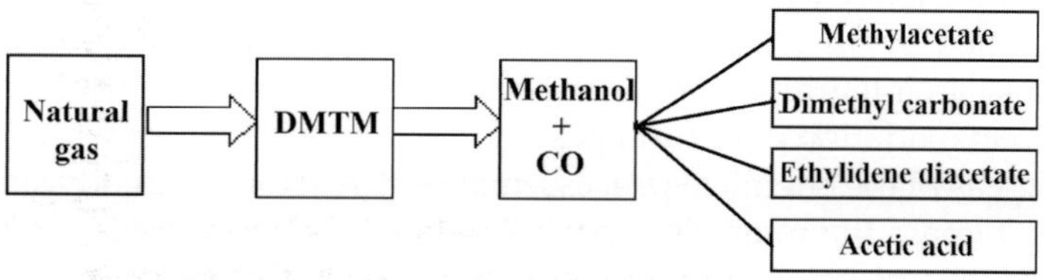

FIGURE 11.22 Schematic diagram of the promising processes for producing chemicals with the use of both major carbon-containing products of the DMTM process. *Based on data of [295].*

The combined production of methanol and power at thermal power plants (TPP) makes it possible to not only simplify and streamline the process of methanol production, but also to significantly reduce NOx emissions during electricity generation. Nitrogen oxides NOx are among the most dangerous anthropogenic emissions, a major component of photochemical smog. For TPP and gas-compressor stations operating on natural gas, they constitute a major hazard to ecosystems. Therefore, the reduction of NOx emissions at TPP is one of the most important global environmental issues. The reduction of NOx emissions by power plants into the atmosphere is highly expensive. Nowadays, only a small part of the power plants in the United States has equipment to remove nitrogen oxides. However, in the future, hundreds of power plants will have to significantly reduce the emission of NOx, with the annual consumption of ammonia for this purpose being estimated at about 1.5 million tons.

In the flue gases of power installations with gas flow rates of 25,000–32,000 m^3/h, nitrogen oxides concentration is 400–800 mg/m^3, whereas the figures for installations with gas flow rates of 35,000–80,000 m^3/h are higher, 700–1500 mg/m^3 [296]. Of the three major kinetic mechanisms of the formation of nitrogen oxides during the combustion of hydrocarbon fuels, thermal (Zeldovich mechanism), prompt (Fenimore mechanism) and fuel NOx, the major role in the combustion of natural gas belongs to the thermal mechanism [297]. This mechanism, described by the two main reactions

$$N_2 + O = NO + N \quad (11.11)$$

$$N + O_2 = NO + O, \quad (11.12)$$

has a strong temperature dependence. The formation of 'thermal' NO decreases sharply as the combustion temperature drops below 1800 K; therefore, reducing the combustion temperature is one of the most effective ways to decrease the concentration of NOx in the flue gas. For this purpose, various methods are used, including the creation of combustion zones with a significant deviation of the fuel-to-oxidant ratio from the stoichiometric value, dilution of the reagents with steam, and partial recirculation of flue gases [296–298]. Reducing the concentration of NOx on average by 50%, entails additional expenditures of 5–50 dollars per kilowatt of the burner unit [297].

The recirculation of combustion products into the combustion zone is one of the most studied and widespread methods of reducing nitrogen oxides, currently widely used in the combustion technology: the admixing of 20% of the recycle gas to the blow air can halve the yield of nitrogen oxides [296]. The maximum flame temperature is reduced thereby by nearly 200 °C. Since the introduction of recycle gas into the combustion chamber results in a decrease in the temperature of the exhaust gas and in a reduction in the efficiency of power plants by about 0.02–0.03% per 1% recycle gas added, it is believed that, under the condition of stable combustion, flue gas recirculation should be limited to 20–30% [296].

As an effective alternative to flue gas recirculation in reducing NOx emissions for all types of power plants fueled by natural gas, it is possible to combine a power installation with the DMTM process into an integrated chemical-power production. Chemical-power technology, combining the direct oxidation of natural gas into methanol with the production of electricity from the outgoing gas of the chemical process, makes it possible to greatly simplify the equipment and, in fact, bypass the main drawbacks of the DMTM process, such as a low conversion per pass and a low selectivity of methanol formation in this process. The simplicity of the equipment and the use of air as the oxidant instead of expensive oxygen make the process economically efficient. At the same time, power plants become able to reduce NOx emissions without additional capital and operational costs for these purposes. Using the outgoing gas from the DMTM process, containing, in addition to methane, a small admixture of other combustible gases (CO and H_2) and about 30% nitrogen (from the air spent for the oxidation) as a fuel for the power plant, makes it possible to reduce NOx emissions due to a lower combustion temperature of the nitrogen-diluted fuel.

According to thermodynamic calculations, for a stoichiometric mixture, a 30% dilution of methane with nitrogen leads to a reduction in the NO concentration by ~20%, whereas a 60% dilution, to a decrease by more than 50%. These results are well consistent with those obtained in experiments with General Electric gas turbines operating on natural-gas diluted with nitrogen. The increase in the molar concentration of N_2 to 48% resulted in an almost twofold decrease in the emission of NOx [299].

Various types of modern power plants can operate on natural gas highly diluted with nitrogen, up to methane content below 50% in electrochemical generators (fuel cells), 40% in spark ignition engines, 30% in conventional gas turbines, 5% in diesel engines, and 1% in gas turbines with catalytic combustion [300].

Thus, the dilution of natural gas with nitrogen in its partial oxidative conversion to methanol can significantly reduce NOx emissions. To further reduce the NOx emissions, to a level of a few percent of the initial concentration, it is advantageous to use the primary (unfractionated) liquid from the DMTM process, containing ~40% methanol and ~5% formaldehyde, introduced into the flue gases at temperatures from 320 to 820 °C. As shown by Zamansky et al [301], the introduction of methanol and some other compounds into flue gases at these temperatures results in the oxidation of nitric oxide to nitrogen dioxide, which then can be removed in scrubbers. The process is mainly controlled by the reactions

$$CH_3OH + OH = CH_2OH + H_2O \quad (11.13)$$

$$CH_2OH + O_2 = CH_2O + HO_2 \quad (11.14)$$

$$HO_2 + NO = OH + NO_2 \quad (11.15)$$

There is a well-pronounced 'temperature window' of the process near 850 K, within which the concentration of NO can be decreased to a few percent of the initial level (Fig. 11.23).

The dependences displayed in Fig. 11.24 shows that the amount of methanol needed to effectively reduce the concentration of NO is approximately equal to the content of NO in the flue gases, i.e., should be about 0.1% of their volume.

The range of initial concentrations of NO in which it is effectively oxidized to NO_2 is quite wide, covering almost the entire range of NO concentrations interesting for the energy industry. The Energy and Environmental Research Corporation (EER, U.S.), specializing in the field of environment and power production, has developed the CombiNOX process, which consists of the three-stage combustion of natural gas, selective noncatalytic oxidation of NO by supplying of methanol in the boiler flue, and purification of the flue gas from NO_2 (and SO_2) in a wet scrubber. According to EER specialists, at a methanol-to-NO molar ratio of 1.5, it is possible to obtain the same effect as when using the selective catalytic reduction (SCR) of NOx with ammonia, but at significantly lower cost. According to estimates, the method developed by EER specialists is three times cheaper than the SCR [302].

Such a process is especially advantageous if methanol is produced directly at the power plant in an integrated process. If natural gas is received directly from a gas-main pipeline at a pressure of 5.5–6.5 MPa, the installation can do without such an expensive unit as a natural gas compressor and without energy expenditures on gas compression. When the

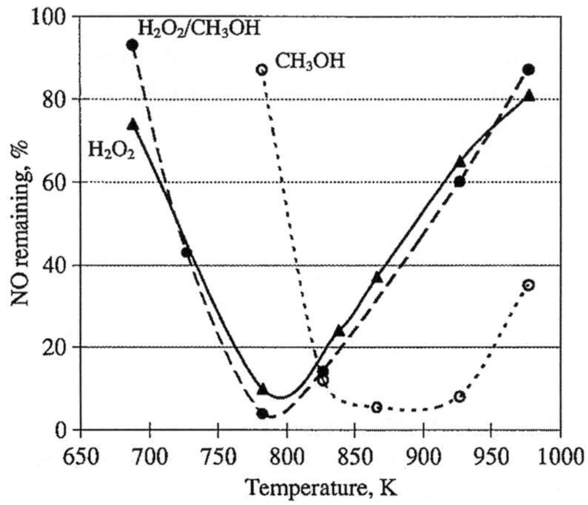

FIGURE 11.23 Temperature window of NO conversion by introducing H_2O_2, CH_3OH, and a 1:1 mixture thereof. $[NO] = 200$ ppm, additive/$[NO] = 1.5$, and $[O_2] = 3.8\%$ [301].

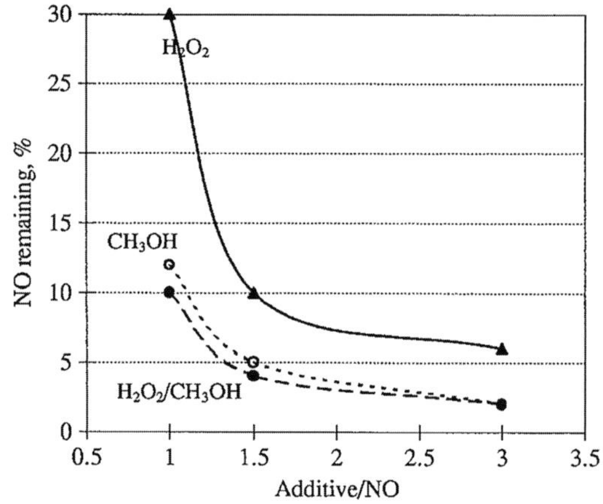

FIGURE 11.24 Effect of the additive/NO molar ratio on the conversion of NO at the optimum temperature of introduction of H_2O_2, CH_3OH, and a 1:1 mixtures thereof. [NO] = 200 ppm, and $[O_2]$ = 3.8% [301].

pressure in the gas pipeline is lower than that, it is expedient to compress the gas to a pressure ~7–8 MPa. After standard oxidation with air and separation of the liquid oxidation products, the high pressure of the discharge gas to be burnt in the power installation can be advantageously released in a turboexpander to drive an air compressor with a power approximately 4–5 times lower than that of the gas flow. The resulting cold gas flow can be advantageously used for the final cooling of the reaction gases before separation, which makes redundant cumbersome air coolers and enhances by ~10% the recovery of liquid oxidation products due to a deeper cooling of the gas–liquid mixture to be separated. A scheme of such integrated production of methanol at TPP is shown in Fig. 11.25.

Given the use of the off-gas of the DMTM process for energy generation and the possibility of utilization of steam produced directly at the TPP, the efficiency of carbon use in such integrated process is expected to be close to 100%. A full use of the reaction heat and calorific value of the exhaust gas actually eliminates the two main drawbacks of the DMTM method associated with the low selectivity of methanol formation and low conversion of gas per pass. At the same time, its benefits, such as one-step nature of the process and relatively low capital and operating costs, make the DMTM-based integrated method of methanol production quite competitive compared with existing methods of production of commercial methanol.

It should be noted that the properties of methanol are similar to petroleum distillate fuel. Currently, there are assurances of the main manufacturers of gas turbines on the possibility of using methanol as a fuel. Methanol can fuel almost any type of boiler installation. The main advantage of methanol compared to the conventional distillate fuel is that it contains no sulphur. Therefore, one additional advantage of integrated chemical-power methanol production at TPP is the possibility of using methanol produced as a clean backup liquid fuel. As a result, the supply of reserve fuel to TPP, for example masout, can be curtailed or completely avoided.

The integrated chemical-power production of methanol can be launched at dozens of power plants fueled by natural gas. By technical-economic assessments, the cost of methanol

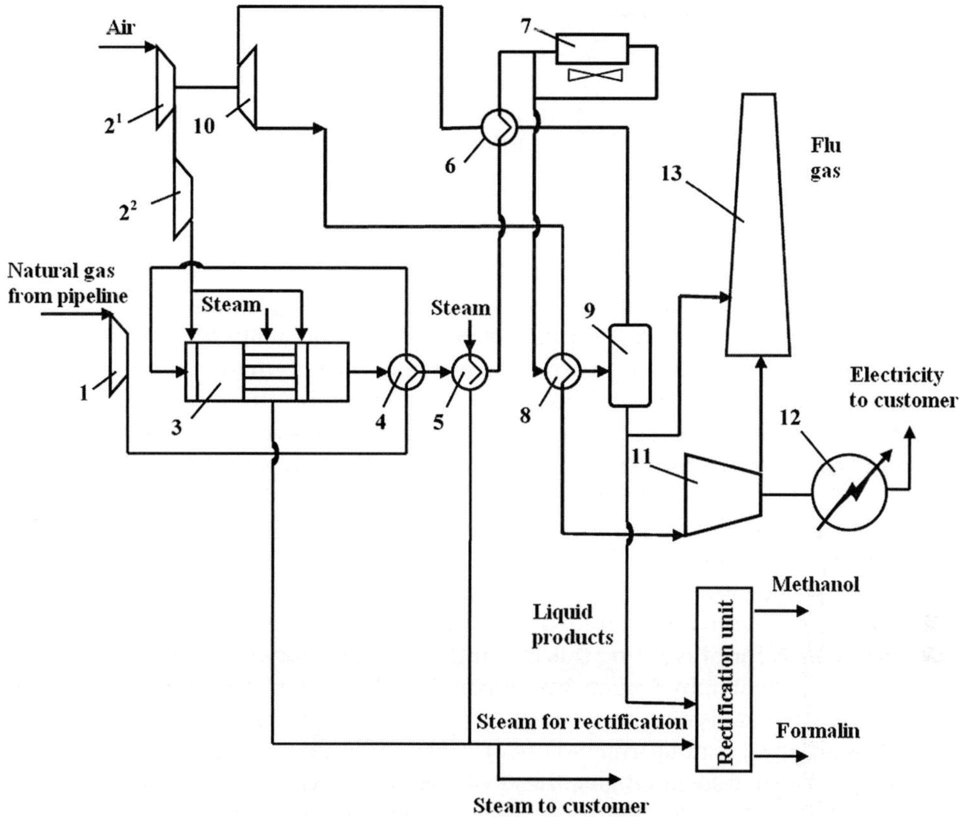

FIGURE 11.25 Flow diagram of the integrated process for the production of methanol and electricity with removal nitrogen oxides from the flue gas: (1) booster compressor, (2^1, 2^2) stages of air compressor, (3) two-sectioned reactor, (4) recuperative heat exchanger, (5) steam generator, (6) heat exchanger, (7) air cooler, (8) end heat-exchanger refrigerator, (9) separator, (10) turboexpander (11) gas turbine, (12) generator, and (13) smokestack.

and specific investment expenditures will be significantly lower than the values typical of modern plants. In this case, the environmental performance of power plants will be improved without additional expenditures, and, at the same time, the reliability of providing them with reserve fuel will be enhanced.

The oxidative prereforming of associated gas can become one of the effective processes of their preparation for traditional catalytic conversion. Real natural gas is a complex mixture of hydrocarbons, the composition of which is specific for each field and changes markedly with time in the course of its exploitation. While the methane content in natural gas is generally greater than 90%, in associated gas it may be less than 70%, with an appropriate proportion of heavier hydrocarbons. The problem of utilization of associated gas and fatty natural gases is one of the greatest challenges in oil and gas production. Every year, over 150 milliard m^3 of various hydrocarbon gases are flared worldwide, and this amount does not reduce in spite of all the efforts.

A large admixture of heavy hydrocarbons complicates the use of natural and associated gas as an energy fuel and chemical raw material. Modern power plants that run on natural gas are intended for the use of dry gas containing not more than 2% C_2+ hydrocarbons, being easily damaged when operating on raw (unrefined) gas. A high content of C_2+ hydrocarbons in natural gas is unfavourable for many catalytic processes, mainly due to catalyst coking or soot formation.

Currently, the vast majority of the chemical processing of natural gas is carried out via its preliminary conversion into syngas, mainly by catalytic steam reforming. This process is sensitive to the content of heavy methane homologues in the gas to be converted. Under typical conditions of steam reforming, the presence of significant concentrations of C_2+ hydrocarbons leads to an abundant sooting and rapid deactivation of the catalyst. One of the practical methods for preventing this is the selective conversion of heavy hydrocarbons under milder conditions. For this purpose, an additional step, prereforming (for example, in the Haldor-Topsoe technology) is introduced. However, prereforming involves significant additional capital and energy costs and, in addition, increases in the consumption of natural gas by about 5%.

Since the joint partial oxidation of C_1-C_4 hydrocarbons significantly reduces the concentration of C_2-C_4 while increasing the concentration of methane (Chapter 10), this process can be used as an oxidizing prereforming step for reducing the content of heavy hydrocarbons. However, at high concentrations of C_2-C_4 hydrocarbons, their oxidation requires the appropriate amount of oxygen, which can lead to a significant heating of the mixture and sooting. The standard method for preventing this is to introduce additional reactants into a mixture to impede this process: water, hydrogen, or oxygen. Just as for methane, for mixtures of hydrocarbons, there are thermodynamically specified minimum concentrations of O_2, H_2, and H_2O at which solid carbon C_s is not formed. For a $0.7CH_4 + 0.1C_2H_6 + 0.2C_3H_8$ (in mole fractions), which simulates a real associated gas, these values, obtained from thermodynamic calculations, are shown in Fig. 11.26. Naturally, they are higher than for pure methane due to a reduction in the H:C ratio in the molecules of heavier hydrocarbons.

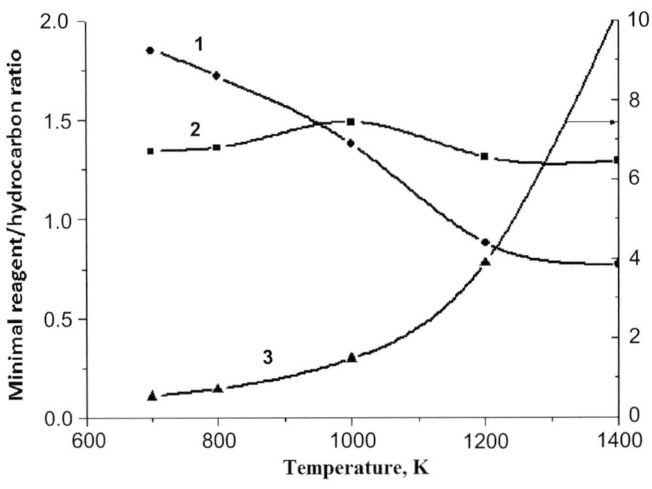

FIGURE 11.26 Minimum ratios of reactants ((1) O_2, (2) H_2O, (3) H_2) for the reaction $0.7CH_4 + 0.1C_2H_6 + 0.2C_3H_8$ + reagent ↔ products at which the formation of C_{solid} is thermodynamically impossible. The pressure in the system was 40 atm [303].

A comparison of the results of thermodynamic equilibrium calculations of the concentration of the oxidation products with the respective experimental data shows that, at relatively low process temperatures and in the absence of the catalyst, the composition of the products is far from equilibrium, being determined by the kinetics of the processes involved. This is especially true for mixtures with low concentration of oxygen and, accordingly, little heating. In the case of oxidation of mixtures of C_1–C_4 hydrocarbons at a higher concentration of oxygen and a heating of up to 640 °C, the concentrations of the main products, methane and CO, approach their equilibrium values, calculated for the fast stage of the process (without the formation of solid product, which takes a longer time.) That the measured concentrations of CO_2 and hydrogen are lower than the thermodynamically calculated is due to an extremely low reactivity of the resultant water. At these temperatures and in the absence of heterogeneous catalysts, the time it takes to reach equilibrium between water, hydrogen, and carbon oxides is by many orders of magnitude longer than the reaction time.

The possibility to avoid sooting over the entire temperature range interesting from the point of view of oxidative prereforming by introducing a 1.5-fold excess of water, which is anyway injected at the stage of steam reforming, makes this process very attractive. When oxidative conversion is carried out under relatively mild conditions, at temperatures below 600 °C, along with carbon oxides, water, and hydrogen (target products of steam reforming), significant amounts of oxygenates, such as methanol, formaldehyde, ethanol, etc. and ethylene are also formed. Experimental studies of the effect of these compounds on the process of steam reforming of natural gas over industrial catalysts have shown that these compounds do not adversely affect the process [303].

Based on experimental studies of the oxidation of methane mixtures with C_2–C_4 hydrocarbons and the demonstrated possibility of selectively reducing the content of C_2–C_4 hydrocarbons in such mixtures (Chapter 10), the authors of [303] proposed a two-step process of conversion of mixtures of methane with alkanes C_2+ to syngas. Proceeding from material balance of the process, a conceptual flow diagram of oxidative prereforming was proposed (Fig. 11.27).

Associated gas or natural gas from the pipeline is compressed with a booster compressor to a pressure of 3 MPa and is directed into the tube space of a gas–gas recuperative heat exchanger located in the second section of the reactor. When the gas heats up to a temperature of superheated vapour, steam in a molar ratio of 1: 1 with respect to the total carbon content in the mixture is also fed here to reduce the maximum temperature of the reaction mixture and to prevent sooting at the partial oxidation stage. In the recuperative heat exchanger, the mixture is heated to 400 °C and fed into the mixer of the first section of the reactor. Oxygen is compressed by the compressor to a pressure of 3 MPa and is directed in two flows into the mixers of the first and second sections of the reactor. The reaction mixture from the second zone of the reactor, at a temperature of ~750 °C and at a pressure up to 2.8 MPa, without subsequent cooling and separation of liquid product, is delivered directly to the stage of steam catalytic conversion, where additional amount of steam required in accordance with technological specifications of the main process is also supplied.

The process may be of practical interest for the steam catalytic conversion of fat natural gases to syngas without preliminary purification from C_2+ hydrocarbons or without the stage their preliminary steam prereforming, making it possible to reduce the energy expenditures on the heating of the steam-hydrocarbon mixture. Furthermore, the process may be

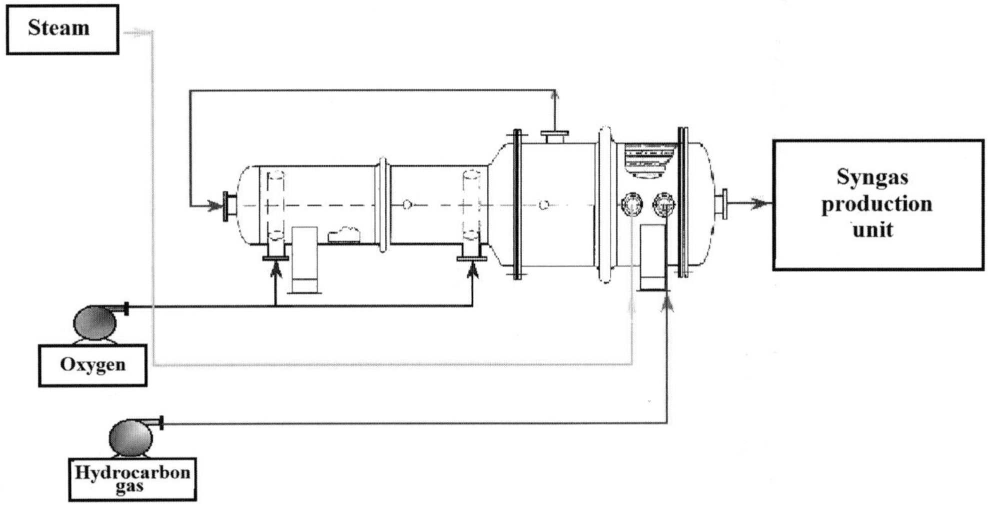

FIGURE 11.27 Conceptual diagram of the oxidative prereforming of C_2–C_4 hydrocarbons in natural gas. (For colour version of this figure, the reader is referred to the online version of this book.)

used for the oxidative purification of hydrocarbon gases from heavy alkanes, after which exhaust gases with a low content of C_2+ hydrocarbons may be directed to standard power plants for generating electricity.

Utilization of coalbed methane for the production of methanol-coal slurries may also be a promising niche for the use of low-scale installations based on the process of direct oxidation. During coal metamorphism, which was accompanied by a low-temperature thermochemical decomposition of organic matter, coal seams produced large amounts of methane. Methane accumulated due to its adsorption in coal, in vertical fractures and cracks of coal seams, disintegrations and cracks between them. One ton of coal typically contains 6.8 m^3 of gas. Since the world's coal resources are approximately 10^4 milliard tons, the gas content of coal deposits is comparable to its content in the gas fields. According to various sources, the coal-bearing strata of the coal basins worldwide contain from 85 to 262 trillion m^3 of natural gas. Therefore, even a moderate gas production from coal seams could make a significant contribution to the world's production of natural gas. In addition, the coalbed methane released to the atmosphere constitutes about 8% of all anthropogenic methane emissions, with its contribution to the anthropogenic greenhouse effect reaching 17% [300].

In the U.S., active coalbed methane extraction has been carried out since the 1950s, and its annual output currently stands at 55 billion m^3, accounting for about 8% of the total natural gas production. The production of coalbed methane is also underway in Canada, Australia, China, India, and other countries. Most of the coalfields in Russia (Kuznetsk, Pechora, Donetsk, Taimyr, Tunguska, etc.) contain gas. The methane content in such highly metamorphosed coal seams increases with the depth of their occurrence, reaching 40–50 m^3/ton. The preliminary degassing of coal seams is a necessary condition for safe operation of the miners and an additional source of methane. The use of coalbed-methane wells is complicated by a low flow rate and its large spread, which can vary from 2 to more than 200 thousand m^3/day, with typical values of 5–15 thousand m^3/day [304].

By the terms of explosion safety (upper and lower explosion limits of methane in air at atmospheric pressure are, respectively, 5% and 15%) and mine-degasification technology, there are two types of coalbed methane: with concentrations of less than 3 vol % (usually not more than 1%, ventilation gases) and above ~25% (degassing gases with a methane concentration of up to 95%). The first kind of gas, due to the very low content of methane in it, is not of interest for the chemical processing and, at best, can be disposed of as low-calorific fuel for catalytic burners or gas turbines with the catalytic combustion of methane [300]. The second type can be used not only as a high-calorie fuel gas, but also is a perspective gas chemical feedstock, for example, for direct oxidation processes. This opens the possibility of a comprehensive utilization of coalbed wastes, including not only methane, but also large amounts of coal waste, fine coal slurry, which accumulates in the course of extraction and enrichment.

Since the 1970s, in Russia and other countries, on a relatively small scale, coal and oil in power plants and furnaces are replaced by water-coal fuel (WCF), which is also used even as a fuel for turbine and diesel power plants [305]. In addition to being a useful method for utilization of waste coal, the WCF makes it possible to transport coal in railway tanks and through pipelines. A lot of experience in this area has already been acquired in Russia, the U.S., and China. The preparation WCF includes the grinding of coal (up to a size of 0–250 μm), separation of a significant part of the mineral component (enrichment for reducing the ash content of the fuel, typically less than 5%), adding a plasticizer (surfactant, 1–2% sodium humate), and homogenization of the mixture. The share of coal in such fuel is typically 60–70%, while its storage time is 25–30 days. Depending on coal used, the heat of combustion of WCF ranges from 21 MJ/kg for black coal to 16 MJ/kg for brown coal.

Coal slurry is metastable colloidal system. Its stability is determined by coal mineral components, mainly finely divided hydrophilic clays and surfactant additives. The expenditures for surfactants and homogenization constitute a significant share of the cost of fuel. Furthermore, in the course of the combustion of WCF, evaporation of each 10% of the water in its composition requires to burn 1% of the present coal. Aside from a relatively low calorific value, WCF tends to freeze during transportation in winter conditions.

In terms of the calorific value, performance, stability, and other properties, another kind of slurry is more promising, namely, coal-methanol slurry. Technology of production, applications, and properties of methanol-coal slurries have been described in a number of patents [306–309]. Methanol-coal slurries with coal particle sizes less than 1500 μm and a specific distribution of the fractions: less than 74 μm (30–65 wt%), less than 10 μm (10–25 wt%), and less than 3 μm (5–15 wt%). Such slurries have good stability and excellent transportation and fuel characteristics. The content of carbon particles ranges from 42.7 to 73.6 wt%. For this purpose, low-grade coals, anthracite, bituminous coal, lignite, and others can be used. The water content may be up to 30 wt%, including the moisture contained in the coal. Wet coals and wastes of coal processes can be directly used. Furthermore, to improve stability and reduce viscosity, surfactants, 0.005–5 wt%, for example, partially hydrolysed salts of lignosulphonic acid, can be introduced. The sources of the latter may be waste cooking oils, wood chips, and other lignin-containing materials in a solution of bisulphite and sulphuric acid, as well as other wastes.

As compared to WCF, the use of methanol-coal slurries reduces transportation costs per unit of energy delivered and eliminates the problem of transportation in winter conditions (freezing point of methanol −98 °C), and allows using low-grade and wet coals. Coal and methanol are mixed well, with many components of coal being partially eluted into methanol, which leads to complex interactions or modifications of both the solid and liquid phases. Methanol effectively penetrates into pores and cracks in coal, making it easier to disintegrate during combustion due to explosive expansion on heating.

Methanol-coal slurries exhibit a high stability and long storage time, more than 2 months even without the periodic stirring, as well as good transport properties due to a sufficiently low viscosity, 50–1000 cP. Their effectiveness as a fuel exceeds that of pulverized fuel. An increase in the percentage of coal in the slurry improves its stability but, simultaneously, increases the viscosity, which makes transportation more difficult. The stability increases with the degree of crushing coal, a factor that reduces, in addition, the wear of pipelines and improves the fuel properties of the emulsion. On the other hand, increasing the proportion of larger coal particles reduces the viscosity, increases the share of coal in the slurry and reduces the energy spent for grinding. Therefore, special attention is paid to ensure an optimal size distribution of the coal particles.

The use of such slurries is particularly effective in the case of coals capable of adsorbing a large amount of methanol, i.e., bituminous and subbituminous coals. Note that methanol-coal slurries can be prepared from methanol containing various impurities formed during its production, including heavier alcohols and water [309].

This combination of the properties of methanol-coal fuel makes the joint utilization of coalbed methane and coal waste on the basis of small-tonnage process of direct oxidation of methane to methanol an attractive option (Fig. 11.28).

One possible scheme of the process is displayed in Fig. 11.29 [310]. Coalbed methane with an admixture of air is introduced into the oxidation unit, where it is boosted to a pressure of 6–7 MPa, heated in an indirect heat exchanger, and fed into the reactor. The liquid condensate separated after cooling the partially oxidized gas mixture, containing up to 40% methanol and 5–8% of other organic compounds, is used, without any pretreatment or rectification, together with crushed and enriched waste coal, to prepare a methanol-coal fuel. The exhausted methane-containing gas, with a methane concentration not less than 25%, is directed to electricity generation for the entire complex and the needs of coal mining.

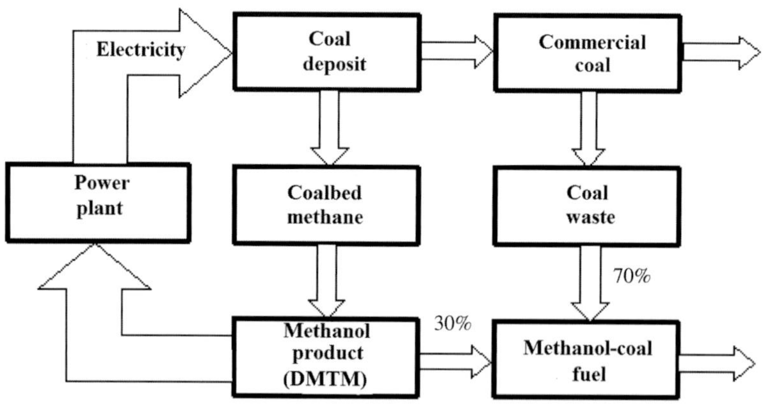

FIGURE 11.28 Flow diagram of the production of methanol-coal fuel from coalbed methane and coal waste.

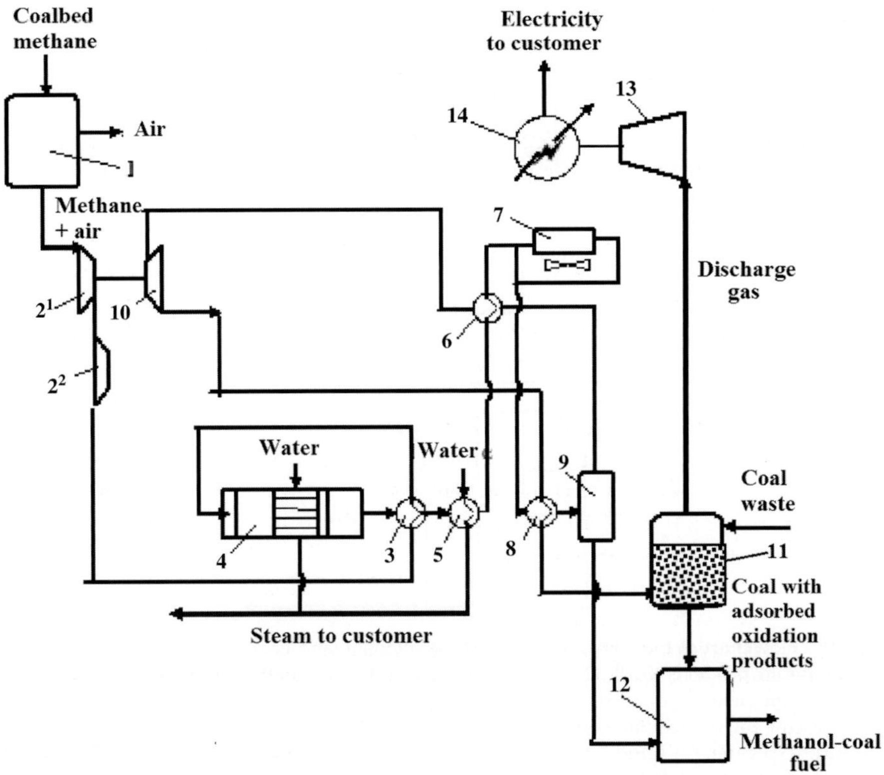

FIGURE 11.29 Flowchart of the production of methanol-coal fuels from converted coalbed methane and coal waste [310]: (1) separator to enrich the air–methanol mixture, (2^1 and 2^2) compressor stages for methane-enriched air mixture, (3) recuperative heat exchanger for heating the methane–air mixture, (4) two-section reactor for the oxidation of methane–air mixture with an intersection heat exchanger, (5) steam generator, (6 and 8) heat exchangers, (7) air cooler, (9) separator, (10) turboexpander, (11) scrubber for absorbing the residual oxidation products from the exhaust gas, (12) cavitation mixer of liquid and solid products, (13) gas turbine, and (14) electricity generator.

The process of production of a methanol-containing condensate via the oxidation of methane with air oxygen is almost identical to the traditional scheme of the direct oxidation of methane to methanol except for the absence of compressed air supply, since coalbed methane contains, as a rule, a sufficient amount of it. Obviously, on the contrary, a methane enrichment unit will be needed to reduce the oxygen concentration in it to a desired methane/oxygen ratio. To partially offset the cost of methane compression, the energy generated by turboexpander is used to drive the first stage of the methane compressor. Furthermore, the exhaust gas direction to energy generation is passed through a layer of finely dispersed coal intended for methanol-coal fuel production to adsorb residual liquid oxidation products untrapped in the separator.

A simpler scheme for the production of methanol-coal fuel without preliminary separation of the liquid methanol product and, consequently, without bulky heat-exchange equipment is presented in Fig. 11.30.

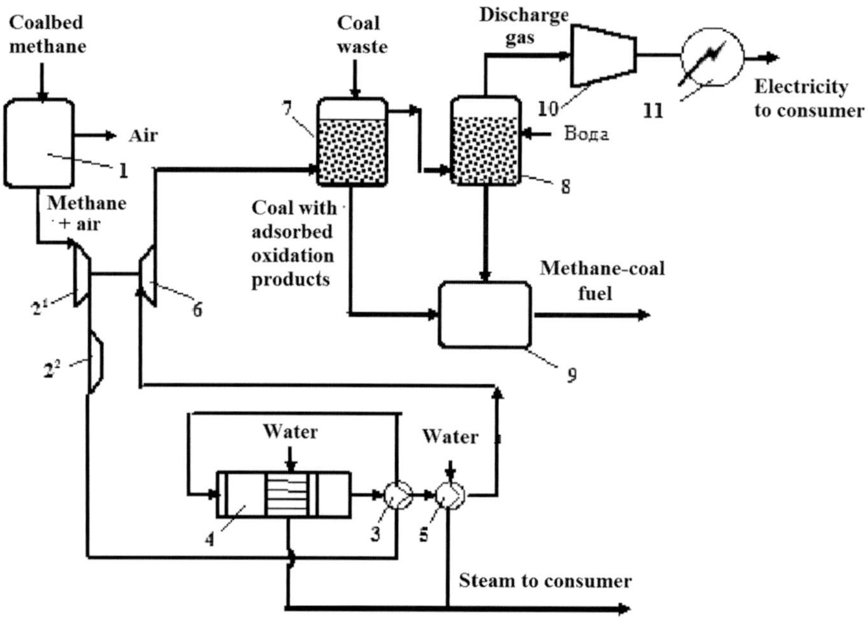

FIGURE 11.30 Flowchart of the conversion of coalbed methane and coal waste to methanol-coal fuel without separation of the liquid products [310]: (1) separator to enrich the air–methanol mixture, (2^1 and 2^2) compressor stages for methane-enriched air mixture, (3) recuperative heat exchanger for heating the methane–air mixture, (4) two-section reactor for the oxidation of methane–air mixture with an intersection heat exchanger, (5) steam generator, (6) turboexpander, (7) scrubber for absorbing the oxidation products, (8) bubble absorber for trapping condensable products, (9) gas turbine, and (10) electricity generator.

In this version, the high pressure of the gas mixture containing methanol and other oxygenates, after precooling in the regenerative heat exchanger and steam generator is immediately used to actuate the turboexpander. Then, the cooled mixture passes through a bed of finely dispersed coal prepared for the production of methanol-coal fuel, where much of oxygenates is adsorbed. The exhaust gases are bubbled through a water layer to saturate it with remaining oxygenates. When the required concentration of methanol is achieved, this aqueous methanol solution is also directed to fuel production.

The properties of the resulting fuel are largely determined by the preparation method. In particular, the high activity of coal in its interaction with methanol and other oxygenates leads to the alkylation of coal and its depolymerization [311]. Therefore, when mechanochemical and especially cavitation technologies are used to produce methanol-coal fuel, one can expect not only the trivial formation of a slurry, but also significant changes in the structure of the original coal and in the fuel composition, appearance of new liquid fraction, and significant enhancement in the stability and storage time of the fuel (without introducing additives).

It is also interesting to examine the interaction of coal with methanol and other oxygenates under supercritical conditions. In this case, the interaction of coal with oxygenates [312] is accompanied by coal hydrolysis with supercritical water to produce liquid and gaseous products, including resins, lower alkanes, and aromatic compounds [313]. This process is expected to give rise to a fundamentally new stable liquid hydrocarbon product with high consumer characteristics.

Another promising area of utilization of coalbed methane (and air oxygen in it) may be associated with its direct interaction with fine coal in supercritical conditions. The hydrogenation of coal by methane (a process reverse to some extent to the natural formation of coal accompanied by methane release) is expected to produce lighter hydrocarbons, including aromatics. The partial oxidation of methane and products formed will also affect the composition of the final liquid product. Given the rapid progress of supercritical processes in chemical technology, one can count on interesting practical results.

Thus, there is a wide spectrum of potential technologies based on the DMTM process for further elaboration and practical implementation. The majority of potential technologies considered in this chapter practically have no viable cost-effective alternatives.

CHAPTER

12

Place of the DMTM among the other Processes of Partial Oxidation of Light Alkanes

PARTIAL OXIDATION OF LIGHT ALKANES AS A BASIS OF THE MOST IMPORTANT GAS-CHEMICAL PROCESSES

A steady increase in the share of light hydrocarbons in the global production of chemicals stimulated by the relative accessibility and abundance of natural gas resources motivates general interest in studying the specifics of gas-chemical processes that distinguishes them from traditional petrochemical processes. Traditional petrochemistry is based on the destructive conversion of large hydrocarbon molecules into smaller, less saturated, and more branched molecules by means of pyrolysis, cracking, dehydrogenation, and isomerization. Gas chemistry, on the contrary, makes use of 'constructive' processes: the transformation of small hydrocarbon molecules, especially methane, to a wide variety of different chemicals, including liquid fuels, polymers, etc. Although destructive processes such as pyrolysis are also used in gas chemistry, their scope of applications is limited because of a small molecular weight and high stability of feed gases. Destructive processes typically require rigid conditions and entail considerable energy costs. The pyrolysis of methane produces only a small set of products, such as soot, hydrogen, and acetylene. The production of other chemicals can only be performed by pyrolyzing heavier (more deficient) methane homologues and can actually be attributed to the domain of traditional petrochemical industry.

The high dissociation energy of the C—H and C—C bonds in the molecules of light alkanes, the major hydrocarbon components of natural gas, makes it extremely difficult to accomplish their chemical activation, which requires, even in the case of catalytic processes, rather high temperatures and, therefore, intense heat fluxes. The most effective and economical way of introducing energy into the reacting mixture of light hydrocarbons is the partial oxidation of feedstock hydrocarbons themselves. In addition, due to the radical-chain character of these processes, it is possible to produce a cross-activation of various parallel reactions.

Oxidative processes with high thermal efficiency account for a large share of the modern chemical industry. By some estimates [314], they yield more than 50% of the world's chemical products. In particular, the major primary product of the production of staple chemicals, such

Direct Methane to Methanol
http://dx.doi.org/10.1016/B978-0-444-63253-1.00012-X

as ammonia, hydrogen, methanol, and synthetic liquid hydrocarbons is syngas produced by oxidative conversion of natural gas. The production of various chemicals, mainly oxygenates, by direct oxidative conversion of natural gas has almost equally long history. Great interest and an unprecedented surge of research activity were caused by the discovery in the 1980s of the possibility of providing a significant yield of olefins in the partial oxidation of methane. In principle, the oxidative conversion of methane and its homologues gives a wide variety of products, up to nanotubes and fullerenes. The output of individual products is usually small; therefore, only in rare instances, oxidative conversion can be considered as a method for their preparation. However, various processes of oxidative conversion of methane and its closest homologues to more valuable products are being actively explored. Generally, along with oxygen, hydrocarbons can be oxidized by other agents, such as halogens or sulphur-containing compounds [315].

Since the slowest step in the chemical conversion of methane is the abstraction of a first hydrogen atom from its molecule, the replacement of a hydrogen atom by any other atom or a functional group greatly facilitates further methane conversion into target products. Therefore, the substitution of one or more hydrogen atoms with other atoms or functional groups is often seen as a self-sufficient goal of research or a technological procedure, because the subsequent operations appear to be straightforward. The methane oxidative functionalization methods occurring completely or largely in the gas phase include the processes of partial oxidation of hydrocarbons at moderate temperatures (intermediate temperature region), from ~300 to ~1200 °C, i.e., below the characteristic temperature of the gas-phase combustion of hydrocarbons, typically well above 1200 °C.

The distinction between combustion and partial oxidation is of fundamental nature and is determined not only by the difference in the temperature of the process, but, above all, in the mechanism. One of the main distinguishing features of the mechanism of the partial oxidation of alkanes in the intermediate-temperature range is an important role played by peroxide compounds and radicals. In the oxidation of methane at the low-temperature end of this range, the most important species are $CH_3OO^•$ radicals and CH_3OOH molecules, whereas at the high-temperature end, the leading role belongs to $HOO^•$ radicals and HOOH molecules. At the same time, the role of such species as O, CH, CH_2, C_2, and excited molecules and radicals, important for combustion processes, is negligible in the partial oxidation.

The different parts of the indicated temperature range in which the oxidation of methane has been most thoroughly studied are dominated by different products (Fig. 12.1). As expected, with increasing temperatures, thermodynamically more stable products become most abundant, with their maximum yields increasing monotonically, reaching almost 100% in the case of syngas. Note, however, that, despite the diversity of the conditions and products, it can be reasonably assumed that the totality of the processes involved in the partial oxidation of methane in the middle-temperature range is described by a single homogeneous–heterogeneous kinetic mechanism [7].

In accordance with the abundance of certain products, the intermediate-temperature range can be in turn divided into three subranges: those of preferential formation of oxygenates (300–600 °C), oxidative coupling of methane (OCM) with predominant formation of C_2 hydrocarbons (600–900 °C), and preferential formation of syngas (>900 °C) (Fig. 12.1). This division corresponds to obvious changes in the kinetic mechanism with the temperature of the process.

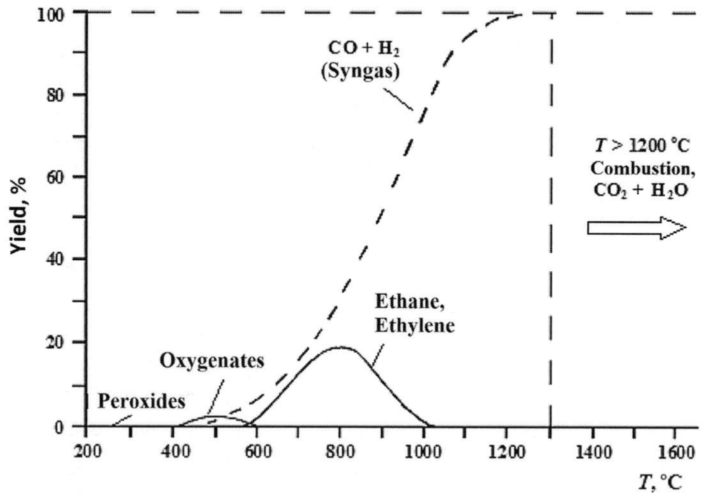

FIGURE 12.1 Temperature ranges of formation and the yields of the main products of methane partial oxidation.

At temperatures below 600 °C, the dominant reaction of the methyl radical $CH_3^{\cdot}$ is its reversible recombination with the oxygen molecule to form the methylperoxy radical,

$$CH_3^{\cdot} + O_2 \leftrightarrow CH_3OO^{\cdot}, \tag{12.1}$$

which plays a key role in the formation of oxygenates. At these temperatures, especially at heightened pressures, the equilibrium in reaction (12.1) is strongly shifted to the right, which gives rise to the subsequent chain of transformations that lead to the preferential formation of oxygenates. At temperatures above 600 °C, the equilibrium in this reaction shifts to the left, so that its role diminishes rapidly (Fig. 12.2, left curve).

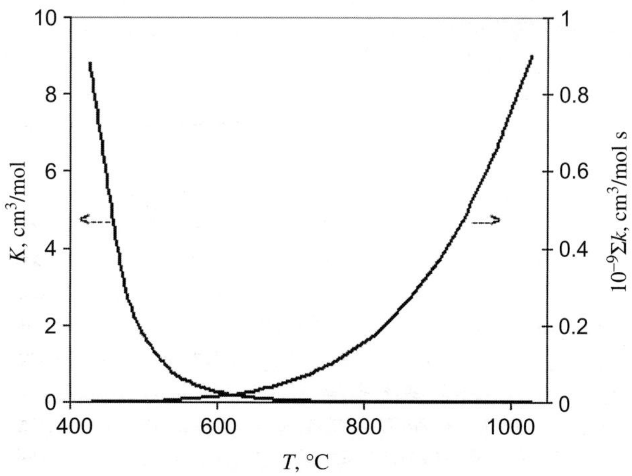

FIGURE 12.2 Temperature range for the oxidative conversion of methane [7]. Left curve - temperature dependence of the equilibrium constant of the reaction 12.1. Right curve - temperature dependence of the sum of rate constants of reactions 12.2 and 12.3.

At temperatures of 600–900 °C, the reactions of endothermic oxidation of the methyl radical,

$$CH_3^{\bullet} + O_2 \rightarrow CH_3O^{\bullet} + O^{\bullet\bullet} \quad (12.2)$$

$$CH_3^{\bullet} + O_2 \rightarrow CH_2O + OH^{\bullet}\, , \quad (12.3)$$

are too slow, despite a significantly higher concentration of oxygen compared to that of methyl radicals (Fig. 12.2, right curve), to compete with the methyl radical recombination

$$CH_3^{\bullet} + CH_3^{\bullet} \rightarrow C_2H_6 \quad (12.4)$$

This reaction leads to the formation of ethane, and further, ethylene, i.e., gives rise to a temperature range of preferential formation of methane coupling products, known as the 'ethylene window'. Above 900 °C, the rates of reactions (12.2) and (12.3) become high enough to ensure a deep oxidation of methane, up to carbon monoxide and hydrogen, i.e., syngas [7].

The effect of temperature on the yield of the oxidation products of heavier homologues of methane, such as butane (Fig. 10.15), is much the same, and only the limits of formation of the respective groups are shifted to lower temperatures.

Thus, despite a great diversity of conditions for the oxidative conversion of methane, there are three basic types of processes and the respective products of oxidative conversion of methane: the direct oxidation of methane to methanol (DMTM), OCM, and partial oxidation to syngas (POM). To gain deeper insights into the prospects of the DMTM, let us, at least briefly, consider the potentialities and perspectives of other processes of oxidative conversion of natural gas.

OXIDATIVE CONVERSION OF METHANE TO SYNGAS

The partial oxidation of methane to syngas (POM),

$$CH_4 + 0.5O_2 \Leftrightarrow CO + 2H_2 \quad \Delta H = -44\ \mathrm{kJ/mol} \quad (12.5)$$

along with steam and, to a lesser extent, carbon dioxide reforming is the basis of modern large-capacity gas-chemical technologies, both as an independent process or as a component of autothermal reforming (ATR). This is a mature large-capacity technology, used, in particular, in Shell's GTL process.

The POM is a weakly exothermic process that requires no additional heat. Developed in the early 1950s [316], by the end of the 1960s, it was used to produce about a quarter of the syngas for ammonia production in the U.S. [93]. The POM is usually conducted at a pressure of 30–100 atm with the use of pure oxygen, which is taken in a small excess over the stoichiometry of reaction (12.5) to provide a more complete conversion and to achieve the desired temperature. This, however, leads to the formation of some amount of deep oxidation products, CO_2 and H_2O. Typical temperatures of the noncatalytic POM that provide a complete conversion of methane and minimize soot formation are 1300–1500 °C, with the thermal efficiency of the process reaching 60–75%.

Despite advances in research and technological applications of methane-steam reforming, the partial oxidation and combined steam-oxygen conversion of methane remain the most popular research areas. In addition to reducing the energy consumption, this method produces a syngas with a H_2/CO ratio of ~2, the most favourable for the Fischer–Tropsch and methanol syntheses. Another attractive aspect of the POM is the possibility of using various types of hydrocarbons, including heavy hydrocarbons, for producing syngas. In this case, however, the H_2/CO ratio is about 1.0, i.e., too low for the production of synthetic hydrocarbons and methanol. It can be modified by converting part of CO into CO_2 and H_2, but it requires a large consumption of steam and reduces the efficiency of the process.

The H_2/CO ratio for syngas produced from natural gas, close to 2, makes the process of partial oxidation of natural gas to syngas very attractive from the point of view of the synthesis of methanol and synthetic hydrocarbons. However, since part of the hydrogen is oxidized to water, the actual H_2/CO ratio is typically 1.5–1.6. The situation can remedied by combining POM with steam reforming, for which the H_2/CO ratio is ~3.

The expected conversion and yield of syngas are estimated based on thermodynamic calculations. However, in the case of a purely homogeneous oxidation of very rich mixtures in actual conditions, it is not always possible to achieve the equilibrium chemical composition of the products. The authors of [317], using a detailed model of the oxidation of C_1–C_2 hydrocarbons, performed a kinetic analysis of the formation of syngas during the homogeneous oxidation of hydrocarbons. Since syngas is produced from mixtures with a low O_2/CH_4 ratio (low oxidant equivalence ratio $\alpha = O_2/2CH_4$), outside the explosion limits under normal conditions, the gas mixture must be preheated and the oxidation should carried out at elevated pressure. Therefore, the calculations were performed at a pressure of $P = 12$ atm for a mixture preheated to $T_0 = 923$ K. Under these conditions, the conversion occurs within ~0.1 s. The dependence of the final concentrations of H_2 and CO on the oxidant equivalence ratio α is displayed in Fig. 12.3. The maximum yield of syngas is achieved at $\alpha \approx 0.4$, a value at which

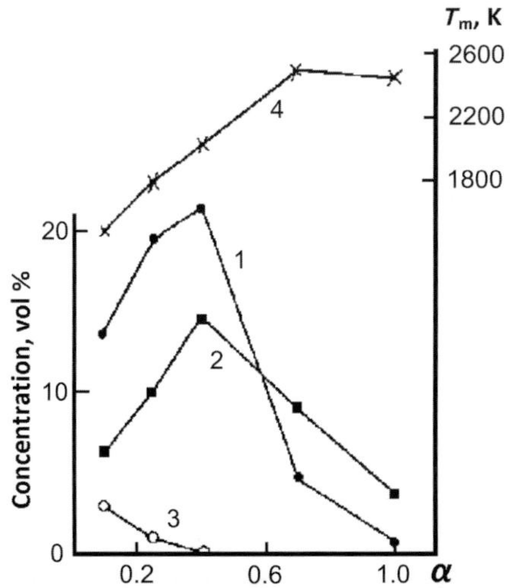

FIGURE 12.3 Dependence of the final concentrations of (1) H_2, (2) CO, and (3) CH_4 and of (4) the maximum temperature T_m on α at $T_0 = 923$ K and $P = 12$ atm [317].

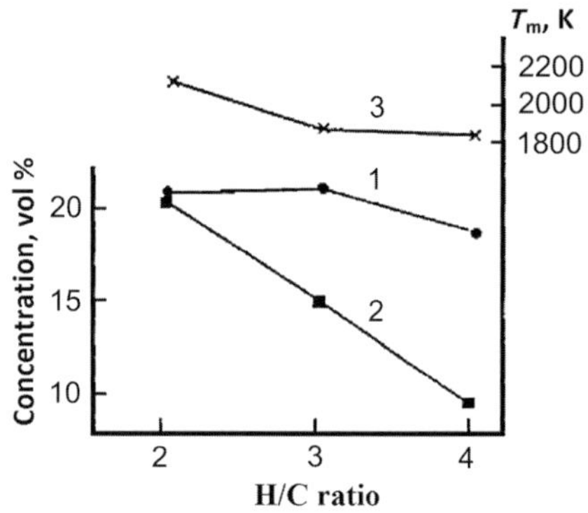

FIGURE 12.4 Dependence of the maximum yields of (1) H_2, (2) CO, and (3) temperature T_m on the H:C ratio in the initial hydrocarbon [317].

the maximum heating of the mixture is well below that for the stoichiometric mixture ($\alpha = 1$). Although the value of α optimal from the standpoint of the thermodynamics and stoichiometry of reaction (12.5) is 0.25, its decrease below $\alpha \approx 0.4$ leads to a rapid decline in the concentrations of H_2 and CO due to a lowering of the process temperature and the corresponding decrease in the methane conversion.

The yield of syngas in the oxidation of heavier hydrocarbons was estimated based on kinetic calculations. The calculations were performed at three values of the H:C ratio for the hydrocarbon, 2, 3, and 4, corresponding to the oxidation of CH_4, C_2H_6, and C_2H_4 at $\alpha = 0.25$ (Fig. 12.4). With decreasing the H:C ratio, the CO yield increases, whereas the yield of H_2 remains approximately constant. It might seem that H_2 yield should decrease with the H:C ratio; however, the simultaneous temperature rise due to the reduction of the bond dissociation energy of the corresponding hydrocarbon compensates for this decrease by increasing the conversion.

These calculations make it possible to estimate the yields of H_2 and CO for the heavier hydrocarbons, such as propane and butane, direct kinetic calculations for which are difficult to perform because of the lack of reliable kinetic models of their oxidation under these conditions. Assuming that the dependences displayed in Fig. 12.4 hold for other hydrocarbons, one estimate the yields of H_2 and CO from propane and butane oxidation by using H:C = 2.67 for propane and H:C = 2.5 for butane. Thus, the main factor determining the yields of H_2 and CO in the homogeneous partial oxidation of hydrocarbons to syngas is the mixture composition. The optimal composition of the mixture and the corresponding maximum yields of the conversion products are determined by the specific conditions of the partial oxidation of the hydrocarbon.

An increase in the total pressure increases the partial pressures of CH_4, CO_2, and H_2O at the equilibrium gas composition; i.e., high pressure is unfavourable for the POM. However, high temperatures achieved in the noncatalytic process compensate for this effect, because selectivities to CO and H_2 increase with the temperature (Fig. 12.5) [318].

According to the above data, the methane–oxygen and methane–air systems produce practically interesting yields of hydrogen and carbon monoxide only at O_2:$CH_4 < 1$, which

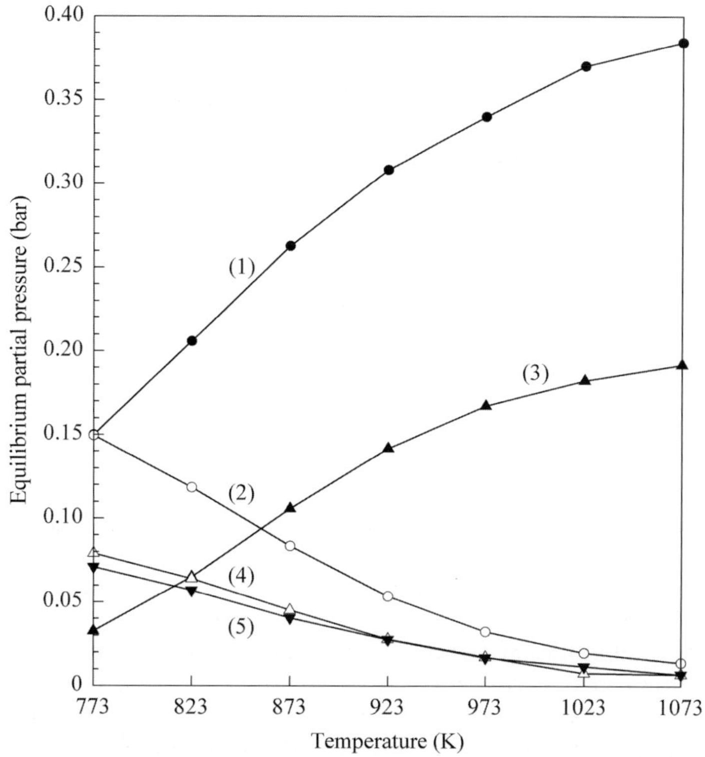

FIGURE 12.5 Equilibrium gas composition for the partial oxidation of methane at O_2:CH_4 = 0.5 and a pressure of 1 bar: (1) H_2, (2) CH_4, (3) CO, (4) CO_2, (5) H_2O [318].

corresponds to a methane-in-air content of more than 17.3 vol %. Under normal conditions, such methane–air mixtures corresponding to the stoichiometric ratio of reaction (12.5) are outside the explosion limits, whereas the oxidation of less rich mixtures with a composition close to the explosion limit yields large amounts of soot. Practically feasible mixtures contain 0.6 molecules of oxygen per methane molecule (with an air equivalence ratio α above 0.3). Thus, for the practical production of syngas, a narrow interval of so-called 'superrich' mixtures is used, in which the methane content exceeds the upper explosion limit for methane–air mixtures under normal conditions (~15%). The ignition and combustion of such mixtures is only possible under special conditions, for example, high initial pressures and/or temperatures. In addition, in the FT processes and others in which syngas is used, oxygen is a catalyst poison, so it is necessary to ensure its essentially complete conversion to a content below ~0.15 vol %. Along with the noncatalytic POM, the catalytic process is considered, the mechanism of which has been studied in recent years in more detail than the other methane conversion reactions.

A number of major world companies rely on the production of syngas and hydrogen by ATR, especially for large-scale productions. ExxonMobil planned to use ATR to produce syngas in its GTL-technology with an output of 154,000 barrels a day. The Haldor Topsoe

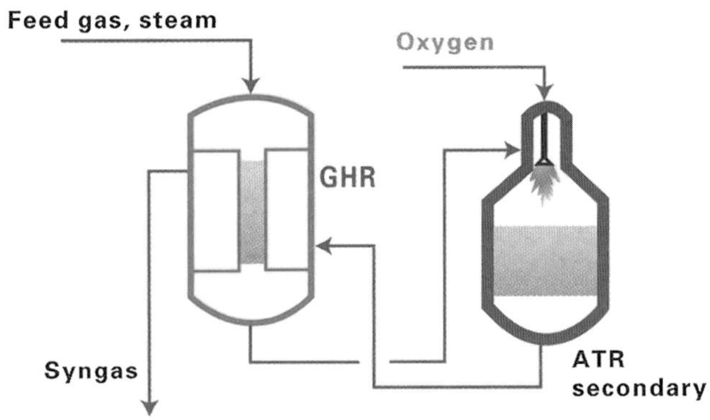

FIGURE 12.6 Combination of ATR reformer with a gas heated reformer (GHR) [319]. (For colour version of this figure, the reader is referred to the online version of this book.)

commercialized the ATR process with a steam-to-carbon ratio of 0.6. A plant with an equivalent capacity of 2 × 8500 barrels of GTL-products a day based on this technology was launched in the South African Republic in 2004 and has shown excellent results. The same technology is also employed at the Oryx GTL plant in Qatar.

The tendency of combining various reactions of methane conversion to syngas to improve the efficiency of the process is currently the dominant. As an example, it is worthwhile to mention a process developed and implemented by the Synetix Company in the early 1990s, in which the ATR was integrated with an advanced gas-heated reformer, capable of utilizing the high-potential heat of reformed gas directly in the process of reforming (Fig. 12.6) [319]. In the first reactor, heated by partial oxidation gases from the second reactor, at 700 °C and 4 MPa, ~75% of the methane is converted by steam into syngas. Next, the gases are fed into the second reactor, wherein the unconverted methane is oxidized by oxygen, with the generated heat being directed to heat the first reactor.

The 'Tandem' process developed in Russia also makes use of two reactors, a feature that allows it to more efficiently combine the processes of steam reforming and partial oxidation [320]. The Syntroleum has developed a process of ATR of natural gas with oxidation by air as a cheaper method of producing syngas at medium-capacity plants with a production capacity in GTL-products beginning from 2000 bbl/day.

The Haldor Topsoe, which contributed enormously to the establishment and development of the ATR method, comes from the fact that, in the nearest years, ATR technology will dominate the production of syngas. Although the process of steam reforming is still considered by many as the most cost-effective option for large installations, ATR supporters point out that the cost of a tubular steam reformer increases almost in proportion to its production capacity. At the same time, the cost of an autothermal reformer with oxygen blowing is determined by economic factors, increasing proportionally to the output to the 0.7 power. Consequently, for high-capacity plants, ATR is preferable [321].

That the scope for improving the classical methods of syngas production is far from being exhausted is seen from the continuous appearance of new promising inventions, for example, one designed at the technological centre of Praxair for catalytic partial oxidation [322], with a

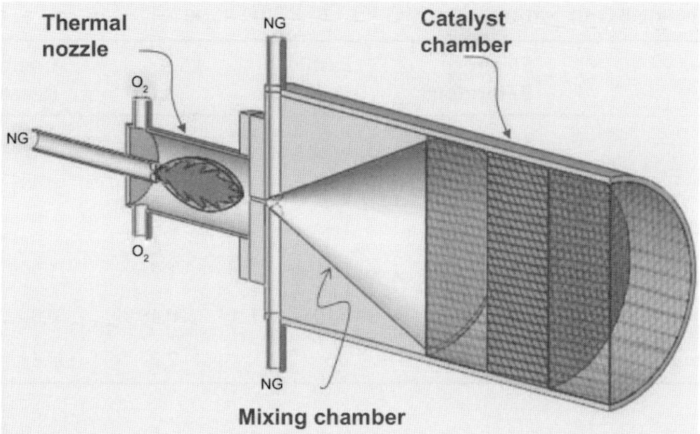

FIGURE 12.7 Schematic of a reactor with heated oxygen supply [322]. (For colour version of this figure, the reader is referred to the online version of this book.)

millisecond time of contact of heated and premixed reactants with a catalyst (Fig. 12.7). Due to a very high rate of injection of oxygen (preheated by burning a small portion of the natural gas) and natural gas through a narrow nozzle into the main chamber with a monolithic catalyst, the gas-phase reaction in the reactor has no time to develop, so the oxidation occurs almost entirely on the catalyst. The conical shape of the mixing chamber ahead of the catalysts provides a rapid and complete mixing of the reactants before they reach the catalyst.

To ensure a more uniform temperature distribution throughout the catalyst and to prevent its overheating or the spreading of the reaction into the gas phase, a three-layer catalyst was used. The first layer consisted of a monolithic foam impregnated with Ru, the second, of pelletized α-alumina impregnated with Ru, and the third, of a pelletized Pt-γ-Al_2O_3 catalyst. The pilot plant demonstrated successful operation at high pressure, producing syngas at pressures up to 18 atm. The conversions of methane and oxygen were 98.0% and 98.9%, respectively, whereas the yields of CO and H_2 were as high as 92.0% and 93.0% respectively. Economic analysis showed certain advantages over the traditional process.

The method of noncatalytic partial oxidation of natural gas to syngas developed by Shell has been implemented at a commercial-scale plant in Bintulu (Malaysia) and then at the biggest gas-chemical complex Perl (Qatar). Notions about the basic parameters and applications of various industrial methods for producing syngas can be drawn from Table 12.1 and Fig. 12.8. The economic efficiency of relatively small-capacity units ($\sim$300,000 m^3 NTP/h) is about the same for all processes. With increasing production capacity, the relative cost of all processes increases linearly, but substantially steeper than for steam reforming. Therefore, at a capacity of 1.5 million m^3 NTP/h, the relative cost of steam reforming is almost two times higher than that of ATR [323].

It should be noted that, due to the high cost of producing syngas, investments into construction of gas-chemical plants for production of GTL-products is significantly higher as compared to oil refineries of similar capacity, ranging from 25,000 to 45,000 dollars per barrel of daily production, while for oil refineries, this value is about 15,000 dollars. The technological complexity and volume of operations of a GTL-plant with a capacity of

TABLE 12.1 Characteristics of Various Processes for Producing Syngas [323]

Parameter	Steam Reforming	Partial Oxidation	ATR	Combined Method (Steam Reforming + ATR)
Steam/carbon	2.7	0.2	1.2	1.45
Oxygen/carbon	–	0.64	0.62	0.48
H_2/CO	4.8	1.9	2.6	3.0
$(H_2 - CO)/(CO + CO_2)$	2.59	1.71	1.74	1.96
Residual content of CH_4, %	3.3	0.15	0.48	0.8
CO_2 content in dry syngas, %	6.9	2.3	7.7	7.6

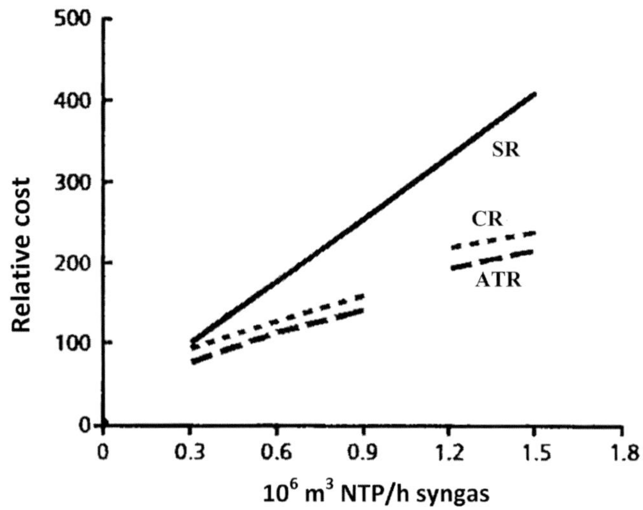

FIGURE 12.8 Dependence of the relative cost of the installation on its syngas production capacity: (SR) steam reforming, (CR) combined reforming, (ATR) autothermal reforming [323].

34,000 bbl/day is equivalent to a refinery with capacity of 100,000 bbl/day [324]. The high costs of syngas production by traditional technology, reaching two thirds of the total costs of the production of methanol (Fig. 12.9) and GTL-products from natural gas is currently the main constraint to wider development of gas chemistry. It is hardly possible to figure on a cost-effective transfer of modern gas-chemical processes onto the level of low-scale processes necessary for the development of spread and remote unconventional natural gas resources.

As follows from Fig. 12.8, the economically viable production of syngas by modern methods is not less than 300,000 m^3/h. With decreasing production capacity, specific investment expenditures (Fig. 12.10), just as the specific operational and all other costs, are soaring.

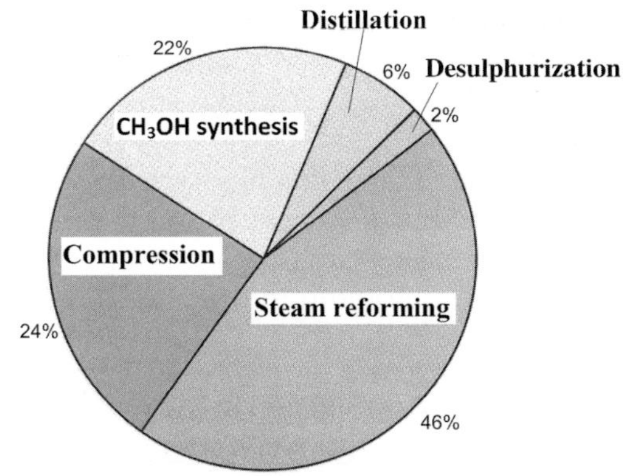

FIGURE 12.9 Share of the different stages in the prime cost of methanol. (For colour version of this figure, the reader is referred to the online version of this book.)

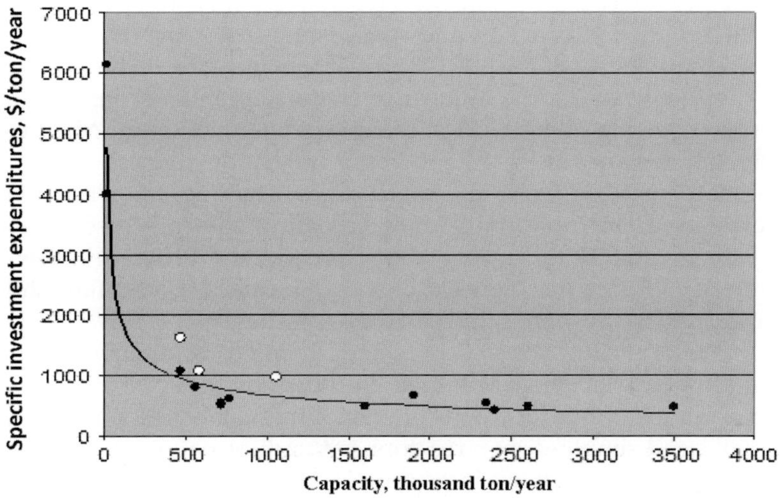

FIGURE 12.10 Dependence of the specific capital costs of modern GTL technologies on their annual capacity [325].

Therefore, without the creation of a fundamentally new technologies for converting natural gas into syngas is difficult to expect a significant role of traditional GTL-technologies in the development of unconventional low-volume sources of natural gas.

This situation motivates a high activity in developing alternative methods for producing syngas. The following is a very brief outline of some of these technologies.

Oxidation on ceramic membranes. The high cost of oxygen is one of the main factors affecting the cost of production of syngas in the POM and ATR processes. Furthermore, the use of oxygen increases explosion hazard. To avoid using expensive air separation units, great efforts have been put in recent years into the development of highly oxygen-permeable ceramic membranes [326]. Their application opens the possibility of combining the processes of air separation and syngas production, thereby eliminating the need to use expensive equipment for oxygen production. Estimates show that such a technology of syngas production could cut the cost by 20–50%. The membranes must be durable, resistant to a highly oxidizing atmosphere, on the one hand, and to a highly reducing atmosphere, on the other, at temperatures above 900 °C. They must also be resistant to poisoning, corrosion, deformation, and withstand large pressure differences.

Microchannel reactors. An interesting trend in the development of methods for the production of syngas is microprocessing technology (MPT) developed by Velocys Inc., which makes it possible to drastically increase the intensity of heat and mass transfer. An MPT reformer consists of a large number of successive layers of microchannels with a diameter of less than 1 mm. Layers in which heat transfer fluid circulates or fuel burns alternate with layers in which the endothermic process of natural gas conversion occurs. Increasing the intensity of heat exchange makes it possible to reduce the size of the apparatus and shorten the reaction time. The latter circumstance is particularly important for steam reforming, because soot formation occurs usually considerably more slowly than the process of reforming. A short residence time in the MPT reactor, less than 10 ms, prevents soot formation, increases the production capacity, and reduces the cost. Operation for several months has demonstrated the stability of the MPT reformer. An increase in the production capacity is achieved by simply increasing the number of layers with microchannels without need to scale the processes occurring in them. The principle of operation of the MPT reactor is applicable to the subsequent stages of the FT process or methanol synthesis [327].

Oxidation of methane at millisecond contact times. Much attention is paid to a relatively new field of research of hydrocarbon oxidation, dealing with processes in which reactive gases contact with a catalyst for a very short time, of the order of a few milliseconds (short-time catalysis). One of the pioneers of these studies, L.D. Shmidt, showed benefits of hydrocarbon oxidation on monolithic catalysts (meshes, metal foam blocks, and blocks with through-channels) at short contact times and, therefore, at high volumetric flow rates, which enable to achieve a very high volumetric production capacity of the reactor [155].

The catalytic oxidation of hydrocarbons on monolithic catalysts for a short time proceeds efficiently in small-sized simple-design reactors, without formation of solid carbon. Characteristic features of these processes are large time ($\sim 10^6$ K/s) and space ($\sim 10^5$ K/cm) gradients of temperature. These autothermal processes occur in nearly adiabatic conditions, since exothermic reactions rapidly heat the catalyst and gas to about 1000 °C, whereas the heat release rate is too high to allow the gas mixture to be effectively cooled through the walls.

A catalyst grid made of a noble metal, for example, Pt-10% Rh, enables to combine very high rates of heating ($\sim$5 ms) on the catalyst with a very rapid subsequent quenching ($\sim$200 μs) of the reactants when they are mixed with the flow passing through the mesh cells. Thus, the reaction, beginning as a catalytic on the surface of the mesh (and maintaining a high temperature due to the catalytic process), then proceeds in the bulk of the reactor as a chain homogeneous reaction. In this case, by adjusting the mesh parameters (diameter of wire, material), it is possible to widely vary the conditions in the gas-phase reaction zone. Depending on the reagents and conditions, the process can yield olefins, oxygenates, or syngas.

Filtration combustion-based processes. In the production of syngas, it is necessary to supply energy into the reaction mixture to initiate endothermic steps and, at the same time, to utilize the heat of the formed products. One approach to solving this duel problem is to carry out the process under conditions of filtration of the reactive gas mixture through a porous inert medium (combustion filtration), combining the oxidizing and steam reforming of methane. The energy consumption of the process is compensated by oxidation of part of the methane, whereas part of the heat released by the reaction is transferred through the solid medium to the incoming gas, so that the reaction zone is maintained at a high temperature. Due to the preheating of the incoming gas through heat exchange with the heated solid medium, the temperature achieved in the combustion zone appears to be higher than the temperature of the adiabatic conversion of a mixture of the same composition. Therefore, these modes are called 'superadiabatic combustion'. Superadiabatic filtration combustion in a porous medium can be used to produce syngas through the conversion of gaseous C_1-C_3 hydrocarbons [328].

Production of syngas by means of power generating technologies. Interesting possibilities of hydrocarbon conversion to syngas are offered by engines operating on hydrocarbon fuels. These can be regarded as combined power-chemical reactors in which both energy and certain chemicals can be produced. The first reports on the production of syngas by partial oxidation of methane in gas engines appeared as long ago as the early 1960s. A method for the gas-phase oxidation of methane to syngas in a modified diesel engine was developed, which acted simultaneously as a chemical reactor. By varying the composition of the methane–air mixture fed into such a reactor, it is possible to prepare syngas of various compositions [329]. The method has been successfully tested using a six-cylinder engine with a gas prechamber-torch ignition, cylinder capacity of 45.8 l, compression of 11.8, and rotation frequency of 500 rpm, operating on a fuel mixture with an oxidant equivalence ratio of 0.42.

Another type of combined power-chemical reactor for the homogeneous oxidation of methane to syngas is a reactor based on a liquid-fuel rocket engine. Such power engines have a uniquely high specific power, 10 times superior to the performance of existing industrial converters. There are reports on tests of a pilot plant based on a rocket engine with a production capacity in terms of methane consumption of 432 kg/h. The composition of the working mixture was varied from $\alpha = 0.35$ to $\alpha = 0.45$. The methane conversion was up to 85% [96]. There are also communications on the production of syngas in a gas turbine [330].

Production of syngas in 3D matrix burners. From the viewpoint of thermodynamics, optimum conditions for syngas production through the partial oxidation of methane are $O_2/CH_4 \approx 0.5$ ($\alpha \approx 0.25$), a temperature of 1073–1273 K, and a low pressure, ~1 atm. However, such a process is difficult to implement in the absence of a catalyst. Therefore, the main technological challenge in the production of syngas by the partial oxidation of natural gas is to ensure a stable conversion process at low values of the oxidant equivalence ratio, within 0.3–0.5. Since mixtures with such α values are well outside the flammability limits under normal conditions, this is not a trivial task.

One of the most effective ways to expand the limits of stable combustion is to reduce heat loss and at any rate partly recover heat of combustion products. A burner unit designed in the form of a geometrically closed cavity (3D matrix) made of a gas-permeable material (Fig. 12.11) is heated to high temperature due to convectional and radiative heat transfer from the flame front stabilized near the surface and from the reaction products to the inner surface of the matrix. Due to an effective heat transfer from the reaction products to the

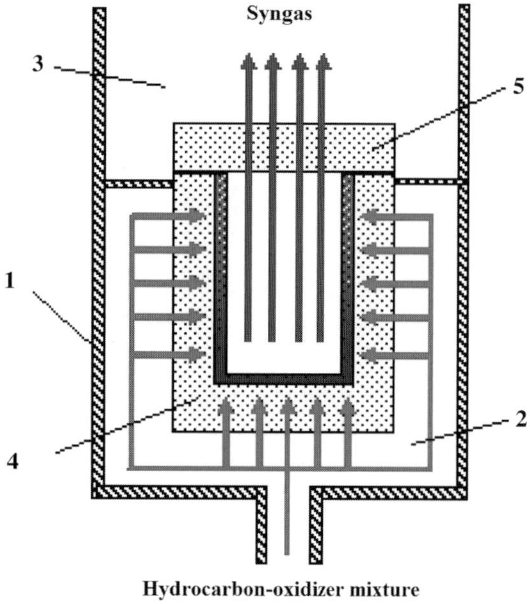

FIGURE 12.11 Schematic diagram of a matrix radiation burner: (1) outer jacket, (2) air—fuel receiver, (3) exhaust gases, (4) sidewalls and bottom of the burner made of honeycomb ceramic, and (5) burner cavity cover made of honeycomb ceramic [332]. (For colour version of this figure, the reader is referred to the online version of this book.)

matrix surface and, through it, to the fresh gas mixture, the latter enters the flame front preheated to nearly the matrix surface temperature. Moreover, the radiation emitted by the flame front occurs almost completely locked in the closed cavity, a feature that drastically reduces radiation loss. The joint action of these factors widened the combustion limits and enables to implement the surface burning of very rich mixtures, which is almost unattainable in other types of burners [331—333].

Such burner units with a closed 3D matrix make it possible to significantly extend the limits of stable combustion of rich mixtures (Fig. 12.12). The possibility of stable conversion in a 3D porous matrix of very rich methane—air mixtures at values $\alpha \leq 0.4$ and a temperature of $\sim 1000\ ^{\circ}C$ to hydrogen and carbon monoxide in concentrations close to the thermodynamically equilibrium values was demonstrated in [331—332].

When air was used as the oxidant, the concentrations of hydrogen and carbon monoxide at the outlet of this converter reached $\sim 20\%$ and $\sim 11\%$, respectively, which is close to the thermodynamically equilibrium values. The possibility of an effective conversion of natural gas into syngas with a nearly optimal ratio of $H_2/CO \approx 2$ was demonstrated. Permeable 3D matrices offer a way of designing relatively simple, compact, and efficient noncatalytic autothermal reformers for the partial oxidation of hydrocarbon gases of different origin and composition to syngas [331—332].

Thus, there are several alternative technologies suggested for low-scale syngas production. However, so far none of them has reached the level of commercialization. A more detailed description of these technologies is given in [6].

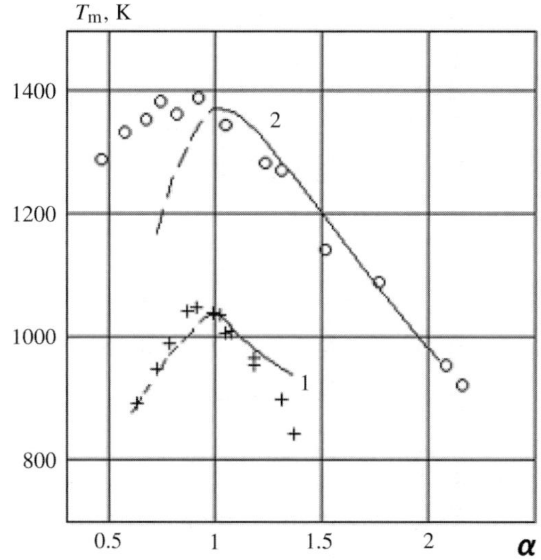

FIGURE 12.12 Temperature of the matrix surface as a function of the air equivalence ratio of the mixture: (1) open flat matrix, (2) geometrically closed 3D matrix. The symbols and curves represent experimental and calculation results, respectively [334].

OXIDATIVE COUPLING OF METHANE

The Oxidative Coupling of Methane (OCM) is the interaction of CH_4 molecules in the presence of O_2 or other oxidizing agent to form C_2-hydrocarbons, ethane and ethylene, according to the gross reaction:

$$2CH_4 + 1/2O_2 \rightarrow C_2H_6 + H_2O \quad (12.6)$$

$$C_2H_6 + 1/2O_2 \rightarrow C_2H_4 + H_2O, \quad (12.7)$$

In some rare cases, heavier hydrocarbons are formed, for example, benzene. The process was discovered in the early 1980s [5,335]. Due to a large number of works, the OCM has become one of the most thoroughly studied reactions of the oxidative conversion of methane, and in particular, its essentially homogeneous—heterogeneous nature has been experimentally demonstrated. Methyl radicals $CH_3^{\cdot}$ formed by the catalytic reaction escape into the bulk of the reactor. In the temperature range optimal for the OCM, 600—950 °C (Fig. 12.2), at an oxygen concentration below 20% and atmospheric pressure, the main reaction of methyl radical is recombination to form ethane (reaction (12.4)).

A huge number of works devoted to the process makes OCM one of the most well-studied catalytic reactions. The main problems of the practical implementation of the OCM are the existence of a kinetic limit of the yield of ethane and ethylene, ~25% according to kinetic estimates [119], and a high heat release by the reaction, which requires designing appropriate reactor systems [335]. So far, despite the great interest in this process, and major international effort, there have been no reports on its practical implementation. In addition, it is not clear what can be an economically justified production capacity of this process, i.e., whether it is suitable for processing small volumes of unconventional gas.

Oxidative coupling of methane to higher hydrocarbons. Unlike the direct coupling of methane to produce higher hydrocarbons,

$$nCH_4 \rightarrow C_nH_{2n+2} + (n-2)H_2 \tag{12.8}$$

the oxidative catalytic coupling of methane is a thermodynamically favourable process. Therefore, in the OCM, along with the formation of C_2-hydrocarbons, further interactions of $CH_3^{\bullet}$ radicals give C_3–C_4 hydrocarbons, the concentration of which may reach several percent.

In a wide range of conditions, the oxidative aromatization of methane,

$$6CH_4 + 4{,}5\ O_2 \rightarrow C_2H_6 + 9H_2O, \tag{12.9}$$

is also thermodynamically favourable; at temperatures above 850 °C, the reaction was observed to occur in the absence of a catalyst as well. The yield of aromatics increases sharply with increasing pressure up to 6 atm, but then becomes virtually independent of pressure. In a $CH_4{:}O_2{:}N_2 = 10{:}1{:}4$ mixture at 1000 °C, the yield of aromatics reached a maximum of 4.9% at a methane conversion of 15.4% and the selectivities of formation of benzene and toluene of 29.2% and 2.6%, respectively. With further increase in temperature, these figures begin to decline. At 950–1000 °C, methane oligomerization to aromatics occurs in the absence of oxygen as well, but the addition of oxygen significantly increases the yield of aromatics. The yield of aromatics was dramatically enhanced by the addition of ethylene to the system. Based on several experiments, it was concluded that benzene is formed according to the scheme $C_1 \rightarrow C_2 \rightarrow C_6$, via the formation of ethylene as an intermediate. The introduction of effective OCM catalysts into the system promotes the formation of aromatics and enhances its yield in this process [336].

It is worthwhile to briefly consider other possible directions of the oxidative and nonoxidative conversion of light alkanes to higher molecular weight compounds. The catalytic aromatization of methane in the absence of oxygen was observed to occur on HZSM-5 zeolite doped with Re and Mo at ~700 °C and 1 atm (dehydroaromatization). However, the maximum methane conversion under these conditions is limited by thermodynamic equilibrium to about 12%. Although the yield of aromatics also reaches about 5%, the catalyst is rapidly deactivated because of carbon deposition. A number of ways to reduce the rate of carbon deposition have been proposed. According to the suggested mechanism, methane aromatization proceeds via the formation of C_2H_4 (C_2H_2) [337]. To demonstrate the feasibility of creating a real technology, Japanese researchers performed studies at a pilot plant in 2000–2006 [338].

The catalytic co-condensation of methane with olefins (cross-condensation, oxidative methylation) was investigated, for example, with ethylene:

$$CH_2{=}CH_2 + CH_4 \rightarrow CH_3CH_2CH_3 \tag{12.10}$$

In this case, at temperatures below 220 °C and a pressure of 0.1 MPa, it is possible to produce propane with an equilibrium yield of 100%. For the co-condensation of propylene with methane,

$$CH_2{=}CH{-}CH_3 + CH_4 \rightarrow CH_3CH_2CH_2CH_3 \tag{12.11}$$

TABLE 12.2 Characteristics of the Processes of Aromatization of C_3–C_4 Paraffins, Including that Developed at the Institute of Catalysis (IC) SB RAS, 'Cyclar' and Z-Forming [340]

Products	IC Process	'Cyclar' (UOP and BP)	Z-Forming (Mitsubishi Oil)
PRODUCT YIELD (% WT)			
Hydrogen	7–8	6.9	3.4
Fuel gas	20–23	29.6	37.1
Total of aromatics	70–72	63.5	59.5
COMPOSITION OF AROMATICS (% WT)			
Benzene (B)	25–27	25	28
Toluene (T)	40–43	43	39
Fraction C_8 (C)	20–23	23	21
Sum of BTC	88–90	91	88
Fraction C_9+	10–12	9	12

the equilibrium yield of butane at 0.1 MPa and 130 °C is 54%, which increases with the pressure [5].

A promising process is the catalytic aromatization of light alkanes C_2–C_4. The thermodynamic probability of this process is high enough at temperatures above 400 °C for butane, >500 °C for propane, and >600 °C for ethane. High-siliceous zeolites of pentasil type (ZSM-5, ZSM-11) catalyze the aromatization of light alkanes with a selectivity of up to 30%. Olefins C_2–C_4, depending on the conditions, can yield a mixture of aliphatic hydrocarbons of gasoline fraction C_6–C_{10} (mainly isoparaffins) or a concentrate of aromatic hydrocarbons, mainly benzene, toluene, and xylenes [339].

Table 12.2 lists the characteristics of the aromatization of C_3–C_4 paraffins developed at the Institute of Catalysis (IC) SB RAS and similar process developed abroad, such as 'Cyclar' (UOP and BP) and Z-Forming (Mitsubishi Oil). In addition to providing higher yields of the target products (70–72% vs. 60–63%), the IC process has a simpler technological scheme. It is carried out in tubular reactors with a fixed catalyst bed, while the other processes are conducted in reactors with a moving catalyst bed. In 2006, the pilot plant with a production capacity of 1000 ton/year [340] was launched into operation.

SELECTIVE OXYCRACKING OF HEAVIER COMPONENTS OF NATURAL GAS

As a promising technology for processing small volumes of natural gas with a high content of heavy methane homologues, the selective oxycracking of these homologues was proposed. The conversion of heavy components of natural gas into lighter, higher-octane compounds less prone to detonation and resin formation makes it possible to use the

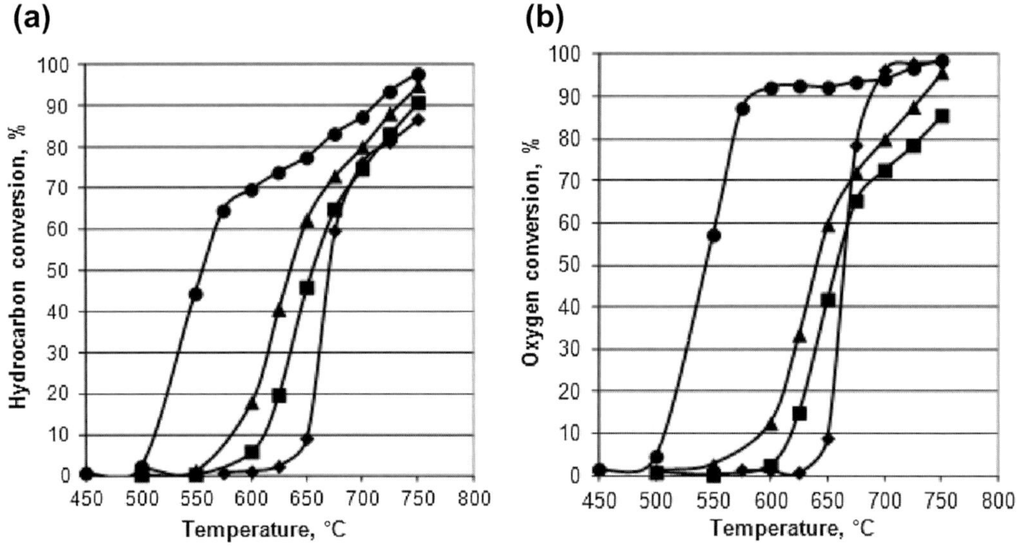

FIGURE 12.13 Temperature dependences of the conversion of (a) hydrocarbons and (b) oxygen in the oxidative conversion of C_2–C_5 alkanes: (◆) C_2H_6, (■) C_3H_8, (▲) n-C_4H_{10}, and (●) n-C_5H_{12}. ($[C_nH_{2n+2}] = 5$ mol %, $[O_2] = 2.5$ mol %, $t_r = 2$ s, balanced by N_2).

produced gas as a fuel for gas-piston and gas-turbine engines [341]. In addition, the possibility of a further catalytic conversion of the resulting products of oxidative cracking into valuable chemical feedstock is discussed.

As was shown in Chapter 10 and in [342–344], the temperature of the gas-phase oxidative conversion of gaseous methane homologues (Fig. 12.13) is substantially below the temperature of oxidative methane conversion, which, under these conditions, exceeds 1000 °C [345,346]. This makes it possible to perform the selective gas phase oxidation of C_3+ alkanes in complex mixtures of hydrocarbon gases, such as natural and associated gas, practically without involving methane into the oxidation process. Moreover, the methane content even increases, since it is one of the main products of oxycracking of heavy hydrocarbons [6].

The main gaseous products of the oxycracking of methane homologues are ethylene, methane, ethane, hydrogen, and carbon monoxide, with ethane and propylene being present in small amounts. The presence of a heterogeneous catalyst significantly accelerates the process at low temperatures, but impedes it at higher temperatures, at which an abrupt transition to the branched-chain-reaction mode occurs (Fig. 12.13). The reactor surface acts in a similar manner, apparently due to the heterogeneous termination of chains. Thus, the presence of a catalyst greatly increases the formation of CO_2 and H_2O because of the deep oxidation of hydrocarbons on the catalyst surface. Therefore, in these conditions, the gas-phase process is more advantageous [343,344].

Oxygen addition markedly increases the rate of conversion of alkanes, particularly at moderate temperatures: above the temperature of the surface catalytic process but below

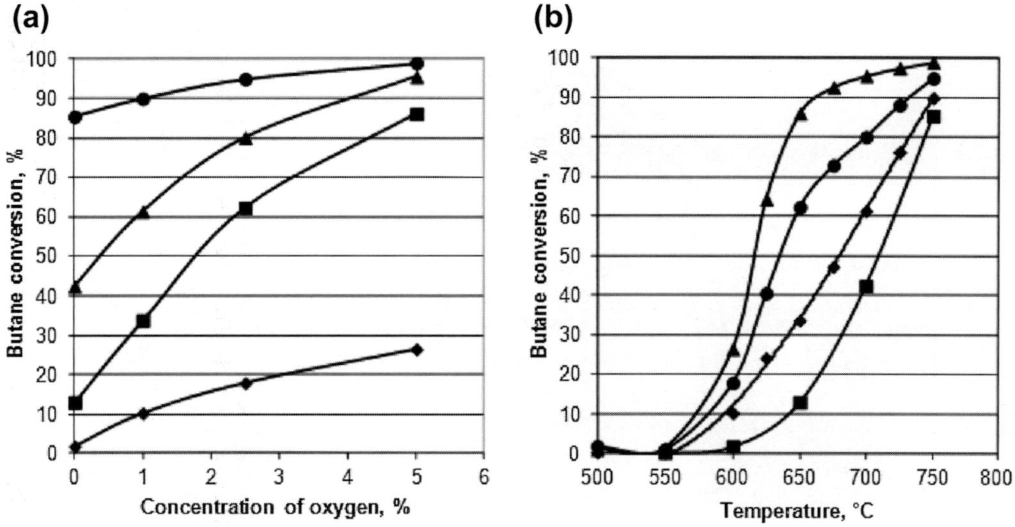

FIGURE 12.14 Conversion of *n*-butane (a) at different temperatures as a function of the initial concentration of oxygen (T = (◆) 600 °C, (■) 650 °C, (▲) 700 °C, and (●) 750 °C) and (b) at different initial concentrations of oxygen as a function of the temperature ($[O_2]$ = (■) 0%, (◆) 1%, (●) 2.5%, and (▲) 5%) ($[n\text{-}C_4H_{10}]$ = 5%, t_r = 2 s, balanced by N_2).

the thermal pyrolysis temperature (Fig. 12.14). Furthermore, oxygen, which is consumed in the gas-phase process simultaneously with hydrocarbons, being present in a greater part of the reactor, significantly inhibits the formation of heavy condensation products, such as C_4H_6 (butadiene) and C_3H_4 (propadiene and methylacetylene), which to some extent may be considered as soot precursors [347].

Selective oxycracking was used to develop a process of oxidative conversion of C_3+ components of associated gas into lighter, high-octane compounds suitable for use as a fuel for gas-piston engines and gas-turbine power plants for local needs of oil and gas fields. The process was successfully pilot-tested, demonstrating the ability to convert up to 90% of C_5+ hydrocarbons and up to 80% of the C_3—C_4 fraction (Fig. 12.15).

This technology can reduce flaring of associated gas at remote oil fields and oil platforms by 20—30% by using part of it to produce energy for local needs. A more interesting prospect is to create technologies capable of effectively processing valuable oxycracking products, such as olefins, CO, and hydrogen, for example, on the basis of the copolymerization of carbon monoxide and ethylene directly in the gas phase [348]. The copolymerization of CO and ethylene, followed by separation of the resulting products would considerably improve the economic attractiveness of the process.

However, despite intense efforts to develop alternative technologies based on syngas production and direct conversion methods, there is currently no economically feasible and industrial-scale tested, small-capacity gas-chemical technologies that could successfully compete with the DMTM process.

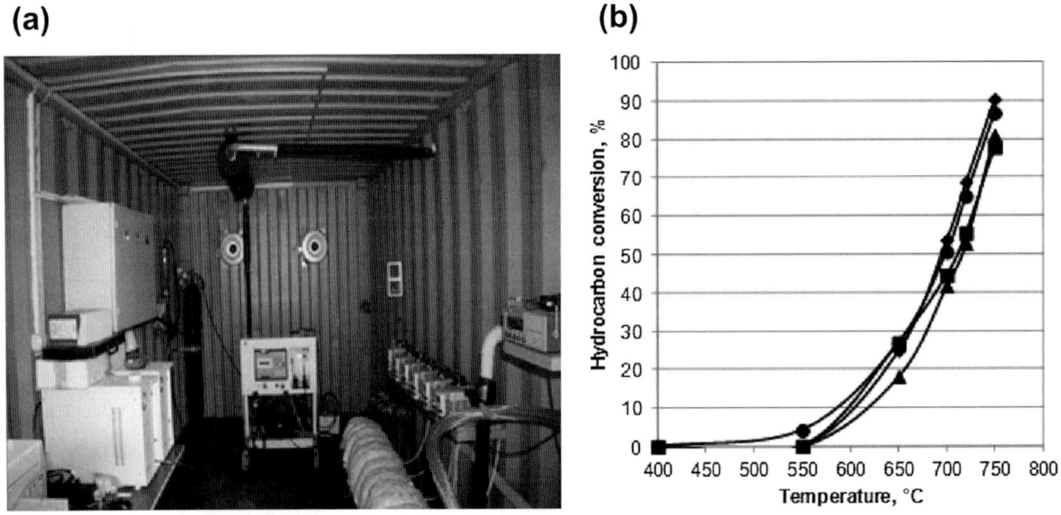

FIGURE 12.15 (a) Pilot unit for the selective oxycracking of heavy gas-phase hydrocarbons and (b) the results of pilot testing of the conversion of typical oil associated gas ((■) C_3H_8, (▲) n-C_4H_{10}, (●) n-C_5H_{12}, (◆) n-C_6H_{14}). The composition of the gas (mol %): CH_4, 81.0; C_2H_6, 1.22; C_3H_8, 17.0; C_4H_{10}, 2.81; C_5H_{12}, 1.50; and n-C_6H_{14}, 1.83. The amount of added air was 15%; $t_r \approx 1$ s. (For colour version of this figure, the reader is referred to the online version of this book.)

GENERAL CONCEPT OF THE MODERN LOW-SCALE GAS CHEMISTRY

The practical prospects of DMTM technology will largely depend on the general trends in the global energy industry and on the role that will played in it by natural gas production and gas chemistry. Today, thanks to new effective technologies of unconventional natural gas production, its resources are constantly growing and the cost of production is decreasing. Whereas even in 2004, the energy cost of natural gas in the U.S. was almost equal to the energy cost of oil, now it is five times lower [349], and this gap is projected to widen in the coming years. This makes natural gas very attractive, not only as an energy source, but also as a petrochemical feedstock. This state of affairs has caused a boom in the petrochemical industry in the United States, where in 2012 the majors companies have launched a number of new projects. For example, the annual production capacity of the ethane pyrolysis plants to be commissioned in the U.S. over the next five years is exceeded 8 million tons.

Note, however, that unconventional sources of natural gas, in addition to obviously higher costs of production, differ from traditional sources by much lower production rate, short duration of active life (the production capacity of shale gas wells falls particularly rapidly during the 1st year (Fig. 12.16)), and spread over huge areas. In addition, a significant portion of these resources is located in remote regions, including Arctic. Even in the United States with its highly developed system of pipelines, more than half of potential resources are out of reach of the gas transportation system.

Given the fact that the probability of discovery of new giant fields of conventional gas decreases exponentially with time, in the coming decades, the global gas industry will

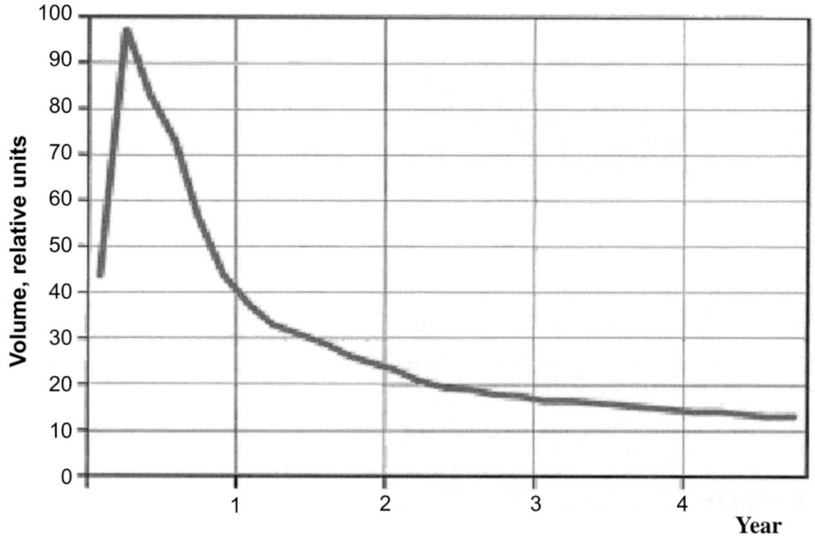

FIGURE 12.16 Shale-gas typical profile. (For colour version of this figure, the reader is referred to the online version of this book.)

increasingly focus on unconventional resources, as well as low-pressure and low-resources fields, associated gas, coalbed methane, and other similar sources. At the same time, in recent years, the trend of the world's leading gas chemistry has been a steady increase in production capacity, which is seen as the main way to improve its economic performance. Upcoming changes in the structure of the gas-chemical feedstock are in stark contrast to the dominant technological trend, which will inevitably lead to the need to introduce certain adjustments.

Equally serious problems may arise in the transportation of natural gas. Nowadays, the dominant means of natural gas transportation are pipelines and shipment in the form of liquefied natural gas (LNG). That these means are profitable only when large volumes are transported over long distances (Fig. 12.17) makes natural gas accessible only to about 20% of the world population. Since most promising sources of natural gas can be attributed to the category of both remote and low-productive, the transportation of nontraditional resources to potential consumers, despite their abundance, can turn out to be a major economic and technical challenge. The use of most of these resources on the basis of the existing technologies of transportation and processing of natural gas is almost unreal. Thus, the new trends in the development of the feedstock base for gas chemistry dictate the need for at least the parallel development of simpler, low-scale, and flexible technologies of transportation and processing of gas resources.

An ideal way to solve both problems is to create relatively compact commercial technologies of conversion of gaseous hydrocarbons to liquid or readily liquefiable hydrocarbon products, such as methanol, dimethyl ether, synthetic liquid hydrocarbons, ethylene etc., since the transportation of liquid products is considerably less expensive than that of gas. Furthermore, in this case, unlike LNG, the customer would receive a more valuable and demanded product of natural gas conversion, with a high added cost.

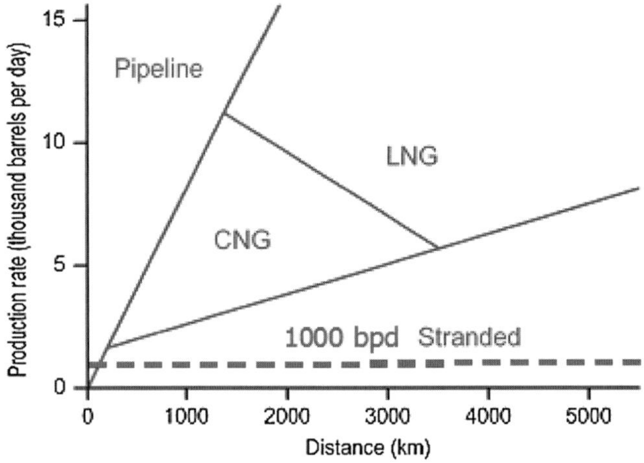

FIGURE 12.17 Areas economically sound use of various technologies for the transportation of natural gas. 1000 bpd is a typical production capacity of the shale gas well. (For colour version of this figure, the reader is referred to the online version of this book.)

It is hard to expect that the development of such processes can be based on 'reverse scaling' of the existing technologies to low capacities, because of complex and energy-intensive processes underlie modern large-scale technologies, for example, those for producing syngas or pyrolysis products. Economic evaluations strongly suggest that a simple reduction of the volume of a production process based on a traditional technology will lead to a sharp increase in the specific investment expenditures (Fig. 12.10), as well as in the specific operational and all other costs.

The reason for this is that, when extrapolated to small production volumes of industrial processes, starting from a certain level, the so-called 'six-tenths rule', which relates the investment expenditures and production capacity as

Expenditures 2/Expenditures 1 = (Production capacity 2/Production capacity 1)$^{0.6}$ [350]

ceases to hold.

This is because the number of units of the processing equipment (apparatuses, valves, measuring devices, etc.) remains the same irrespective of the production capacity, so that after a certain threshold, a further decrease in the production capacity has little effect on the cost (Fig. 12.18). Therefore, the development of low-yield resources of natural and associated gas is not possible based on the 'scaling back' of existing large-capacity processes: it requires fundamentally new technological solutions.

Thus, it is necessary to develop new flexible technologies capable of producing more valuable and easily transportable products immediately on the site of production of gaseous hydrocarbons. Such technologies should be implemented in low-scale gas-chemical plants that meet the following basic requirements [351]:

- Fully factory-manufactured equipment without on-site construction-and-assembling operations.

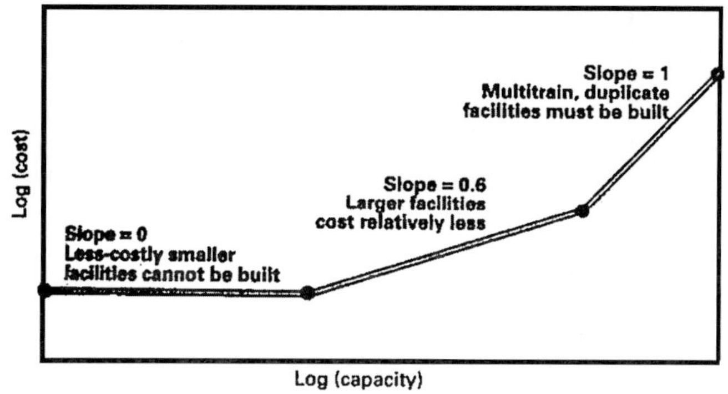

FIGURE 12.18 Logarithmic dependence of the investment expenditure on the production capacity [350].

- Modular configuration of the equipment adjustable to the gas well output.
- The equipment should be easy to dismantle and move after exhaustion of the deposit.
- Technological modules should be versatile, suitable for being assembled as standard equipment irrespective of the manufacturer.
- The availability of gas pretreatment modules designed for a wide range of parameters of the produced gas, such as volume, composition, content of impurities, presence of moisture, etc.
- An independent power supply at the expense of the gas produced.
- A high degree of automation, minimal service staff.
- Minimum number of technological stages for obtaining a single easily transportable product for further processing at special plants.

As one of the most promising directions for designing such technologies, the DMTM process can regain its practical value. In fact, it is necessary to create a new generation of technologies intended for processing the resources of unconventional sources of hydrocarbon gases with low production, short living time, remote and widely dispersed.

CHAPTER

13

Conclusions

Currently, the method of partial oxidation of hydrocarbon gases to oxygenates does not exist as an industrial technology [351], a circumstance that predetermines the basic problems of its further progress. Many of the technological, engineering, and scientific problems has not been solved not because they are intractable, but simply because there has been no demand for such solutions. For comparison, in recent years, leading oil and gas companies have invested milliards of dollars in the development of various aspects of the production of syngas and accompanying processes; this area employs tens of thousands of highly skilled professionals. It is not surprising, therefore, that modern methods of syngas production are the technologically most advanced chemical processes.

There is no doubt that as soon as a new wave of industrial applications of technology of partial oxidation begins to gain strength, most of the technical issues that impede or limit the capabilities of this technology will be successfully resolved. That this technology will experience a renaissance is beyond doubt, since the need for simple low-scale methods for processing of hydrocarbon gases is great, and in this respect, the DMTM has no economically and technologically feasible alternatives. The small deposits of natural gas, with reserves of less than 7 milliard m^3, the number of which in the world exceeds 25,000, account for 12% of all resources. In addition, at least 10% of the gas in the Russia, and in some countries up to 100%, is associated gas [352].

As with any technology benefiting from using large volumes of oxygen or enriched air, for the development of DMTM technology, especially the establishment of autonomous low-capacity plants for complete processing of associated gas in order to eliminate hundreds of flares, of great importance would be a progress in the membrane and adsorption separation of air. To make membrane air-separation plants practically feasible in the processing of associates gas, it is necessary to enhance their productivity, reduce the cost, and increase the concentration of oxygen produced. Even today, the existence of commercially available membrane systems for enriched air production with an oxygen concentration of ~40% and a capacity of tens of thousands of cubic metres per hour, as well as the availability of pressure swing adsorption units for the production of 93–95% pure oxygen can solve many problems.

Another area of application of membrane technology to the process of partial oxidation could be solving the inverse problem: the separation of nitrogen or depleted nitrogen-containing gas from the recycling flow. In technological sense, it is a simpler task, since the permeability of the membrane with respect to nitrogen is higher than for methane, so

http://dx.doi.org/10.1016/B978-0-444-63253-1.00013-1

nitrogen is more readily separated by the membrane method than methane. Depleted nitrogen-containing methane gas can either be reinjected into the reservoir to increase its productivity, or can be used as a clean fuel for local needs, with its combustion producing very low emission of nitrogen oxides.

A significant influence on the progress in DMTM circulation technology will be produced by creating effective methods for removing CO from the circulation gases. However, a more attractive prospect of drastically increasing the selectivity of the process is to use resulting CO, for example, as a carbonylation agent for producing from methanol acetic acid or other products by means of the processes discussed in Chapter 11 or similar ones. In this regard, the catalytic carboxylation of alkanes, particularly those present in the feed gas, and heavy reaction products of oxidation of methane homologues (which more easily enter into such reactions with the formation of acids and other liquid products) may also prove to be of considerable interest [152,349,353].

The possibility of catalytic hydrogenation of formaldehyde directly in the gas flow, considered in a number of works, is also noteworthy for increasing the methanol yield [68,188]. Of certain interest, especially for designing simple technologies of production of methanol for inhibiting hydrate formation in gas fields, are simple methods for the conversion of diluted formalin (vat residue) into products consumed locally or readily transportable, such as paraformaldehyde or urotropin. In some cases, it may be advantageous to combine the DMTM process with the subsequent conversion of methanol formed into products that can be used as engine fuel for local needs. It may turn out to be promising to produce methanol by the DMTM and then use it to synthesize dimethyl ether or heavier alcohols [354], to increase the octane number of petroleum fuels by alkylating or etherifying them with methanol [355], to modify biodiesel fuel by etherifying vegetable oils with methanol [356].

Of considerable interest is the possibility of extracting formed oxygenates from the gas flow without using cumbersome and energy-intensive processes of heat transfer and separation, for example by adsorption in a moving adsorbent bed ('chromatographic reactor' [106]), in cyclic adsorbers, or by chemical bonding. Should it become possible to remove oxygenates without using cycles of complete cooling and subsequent heating of the reaction mixture, the problem of low conversion per passage of gas through the reactor would be mostly solved.

There are some other interesting ideas, such as the return of part of the active products into the initial phase of the process for its intensification, the intensification of the process with simultaneous reduction in gas heating by introducing a solid heat carrier (catalyst) circulating in the reactor [357], or a partial recycling of the hot-gas flow containing active reaction products immediately in the reactor.

A great incentive to expand the scope of applications of this technology would be its use to produce promising methanol-coal suspension fuels from crude methanol and coal-mine dumps. This would solve the problem of huge waste piles of coal and at the same time produce a large amount of nonoil liquid fuel, capable of replacing fuel oil and even heavy diesel fuel.

The most interesting prospects, however, are associated with a further study of the mechanism of the oxidation of hydrocarbons, especially in nonsteady and oscillatory regimes, which provide an increase in the selectivity to methanol and in the direct production of valuable products, such as olefin oxides. There is an evidence of promotion of the formation of olefins and their oxides by small additives of olefins themselves. It is equally interesting to study the possibility of controlling this nonlinear process by means of short-time catalytic or physical impacts.

Now we are at the very beginning of the new stage of development of DMTM technology and its applications, and it is difficult to predict what will be the real technological processes created on its basis. However, with a high degree of confidence, it is possible to predict a significant role of the DMTM technology in the development of world's natural gas resources.

References

[1] Arutyunov VS, Basevich VYa, Vedeneev VI. Modern state of direct partial oxidation of natural gas to methanol. Ind Eng Chem Res 1995;34:4238–43.

[2] Arutyunov VS, Basevich VYa, Vedeneev VI. Direct high pressure gas-phase oxidation of natural gas to methanol and other oxygenates. Russ Chem Rev 1996;65:197–224.

[3] Arutyunov VS. Recent results on fast flow gas-phase partial oxidation of lower alkanes. J Nat Gas Chem 2004;13:10–22.

[4] Arutyunov VS, Krylov OV. Oxidative conversion of methane. Russ Chem Rev 2005;74:1111–37.

[5] Arutyunov VS, Krylov OV. Oxidative conversion of methane. Moscow: Nauka; 1998. pp. 361. [in Russian].

[6] Arutyunov VS. Oxidative conversion of natural gas. Moscow: KRASAND; 2011. pp. 640. [in Russian].

[7] Sinev M, Arutyunov V, Romanets A. Kinetic models of C_1–C_4 alkane oxidation as applied to processing of hydrocarbon gases: principles, approaches and developments. In: Marin GB, editor. Advances in chemical engineering, 32. Elsevier; 2007. p. 171–263.

[8] Bone WA, Wheeler RV. J Chem Soc 1902;81:535–49.

[9] Bone WA, Wheeler RV. J Chem Soc 1903;83:1074–87.

[10] Bone WA, Allum RE. The slow combustion of methane. Proc R Soc London 1932;A134:578–91.

[11] Walker JC, Malakoff HL. Basic oxygenated chemicals from C_1, C_2 and C_3 paraffins and olefins. Oil Gas J 1946;45(33):59–65.

[12] Bludworth JE. First tonnage oxygen plant ties in with petrochemical manufacture. Pet Process 1949;4(4):377–9.

[13] Shtern VYa. Gas phase oxidation of hydrocarbons. Oxford. London. New-York: Pergamon Press; 1964.

[14] Yoshikawa K. Oxidation of methane. II. The homogeneous oxidation of methane under high pressures. Bull Inst Phys Chem Res (Japan) 1931;10:305. Chem Abstr 1931;25:4842.

[15] Pichler H, Reder DR. Über die bildung von methylalkohol bei der partiellen oxydation von methan und athylen unter hohen druck. Angew Chem 1933;11:161–5.

[16] Newitt DM, Huffner AE. The formation of methyl alcohol and formaldehyde in the slow combustion of methane at high pressures. Proc R Soc London 1932;A134:591–604.

[17] Newitt DM, Szego P. Slow oxidation at high pressures. Proc R Soc London 1934;A147:555–71.

[18] Newitt DM. The oxidation of hydrocarbons at high pressures. Chem Rev 1937;21:299–317.

[19] Patent of France 352687; 1905.

[20] Patent of Canada 291411; 1929.

[21] Patent US 1776771; 1930.

[22] Newitt DM, Bloch AM. The slow combustion of ethane at high pressures. Proc R Soc 1933;A140:426–39.

[23] Newitt DM, Schmidt WG. The oxidation of propane. Part II. The products of the slow oxidation at high pressure. J Chem Soc 1937:1665–9.

[24] Newitt DM, Gardner JB. The initial formation of alcohols during the slow combustion of methane and ethane at atmospheric pressure. Proc R Soc 1936;A154:329–35.

[25] Luckett GA, Mile B. The formation of methanol during the gas-phase oxidation of methane. Combust Flame 1976;26:299–302.

[26] Wiezevich PJ, Frolich PK. Direct oxidation of saturated hydrocarbons at high pressures. Ind Eng Chem 1934;26:267.

[27] Paris A. Mild oxidation of methane under pressure. Chim Ind April 1934:411–20. Chem Abstr 1934;28:5806.

[28] Asinger F. Chemie und technologie der paraffin-kohlenwasserstoffe. Berlin: Akademie-Verlag; 1958.

[29] Anisonian AA, Gudkov SF, Enikolopian NS, Kle'menov NA, Markevich AM, Nalbandian AB, et al. Production of formaldehyde by direct oxidation of natural gas by air. Gazov Prom-st 1957;6:32–40 [in Russian].

[30] Semenov NN. Chemical kinetics and chain reactions. Oxford: Clarendon Press; 1935.

[31] Gesser HD, Hunter NR, Prakash CB. The direct conversion of methane to methanol by controlled oxidation. Chem Rev 1985;85:235–44.
[32] Gesser HD, Hunter NR. The direct conversion of methane to methanol (DMTM). In: Wolf EE, editor. Direct methane conversion by oxidative processes: fundamental and engineering. N.Y.: Van Nostrad; 1992. p. 402–25.
[33] Garcia EY, Loffer DA. Partial oxidation of methane. Lat Am J Chem Eng Appl Chem 1984;14:267–84.
[34] Foster NR. Direct catalytic oxidation of methane to methanol – a review. Appl Catal 1985;19:1–11.
[35] Pitchai P, Klier K. Partial oxidation of methane. Catal Rev Sci Eng 1986;28:13–88.
[36] Sinev MYu, Korchak VN, Krylov OV. The mechanism of the partial oxidation of methane. Russ Chem Rev 1989;58:22–34.
[37] Brown MJ, Parkyns ND. Progress in the partial oxidation of methane to methanol and formaldehyde. Catal Today 1991;8:305–35.
[38] Mackie JC. Partial oxidation of methane: the role of gas phase reaction. Catal Rev Sci Eng 1991;33:169.
[39] Edwards JH, Foster NR. The potential for methanol production from natural gas by direct catalytic partial oxidation. Fuel Sci Technol Int 1986;4:365–90.
[40] Brockhaus R, Franke HJ. Process for the manufacture of formaldehyde and methanol by partial oxidation of methane. US Patent 4 243 613; Jan 6, 1981.
[41] Zhang Q, He D, Zhang X, Zhu Q. Methanol production from partial oxidation of methane in a specially designed reactor. ACS Fuel Chem Div Prepr 2002;47(1):334.
[42] Yarlagadda PS, Morton L, Hunter NR, Gesser HD. Direct conversion of methane to methanol in a flow reactor. Ind Eng Chem Res 1988;27:252–6.
[43] Feng W, Knopf FC, Dooley KM. Effects of pressure, third bodies, and temperature profiling on the non-catalytic partial oxidation of methane. Energy Fuels 1994;8:815–22.
[44] Gesser HD, Hunter NR, Morton L. Direct conversion of natural gas to methanol by controlled oxidation. US Patent 4 618 732; Oct 21, 1986.
[45] Burch R, Squire GD, Tsang SC. Direct conversion of methane into methanol. J Chem Soc Faraday Trans I 1989; 85:3561–8.
[46] Lott JL, Sliepcevich CM. Partial oxidation of methane at high pressures. Ind Eng Chem Process Des Dev 1967; 6:67–74.
[47] Bauerle GL, Lott JL, Sliepcevich CM. Oxidation of methane at elevated pressures. II. A reaction mechanism. J Fire Flammability 1974;5:190–202.
[48] Tripathy N. Slow combustion of methane at elevated pressures. Isr J Chem 1975;13:190–6.
[49] Oberdorfer PE, Winch RF. Chemicals from methane in a high compression engine. Ind Eng Chem 1961;53: 41–4.
[50] Torchian HG, Mantashian AH, Nalbandian AB. Oxidation of methane under adiabatic compression conditions. Arm Khim Zh 1974;27:271–8 [in Russian].
[51] Furman MS, Shestakova AD. Oxidation of methane under pressure. Tr Gos Inst Azot Promyshl 1956;6:98–108 [in Russian].
[52] Han L-B, Tsubota S, Kobayashi T, Haruta M. Formation of methanol by the gas phase partial oxidation of methane under normal pressures. J Chem Soc Chem Commun 1995:93–4.
[53] Omata K, Fukuoka N, Fujimoto K. Methane partial oxidation to methanol 1. Effects of reaction conditions and additives. Ind Eng Chem Res 1994;33:784–9.
[54] Melvin A. Spontaneous ignition of methane–air mixtures at high pressure. I. The ignition delay preceding explosion. Combust Flame 1966;10:120–8.
[55] Atroshchenko VI, Gavrya NA, Shchedrinskaya ZM. On direct production of formaldehyde and methanol from natural gas. Tr Khark Polit Inst 1962;39:11–8 [in Russian].
[56] Atroshchenko VI, Shchedrinskaya ZM, Gavrya NA. Catalysts for the process of natural gas oxidation to formaldehyde and methanol. Zh Prikl Khim 1965;38:643–9 [in Russian].
[57] Baldwin TR, Burch R, Squir GD, Tsang SC. Influence of homogeneous gas phase reactions in the partial oxidation of methane to methanol and formaldehyde in the presence of oxide catalysts. Appl Catal 1991;74: 137–52.
[58] Parmaliana A, Frusteri F, Mezzapica A, Scurrell MS, Giordano N. Novel high activity catalysts for partial oxidation of methane to formaldehyde. J Chem Soc Chem Commun 1993:751–3.

[59] Bafas IC, Constantinou IE, Vayenas CG. Partial oxidation of methane to formaldehyde with 50% yield in a continuous recycle reactor separator (CRRS). Chem Eng J 2001:109–15.
[60] Irusta S, Lombardo EA, Miro EE. Effect of NO and solids on the oxidation of methane to formaldehyde. Catal Lett 1994;29:339–48.
[61] Zhang J, Burkle-Vitzthum V, Marquaire PM, Wild G, Commenge JM. Direct conversion of methane in formaldehyde at very short residence time. Chem Eng Sci 2011;66:6331–40.
[62] Zalc JM, Green WH, Iglesia E. NOx-mediated homogeneous pathways for the synthesis of formaldehyde from CH_4-O_2 mixtures. Ind Eng Chem Res 2006;45:2677–88.
[63] Vedeneev VI, Goldenberg MYa, Gorban' NI, Teitelboim MA. Quantitative model of the oxidation of methane at high pressures. I. Description of model. Kinet Catal 1988;29:1–8.
[64] Vedeneev VI, Goldenberg MYa, Gorban' NI, Teitelboim MA. Quantitative model of the oxidation of methane at high pressures. II. Mechanism of autoacceleration. Kinet Catal 1988;29:8–14.
[65] Vedeneev VI, Goldenberg MYa, Gorban' NI, Teitelboim MA. Quantitative model of the oxidation of methane at high pressures. III. Mechanism of formation of reaction products. Kinet Catal 1988;29:1121–6.
[66] Vedeneev VI, Goldenberg MYa, Gorban' NI, Teitelboim MA. Quantitative model of the oxidation of methane at high pressures. IV. Ignition delays. Kinet Catal 1988;29:1126–33.
[67] Lødeng R, Lindvag OA, Søraker P, Roterud PT, Onsager OT. Experimental and modeling study of the selective homogeneous gas phase oxidation of methane to methanol. Ind Eng Chem Res 1995;34:1044–59.
[68] Takemoto T, He D, Teng Y, Tabata K, Suzuki E. The optimization of methanol yield in direct selective oxidation of methane with O_2 and NO in the presence of $Cu-ZnO/Al_2O_3$. J Mol Catal A Chem 2002;179:279–86.
[69] Arutyunov VS, Vedeneev VI, Klimovetskaya SYu, Leonov VE, Pavlii LV. Influence of pressure on the formation of products of partial oxidation of methane. Theor Found Chem Eng 1994;28:563–8.
[70] Arutyunov VS. On catalysis of homogeneous–heterogeneous processes of oxidizing hydrocarbon conversion. Catal Ind 2003;(3):3–9 [in Russian].
[71] Vedeneev VI, Arutyunov VS, Krymov NYu, Cherbakov PM, Sedykh AD. Some features of methane oxidation at high pressures. Catal Today 1992;13:613–6.
[72] Chou T-C, Albright LF. Partial oxidation of methane in glass and metal tubular reactors. Ind Eng Chem Process Des Dev 1978;17:454–9.
[73] Mahajan S, Menzies WR, Albright LF. Partial oxidation of light hydrocarbons. 1. Major differences noted in various tubular reactors. Ind Eng Chem Process Des Dev 1977;16:271–4.
[74] Pat. SU 1145014; Dec 23, 1981.
[75] Pat. SU 1469788; June 19, 1986.
[76] Budymka VF, Egorov SA, Gavrya NA, Mochaev AS, Khomenko GA, Leonov VE. Effect of the process parameters on the gas-phase partial oxidation of natural gas. Khim Prom-st 1987;6:10–1 [in Russian].
[77] Morton LA, Gesser HD, Hunter NR. The partial oxidation of CH_4 to CH_3OH at high pressure in a packed reactor. Fuel Sci Tech Int 1991;9:913–33.
[78] Vedeneev VI, Simchenko VP, Shcherbakov PM. Production of methanol in the stepwise oxidation of natural gas. Gazov Prom-st 1995;11:30–1 [in Russian].
[79] Gesser HD, Hunter NR, Shigapov AN. Some characteristics of the partial oxidation of CH_4 to CH_3OH at high pressures. In: Bhasin MM, Slocum DW, editors. Methane and alkane conversion chemistry. N.Y.: Plenum Press; 1995. p. 271–86.
[80] Onsager OT, Lødeng R, Søraker P. The homogeneous gas phase oxidation of methane and the retarding effect of basic/inert surfaces. Catal Today 1989;4:355–63.
[81] Walsh DE, Martenak DJ, Han S, Palermo RE. Direct oxidative methane conversion at elevated pressure and moderate temperatures. Ind Eng Chem Res 1992;31:1259–62.
[82] Walsh DE, Martenak DJ, Han S, Palermo RE, Michaels JN, Stern DL. Pressure, temperature, and product yield relationships in direct oxidative methane conversion at elevated pressure and moderate temperatures. Ind Eng Chem Res 1992;31:2422–5.
[83] Chun J-W, Anthony RG. Partial oxidation of methane, methanol, and mixtures of methane and methanol, methane and ethane, and methane, carbon dioxide, and carbon monoxide. Ind Eng Chem Res 1993;32:788–95.
[84] Casey PS, McAllister T, Foger K. Selective oxidation of methane to methanol at high pressures. Ind Eng Chem Res 1994;33:1120–5.

[85] Chellappa AS, Fuangfoo S, Viswanath DS. Homogeneous oxidation of methane to methanol: effect of CO_2, N_2, and H_2 at high oxygen conversions. Ind Eng Chem Res 1997;36:1401–9.
[86] Zhang Q, He D, Li J, Xu B, Han Z, Liang Y, et al. Comparatively high yield methanol production from gas phase partial oxidation of methane. Appl Catal A 2002;224:201–7.
[87] Boomer EH, Thomas V. The oxidation of methane at high pressures. III. Experiments using pure methane and principally copper as catalyst. Can J Res Sect B 1937;15:414–33.
[88] Foulds GA, Gray BF, Miller SA, Walker GS. Homogeneous gas-phase oxidation of methane using oxygen as oxidant in an annular reactor. Ind Eng Chem Res 1993;32:780–7.
[89] Arutyunov VS, Basevich VYa, Vedeneev VI, Romanovich LB. Kinetic modeling of direct gas-phase methane oxidation to methanol at high pressures. Kinet Catal 1996;37(1):20–7.
[90] Yarlagadda PS, Morton LA, Hunter NR, Gesser HD. Temperature oscillations during the high-pressure partial oxidation of methane in a tubular flow reactor. Combust Flame 1990;79:216–7.
[91] Arutyunov VS, Rudakov VM, Savchenko VI, Sheverdenkin EV, Sheverdenkina OG, Zheltyakov AYu. Partial alkane oxidation at high pressures: methane oxidation in stainless steel and quartz reactors. Theor Found Chem Eng 2002;36:472–6.
[92] Fujimoto K, Sekine Y. Low temperature direct oxidation of methane to methanol. In: De Pontes M, Espinoza RL, Nicolaides CP, Scurrell MS, editors. Natural gas conversion IV. Studies in surface science and catalysis, vol. 107. Elsevier Science B.V; 1997. p. 63–6.
[93] Sitting M. Combining oxygen and hydrocarbons for profit. Houston, Texas: Gulf Pub. Co.; 1962 (Sittig M. Processes in the oxidation of hydrocarbon raw material. Moscow: Khimiya; 1970 [in Russian]).
[94] Vedeneev VI, Arutyunov VS, Basevich VYa, Goldenberg MYa, Teitelboim MA, Krymov NYu. Role of pressure in homogeneous and/or catalytic oxidation of methane. Catal Today 1994;21:527–32.
[95] Furman MS. Oxidation of methane under pressure. Khim Prom 1946;(1–2):24–6 [in Russian].
[96] Kolbanovskii YuA, Bilera IV, Rossikhin IV, Borisov AA, Troshin KYa. Single-stage conversion of associated petroleum gas and natural gas to syngas in combustion and auto-ignition processes. Russ J Gen Chem 2011;81: 2594–603.
[97] Simchenko VP, Shcherbakov PM, Vedeneev VI, Arutyunov VS. The kinetics of methane oxidation at high pressures. Theor Found Chem Eng 1999;33:362–4.
[98] Foral MJ. The noncatalytic partial oxidation of natural gas to methanol. In: Preprints, symposium on natural gas upgrading II, 203d National meeting of the American chemical society, Division of petroleum chemistry, San Francisco, CA. Washington, DC: American Chemical Society; April 5–10, 1992. p. 34–40.
[99] Foulds GA, Gray BF. Homogeneous gas-phase oxidation of methane to methanol and formaldehyde. Fuel Process Technol 1995;42:129–50.
[100] Arutyunov VS, Vedeneev VI, Klimovetskaya SYu, Leonov VE, Pavlii LV. Effect of hydrogen, carbon monoxide, and nitrogen on partial oxidation of methane. Theor Found Chem Eng 1995;29:63–7.
[101] Foulds GA, Charlton BG, Le BT, Jones JC, Gray BF. The use of a jet-stirred continuously stirred tank reactor (CSTR) to study the homogeneous gas phase partial oxidation of methane to methanol. In: De Pontes M, Espinoza RL, Nicolaides CP, Scurrell MS, editors. Natural gas conversion IV. Studies in surface science and catalysis, vol. 107. Elsevier Science B.V; 1997. p. 3–8.
[102] Arutyunov VS, Basevich VYa, Vedeneev VI, Parfenov Yu V, Sokolov OV. Dependence of the kinetic of gas phase methane oxidation at high pressures on the concentration of oxygen and on temperature. Russ Chem Bull 1996;45:45–8.
[103] Patent RU 2200731 from 10.10.2001.
[104] Gray BF, Griffiths JF, Foulds GA, Charlton BG, Walker GS. The relevance of thermokinetic interactions and numerical modeling to the homogeneous partial oxidation of methane. Ind Eng Chem Res 1994;33:1126–35.
[105] Sokolovskii VD, Coville NJ, Parmaliana A, Eskendirov I, Makoa M. Methane partial oxidation. Challenge and perspective. Catal Today 1998;42:191–5.
[106] Bjorklund MC, Carr RW. Enhanced methanol yields from the direct partial oxidation of methane in a simulated countercurrent moving bed chromatographic reactor. Ind Eng Chem Res 2002;41:6528–36.
[107] Basevich VYa, Vedeneev VI, Arutyunov VS. Radial and axial diffusion at gas phase oxidation of methane to methanol. Theor Found Chem Eng 1996;30:579–84.
[108] Sheverdenkin EV, Arutyunov VS, Rudakov VM, Savchenko VI. Kinetics of partial oxidation of alkanes at high pressures: oxidation of ethane and methane–ethane mixtures. Theor Found Chem Eng 2004;38:311–5.

[109] Vedeneev VI, Goldenberg MYa, Gorban' NI, Teitelboim MA. Mechanism of the initial stage of oxidation of methane at high pressures. Khim Fiz 1987;6:626–32 [in Russian].

[110] Vedeneev VI, Arutyunov VS, Basevich VYa, Goldenberg MYa, Teitelboim MA, Krymov NYu. Role of pressure in homogeneous and/or catalytic oxidation of methane. Catal Today 1994;21:527–32.

[111] Chun J-W, Anthony RG. Free radical kinetic model for homogeneous oxidation of methane to methanol. Ind Eng Chem Res 1993;32:796–9.

[112] Hunter TB, Wang H, Litzinger TA, Frenklach M. The oxidation of methane at elevated pressures: experiments and modeling. Combust Flame 1994;97:201–24.

[113] Rasmussen CL, Jacobsen JG, Glarborg P. Experimental measurements and kinetic modeling of CH_4/O_2 and $CH_4/C_2H_6/O_2$ conversion at high pressure. Int J Chem Kinet 2008;40:778–807.

[114] Rasmussen CL, Glarborg P. Direct partial oxidation of natural gas to liquid chemicals: chemical kinetic modeling and global optimization. Ind Eng Chem Res 2008;47:6579–88.

[115] Knyazev VD, Arutyunov VS, Vedeneev VI. The mechanism of $O(^3P)$ atom reaction with ethylene and simple olefins. Int J Chem Kinet 1992;24:545–61.

[116] Krylov OV. The partial catalytic oxidation of methane to give oxygen-containing compounds. Russ Chem Rev 1992;61:1118–29.

[117] Tabata K, Teng Y, Takemoto T, Suzuki E, Bañares MA, Peña MA, et al. Activation of methane by oxygen and nitrogen oxides. Catal Rev 2002;44:1–58.

[118] Vedeneev VI, Arutyunov VS, Bacevich VYa. Simulation of the gas-phase oxidation of methane: the kinetic limit of the yield of ethane and ethylene in the oxidative coupling of methane. Khim Fiz 1997;16:76–91 [in Russian].

[119] Arutyunov VS, Basevich VYa, Vedeneev VI, Krylov OV. On the role of catalyst in high temperature reactions of methane oxidation. Kinet Catal 1999;40:382–7.

[120] Wang YH, Hermann RO, Klier K. Dissociative adsorption of methane on Pd(679) surface. Surf Sci 1992;279: 33–48.

[121] Arutyunov VS. Recent results on fast flow gas-phase partial oxidation of lower alkanes. J Nat Gas Chem 2004; 13:10–22.

[122] Slovetskaya KI, Kustov LM. High-temperature methane oxidation over metallic monolith-supported zeolite catalysts containing Mn, Co, and Pd ions. Russ Chem Bull Int Ed 2003;52:1933–9.

[123] Bobrova II, Chesnokov VV, Bobrov NN, Zaikovskii VI, Parmon VN. New data on gas-phase radical reactions in the steam reforming of methane in the presence of catalysts: I. Nickel catalysts. Kinet Catal 2000;41: 19–24.

[124] Andrae J, Björnbom P, Edsberg L. Numerical studies of wall effects with laminar methane flames. Combust Flame 2002;128:165–80.

[125] Bobrova II, Bobrov NN, Simonova LG, Parmon VN. Direct catalytic oxidation of methane to formaldehyde: new investigation opportunities provided by an improved flow circulation method. Kinet Catal 2007;48: 676–92.

[126] Davis MB, Pawson MD, Veser G, Schmidt LD. Methane oxidation over noble metal gauzes: an LIF study. Combust Flame 2000;123:159–74.

[127] Arutyunov VS. The role of pressure in partial oxidation of methane. Russ Chem Bull 2002;51:2170–5.

[128] Tyulenin YuP, Savkin VV, Sinev MYu, Korchak VN. The order of product formation in the partial oxidation of methane to syngas. Kinet Catal 2002;43:847–53.

[129] Tulenin YuP, Sokolov OV, Sinev MYu, Savkin VV, Arutyunov VS, Korchak VN. Interaction between heterogeneous and homogeneous processes in gas-phase methane oxidation by supported platinum catalysts. EvropCat Symposium 5. Alkane activation; Sept 2001. Abstracts. Book 4. 5-P-12.

[130] Tulenin YuP, Sinev MYu, Savkin VV, Korchak VN. Dynamic behavior of Ni-containing catalysts during partial oxidation of methane to synthesis gas. In: Bao X, Xu Y, editors. Proceedings of the 7th natural gas conversion symposium, June 6–10, Dalian, China. Studies in surface science and Catalysis. Natural gas conversion VII, vol. 147. Amsterdam: Elsevier; 2004. p. 151–6.

[131] Hansen PL, Wagner JB, Helveg S, Rostrup-Nielsen JR, Clausen BS, Topsoe H. Atom-resolved imaging of dynamic shape changes in supported cooper nanocrystals. Science 2002;295:2053–5.

[132] Helveg S, Lopez-Cartes C, Sehested J, Hansen PL, Clausen BS, Rostrup-Nielsen JR, et al. Atomic-scale imaging of carbon nanofibre growth. Nature 2004;427:426–9.

[133] Kasztelan S, Moffat JB. Partial oxidation of methane by oxygen over silica. J Chem Soc Chem Commun 1987:1663–4.

[134] Kastanas GN, Tsigdinos GA, Schwank J. Selective oxidation of methane over vicor glass, quartz glass and various silica, magnesia and alumina surfaces. Appl Catal 1988;44:33–51.

[135] Nersessian LA, Vardanyan IA, Kegeyan EM, Margolis LYa, Nalbandyan AB. Heterogeneous–homogeneous oxidation of methane. Dokl Akad Nauk SSSR 1975;220:605–7 [in Russian].

[136] Nersesyan LA, Vardanian IA, Margolis LYa, Nalbandyan AB. Dokl Akad Nauk SSSR 1973;220:605 [in Russian].

[137] Arena F, Frusteri F, Parmaliana A, Giordano N. A temperature programmed reaction method for catalysts testing in the partial oxidation of methane to formaldehyde. J Catal 1993;143:299–303.

[138] Amiridis MD, Rekoske JE, Dumesic JA, Rudd DF, Spencer ND, Pereira CJ. Simulation of methane partial oxidation over silica-supported MoO_3 and V_2O. AIChE J 1991;37:87–97.

[139] Ozturk S, Onal I, Senkan S. Partial oxidation of methane on the SiO_2 surfaces. A quantum chemical study. Ind Eng Chem Res 2000;39:250–8.

[140] Hunter NR, Gesser HD, Morton LA, Yarlagadda PS, Fung DPC. Methanol formation at high pressure by the catalyzed oxidation of natural gas and by the sensitized oxidation of methane. Appl Catal 1990;57:45–54.

[141] Thomas DJ, Willi R, Baiker A. Partial oxidation of methane: the role of surface reactions. Ind Eng Chem Res 1992;31:2272–8.

[142] Alzueta MU, Bilbao R, Finestra M. Methanol oxidation and its interaction with nitric oxide. Energy Fuels 2001;15:724–9.

[143] Birtill JJ, Ridley P, Liddle S, Nunney TS, Raval R. Degradation of methanol and methylamines to carbon over heated alloy surfaces. Ind Eng Chem Res 2001;40:553–7.

[144] Zhang Q, He D, Zhu Q. Direct partial oxidation of methane to methanol: reaction zones and role of catalyst location. J Nat Gas Chem 2008;17:24–8.

[145] Cheng WH. Selectivity in methanol oxidation over silica supported molybdena. Catal Lett 1996;36:87.

[146] Yao S, Yang F Ou, Shimumora S, Sakurai H, Tabata K, Suzuki E. A kinetic study of methanol oxidation over SiO_2. Appl Catal A 2000;198:43–50.

[147] Boomer EH, Naldrett N. The oxidation of methane at high pressures. IV. Experiments using pure methane and copper, silver, zinc, nickel and monel metal as catalysts. Can J Res Sect B 1947;25:494–501.

[148] Gavrya NA, Atroshchenko VI, Shchedrinskaya ZM. Technological and constructive factors in the oxidation gas. Vestn Khark Polytech Inst 1968;(34):53–6 [in Russian].

[149] Rytz DW, Baiker A. Partial oxidation of methane to methanol in a flow reactor at elevated pressure. Ind Eng Chem Res 1991;30:2287–92.

[150] Chun J-W, Anthony RG. Catalytic oxidation of methane to methanol. Ind Eng Chem Res 1993;32:259–63.

[151] Walker GS, Lapszewicz JA, Foulds GA. Partial oxidation of methane to methanol – comparison of heterogeneous catalyst and homogeneous gas phase reactions. Catal Today 1994;21:519–26.

[152] Otsuka K, Wang Y. Direct conversion of methane into oxygenates. Appl Catal A 2001;A222:145–61.

[153] Zhang Q, He D, Zhu Q. Recent progress in direct partial oxidation of methane to methanol. J Nat Gas Chem 2003;12:81–9.

[154] Barbero JA, Alvarez MC, Bañares MA, Peña MA, Fierro JLG. Breakthrough in the direct conversion of methane into C_1-oxygenates. Chem Commun 2002:1184–5.

[155] Schmidt LD. Modeling millisecond reactors. In: Iglesia E, Fleish TH, editors. Proceedings of the 6th natural gas conversion symposium. Girdwood, Alaska, 2001. Studies in surface science and catalysis. Natural gas conversion VI, vol. 136. Amsterdam-London-New York-Oxford-Paris-Shannon-Tokyo: Elsevier Science B.V; 2001.

[156] Dagaut P, Cathonnet M, Boettner J-C. Chemical kinetic modeling of the supercritical-water oxidation of methanol. J Supercrit Fluids 1996;9:33–42.

[157] Pietruszka B, Heintze M. Methane conversion at low temperature: the combined application of catalysis and non-equilibrium plasma. Catal. Today. 2004;90(1–2):151–8.

[158] Takemoto T, Tabata K, Teng Y, Nakayama A, Suzuki E. The effect of reaction pressure on the production of methanol trough the selective oxidation of methane in CH_4–O_2–NOx (x = 1 or 2). Appl Catal 2001;A205:51–9.

[159] Sokolik AS. Self-ignition, flame, and detonation in gases. Moscow: USSR Academy of Sciences; 1960 [in Russian].
[160] Lewis B, Elbe G. Combustion, flames, and explosions of gases. Orlando: Acad. Press; 1987.
[161] Vanpee M. Sur les flames froides du methane. C R Acad Sci Paris 1956;243:804–6.
[162] Sokolov OV, Arutyunov VS, Basevich VYa, Vedeneev VI. Observation of cool-flame phenomena during self-ignition of methane–oxygen mixtures. Kinet Catal 1995;36:317–8.
[163] Arutyunov VS, Basevich VYa, Vedeneev VI, Sokolov OV. Discovery of the negative temperature coefficient in the reaction of methane oxidation. Kinet Catal 1995;36:458–9.
[164] Sokolov OV, Parfenov YuV, Arutyunov VS, Basevich VYa, Vedeneev VI. Study of cool-flame phenomena during self-ignition of methane–oxygen mixtures. Russ Chem Bull 1996;45(10):2316–20.
[165] Vedeneev VI, Romanovich LB, Bacevich VY, Arutyunov VS, Sokolov OV, Parfenov YuV. Experimental investigation and kinetic modeling of the negative temperature coefficient of the reaction rate in rich propane–oxygen mixtures. Russ Chem Bull 1997;46:2006–10.
[166] Bacevich VYa, Vedeneev VI, Arutyunov VS. Theoretical study of cool-flame phenomena in the self-ignition methane–oxygen mixtures. Khim Fiz 1994;13:157–63 [in Russian].
[167] Basevich VYa, Frolov SM. Kinetics of 'blue' flames in the gas-phase oxidation and combustion of hydrocarbons and their derivatives. Russ Chem Rev 2007;76:867.
[168] Barckholtz TA. Modeling the negative temperature coefficient in the low temperature oxidation of light alkanes. Prepr Pap Am Chem Soc Div Fuel Chem 2003;48(2):518–9.
[169] Zhu G, Gesser HD, Hunter NR. The O_3 sensitized partial oxidation of CH_4 to CH_3OH. Studies in surface science and catalysis. In: Curry-Hyde HE, Howe RF, editors. Natural gas conversion III. Proceedings of the third natural gas conversion symposium, Sydney, vol. 81. Amsterdam: Elsevier; July 4–9, 1993.
[170] Eskendirov I, Coville NJ, Parmaliana A, Sokolovski VD. Direct oxidative conversion of methane into higher hydrocarbons and oxy-products in the presence of hydrogen peroxide. In Natural gas conversion IV. Studies in surface science and catalysis, vol. 107. Elsevier Sci. B.V; 1997. 301–306.
[171] Davis MB, Schmidt LD. The seeding of methane oxidation. Combust Flame 1999;119:182–8.
[172] Layng TE, Soukop R. Ind Eng Chem Res 1928;20:1052.
[173] Bibb CH, Lucas HJ. Air oxidation of hydrocarbons catalyzed by nitrogen oxides. Ind Eng Chem 1929;21: 633–5.
[174] Paushkin YaM, Adelson SV, Vishnyakova TP. Technology of petrochemical synthesis. Part I. Moscow: Chemistry; 1973 [in Russian].
[175] Urizko VI, Polyakov MV. Effect of the wall surface on the oxidation of methane to formaldehyde. Dokl Akad Nauk SSSR 1954;95:1239–41 [in Russian].
[176] Enikolopyan NS, Konoreva GP. Homogeneous catalysis in the gas-phase oxidation of hydrocarbons. Russ Chem Bull 1960;9:389–95.
[177] Maizus ZK, Emanuel NM. Macroscopic step catalysed nitrogen oxides propane oxidation. Dokl Akad Nauk SSSR 1954;95:1009–11 [in Russian].
[178] Fedorova TV, Ballod AP, Topchiev AV, Stern VYa. On the kinetic mechanism of the interaction of methane with nitrogen dioxide. Dokl Akad Nauk SSSR 1958;123:860–3 [in Russian].
[179] Ballod AP, Molchanova SI, Topchiev AV, Fedorova TV, Stern VYa. Three types of kinetic curves for the interaction of methane and propane with nitrogen dioxide. Dokl Akad Nauk SSSR 1958;123:464–7 [in Russian].
[180] Bromly JH, Barnes FJ, Muris S, You X, Haynes BS. Kinetic and thermodynamic sensitivity analysis of the NO-sensitised oxidation of methane. Combust Sci Technol 1996;115:259–96.
[181] Hori M, Matsunaga N, Marinov N, Pitz W, Westbrook C. An experimental kinetic calculation of the promotion effect of hydrocarbons on the NO–NO2 conversion in a flow reactor. Twenty-seventh symposium (International) on combustion/The Combustion Institute. 1998. p.389–396.
[182] Taylor PH, Cheng Li, Dellinger B. The influence of nitric oxide on the oxidation of methanol and ethanol. Combust Flame 1998;115:561–7.
[183] Otsuka K, Takahashi R, Amakawa K, Yamanaka I. Partial oxidation of light alkanes by NOx in the gas phase. Catal Today 1998;45:23–8.
[184] Otsuka K, Takahashi R, Yamanaka I. Oxygenates from light alkanes catalyzed by NOx in the gas phase. J Catal 1999;185:182–91.

[185] Teng Y, Ouyang F, Dai L, Karasuda T, Sakurai H, Tabata K, et al. High yield methanol formation in CH_4-O_2-NO gaseous selective oxidation at 1 atm. Chem Lett 1999:991–2.

[186] Teng Y, Sakurai H, Tabata K, Suzuki E. Methanol formation from methane partial oxidation in in CH_4-O_2-NO gaseous phase at atmospheric pressure. Appl Catal 2000;A190:283–9.

[187] Takemoto T, Tabata K, Teng Y, Yao S, Nakayama A, Suzuki E. Optimization of C_1-oxygenates for the selective oxidation of methane in a gas-phase reaction of CH_4-O_2-NO at atmospheric pressure. Energy Fuels 2001;15: 44–51.

[188] Takemoto T, He D, Teng Y, Nakayama A, Tabata K, Suzuki E. Enhancement of methanol selectivity in products of direct selective oxidation of methane in CH_4-O_2-NO with $Cu-ZnO/Al_2O_3$. J Catal 2001;198:109–15.

[189] Han L-B, Tsubota S, Haruta M. Effect of the addition of nitrogen dioxide on the gas phase partial oxidation of methane with oxygen under normal pressures. Chem Lett 1995:931–2.

[190] Yamaguchi Y, Teng Y, Shimomura S, Tabata K, Suzuki E. Ab initio study for selective oxidation of methane with NOx ($x = 1, 2$). J Phys Chem A 1999;103:8272–8.

[191] Tabata K, Teng Y, Yamaguchi Y, Sakurai H, Suzuki E. Experimental verification of theoretically calculated transition barriers of the reactions in a gaseous selective oxidation of $CH_4-O_2-NO_2$. J Chem Phys 2000;104: 2648–54.

[192] Faravelli T, Frassoldati A, Ranzi E. Kinetic modeling of the interaction between NO and hydrocarbons in the oxidation of hydrocarbons at low temperatures. Combust Flame 2003;132:188–207.

[193] Banares MA, Cardoso JH, Hutchings GJ, Bueno JMC, Fierro JLG. Selective oxidation of methane to methanol and formaldehyde over V_2O_5/SiO_2 catalysts. Role of NO in the gas phase. Catal Lett 1998;56:149–53.

[194] Sen A, Lin M. NOx-catalyzed partial oxidation of methane and ethane to formaldehyde by dioxygen. In: Iglesia E, Fleish TH, editors. Proceedings of the 6th natural gas conversion symposium. Girdwood, Alaska, 2001. Studies in surface science and catalysis. Natural gas conversion VI, vol. 136. Amsterdam-London-New York-Oxford-Paris-Shannon-Tokyo: Elsevier Science B.V; 2001. p. 227.

[195] Rasmussen CL, Rasmussen AE, Glarborg P. Sensitizing effects of NOx on CH_4 oxidation at high pressures. Combust Flame 2008;154:529–45.

[196] Gudkov SF, Fedulova VN. Partial oxidation of methane–propane mixtures. Gazov Prom 1957;(10):32–9 [in Russian].

[197] Fischer F. Brennst Chem 1921;9:308.

[198] Hirota K. Researches on the chemical reaction by use of the thermal diffusion apparatus of Clusius and Dickel. I. Thermal polymerisation of methane. Bull Chem Soc Jpn 1941;16(8):274–8.

[199] Yamaguchi J, Saito C. Selective synthesis of ethylene by dehydrogenative coupling of methane by use of thermal diffusion column. Bull Chem Soc Jpn 1989;61:2649–50.

[200] Gesser HD, Morton LA. The dehydrogenative coupling of methane to form higher hydrocarbons in a hot wire thermal diffusion column. Catal Lett 1991;11:357–63.

[201] El Behery RE, Mohamad AA, Kamal MM. Combustion enhancement of gas flare using acoustical excitation. Combust Sci Technol 2005;177:1627–59.

[202] El'piner IE, Sokolovskaya AV. On the oxidation of aqueous solutions of methane and ethylene in an ultrasonic field. Zh Fiz Khim 1971;45:3071–2.

[203] Ogura K, Kataoka M. Photochemical conversion of methane. J Mol Catal 1988;43:371–9.

[204] Baughum SL, Oldenborg RC, Danen WC, Streit GE, Rofer C. Laser initiated chain reactions in the partial oxidation of methane. In: Proceedings of the 1986 Quebec international symposium on optical and opto-electric applied sciences and engineering. Quebec City, Canada. SPIE proceedings, vol. 669. Laser applications in chemistry, 1986. p. 81–89.

[205] Tepermeister I, Smith J, Streit J, Oldenborg R, Finch F, Danen W, et al. Methane to methanol: reactor design and process evaluation. (LA-11396-MS. DE89 003381).

[206] Ponomarev AV, Ershov BG, Makarov IE. Electron beam processing of gas: GTL-synthesis of high-octane fuel. Gazokhimiya 2009;(1):46–51 [in Russian].

[207] Shepelev SS, Gesser HD, Hunter NR. Light paraffins oxidative conversion in a silent electric discharge. Plasma Chem Plasma Process 1993;13:479.

[208] Huang J, Suib SL. Dimerization of methane trough microwave plasmas. J Phys Chem 1993;97:9403–7.

[209] Huang J, Badani MV, Suib SL, Harrison JB, Kablauoi M. Partial oxidation of methane through microwave plasmas. Reactor design to control free-radical reactions. J Phys Chem 1994;98:206–10.

[210] Fedoseyev VI, Aristov YuI, Tanashev YuYu, Parmon VN. Methane pyrolysis under pulsed microwave radiation in the presence of solid catalysts. Kinet Catal 1996;37:808–11.
[211] Zhou LM, Xue B, Kogelschtz U, Eliasson B. Partial oxidation of methane to methanol with oxygen or air in a nonequilibrium discharge plasma. Plasma Chem Plasma Process 1998;18:375–93.
[212] Pushkarev AI, Remnev GE, Yezhov VV. Non-equilibrium plasma chemical conversion of methane. Gorenie I Plazmochimiya 2005;3:106–22 [in Russian].
[213] Larkin DW, Lobban LL, Malinson RG. Production of organic oxygenates in the partial oxidation of methane in a silent electric discharge reactor. Ind Eng Chem Res 2001;40:1594–601.
[214] He M, Zou J, Li Y, Eliasson B, Liu C. Oxygenate synthesis from methane and carbon dioxide using dielectric-barrier discharges. ACS Fuel Chem Div Prepr 2002;47(1):328–9.
[215] Wang Yu, Liu C. Oxygenates synthesis using methanol plasmas. Prepr Pap Am Chem Soc Div Fuel Chem 2005;50(2):641.
[216] Bugaev SP, Kozyrev AV, Kuvshinov VA, Sochugov NS, Khryapov PA. Stimulated condensation of the products plasmochemical oxidative conversion of lower hydrocarbons. Dokl Ross Akad Nauk 1997;354: 200–2 [in Russian].
[217] Nozaki T, Hattori A, Okazaki K. Partial oxidation of methane using a microscale non-equilibrium plasma reactor. Catal Today 2004;98:607–16.
[218] Sergio R, Thornton JD. Synthesis of formaldehyde and methanol from methane and water in an electrical discharge two-phase reactor. J Appl Chem 1967;17:325–8.
[219] Savage PE, Gopalan S, Mizan TI, Martino CJ, Brock E. Reactions at supercritical conditions: applications and fundamentals. AIChE J 1995;41:1723–78.
[220] Savage PE. Organic chemical reactions in supercritical water. Chem Rev 1999;99:603–21.
[221] Arutyunov VS, Zalepugin D Yu, Sinev M Yu. Oxidative conversion of light alkanes in supercritical conditions: first results, problems, and prospects. SKF-TP 2007;2:5–29 [in Russian].
[222] Sato T, Watanabe M, Smith Jr RL, Adschiri T, Arai K. Analysis of the density effect on partial oxidation of methane in supercritical water. J Supercrit Fluids 2004;28:69–77.
[223] Vohringer-Martinez E, Hansmann B, Hernandez H, Francisco JS, Troe J, Abel B. Water catalysis of a radical-molecule gas-phase reaction. Science 2007;315:497–501.
[224] Hirth Th, Franck EU. Oxidation and hydrothermolysis of hydrocarbons in supercritical water at high pressures. Ber Bunsenges Phys Chem 1993;97:1091–8.
[225] Lee JH, Foster NR. Direct partial oxidation of methane to methanol in supercritical water. J Supercrit Fluids 1996;9:99–105.
[226] Ksandopulo GI, Dubinin VV. Gas-phase chemistry of combustion. Moscow: Khimiya; 1987 [in Russian].
[227] Dixon CN, Abraham MA. Conversion of methane to methanol by catalytic supercritical water oxidation. J Supercrit Fluids 1992;5:269–73.
[228] Savage PE, Li R, Santini JTJ. J Supercrit Fluids 1994;7:135.
[229] Steeper RR, Rice SF, Kennedy IM, Aiken JD. Kinetic measurements of methane oxidation in supercritical water. J Phys Chem 1996;100:184–9.
[230] Savage PE, Rovira J, Stylski N, Martino CJ. Oxidation kinetics for methane/methanol mixtures in supercritical water. J Supercrit Fluids 2000;17:155–70.
[231] Armbruster U, Martin A, Krepel A. Partial oxidation of propane in sub- and supercritical water. J Supercrit Fluids 2001;21:233–43.
[232] Suresh AK. Isobutane oxidation in the liquid and supercritical phases: comparison of features. J Supercrit Fluids 1998;12:165–76.
[233] Bröll D, Kaul C, Krämer A, Richter T, Jung M, Vogel H, et al. Angew Chem Int Ed 1999;38:2998.
[234] Moshkina RI, Polyak SS, Sokolova NA, Masterovoi IF, Nalbandyan AB. Study of the ethane oxidation reaction by the kinetic tracer method. Int J Chem Kinet 1980;12:315–38.
[235] Gesser HD, Hunter NR, Morton LA, Yarlagadda PS. Direct conversion of ethane to ethanol by the homogeneous partial oxidation at high pressures. Energy Fuels 1991;5:423–7.
[236] Shigapov AN, Hunter NR, Gesser HD. The direct oxidation of ethane to alcohols at high pressures. Catal Today 1998;42:311–4.
[237] Arutyunov VS, Basevich VYa, Vedeneev VI, Parfenov YuV, Sokolov OV, Leonov VE. Oxidation of methane–ethane mixtures into alcohols under enhanced pressures. Theor Found Chem Tech 1998;32:152–8.

[238] Arutyunov VS, Basevich VYa, Vedeneev VI, Parfenov YuV, Sokolov OV, Leonov VE. Oxidation of methane–ethane mixtures into alcohols under enhanced pressures. Catal Today 1998;42:241–5.
[239] Kooijman PL. The preparation of peroxides by the oxidation of propane. Recueil 1947;66(1/2):5–23.
[240] Satterfield CN, Wilson RE. Partial oxidation of propane. The role of hydrogen peroxide. Ind Eng Chem 1954; 46:1001–7.
[241] Badrian AS, Furman MS. Effect of pressure on the formation of intermediate propane oxidation products. Dokl Akad Nauk SSSR 1956;108:861–3 [in Russian].
[242] Badrian AS, Furman MS. Kinetics and mechanism of the oxidation of propane under pressure. Tr GIAP 1957; 7:79–100 [in Russian].
[243] Livshits VD. Effect of the solid phase of "catalysts" on the oxidation of propane. Tr GIAP 1959;(9):216–22 [in Russian].
[244] Livshits VD, Sokolova NP, Beskova AP. Oxidation of propane under pressure. Tr GIAP 1959;(9):223–37 [in Russian].
[245] Hoffman JS, Lee W, Litzinger TA, Santavicca DA. Oxidation of propane at elevated pressures: experiments and modeling. Combust Sci Technol 1991;77:95–125.
[246] Koert DN, Miller DL, Cernansky NP. Experimental studies of propane oxidation through the negative temperature coefficient region at 10 and 15 atmospheres. Combust Flame 1994;96:34–49.
[247] Dagaut P, Cathonnet M, Boettner J-C. Kinetic modeling of propane oxidation and pyrolysis. Int J Chem Kinet 1992;24:813–37.
[248] Mitchell RL. Make petrochemicals from air and LPG. Pet Refin July 1956:179–82.
[249] Emanuel NM, Denisov ET, Maizus ZK. Chain reactions of oxidation of hydrocarbons in the liquid phase. Moscow: Nauka; 1974 [in Russian].
[250] Quon D, Dalla Lana I, Govier GW. The vapor phase partial oxidation of *n*-butane – effect of pressure, reaction time, and inlet gas composition. Can J Chem 1954;32:880–95.
[251] Furman MS, Badrian AS, Gol'tyaeva NA, Savchuk SN. Gas-phase oxidation of *n*-butane under pressure. Gazov Prom 1958;(10):36–43 [in Russian].
[252] Furman MS, Badrian AS, Gol'tyaeva NA. Oxidation of *n*-butane under pressure. Tr GIAP 1959;(9):238–47 [in Russian].
[253] Wilk RD, Cohen RS, Cernansky NP. Oxidation of *n*-butane: transition in the mechanism across the region of negative temperature coefficient. Ind Eng Chem Res 1995;34:2285–97.
[254] Kwok ESC, Arey J, Atkinson R. Alkoxy radical isomerization in the OH radical initiated reaction of C_4–C_8 *n*-alkanes. J Phys Chem 1996;100:214–9.
[255] Warth V, Stef N, Glaude PA, Battin-Leclerc F, Scacchi G, Come GM. Computer-aided derivation of gas-phase oxidation mechanisms: application to the modeling of the oxidation of *n*-butane. Combust Flame 1998;114: 81–102.
[256] Minetti R, Ribaucour M, Carlier M, Fittschen C, Sochet LR. Experimental and modeling study of oxidation and autoignition of butane at high pressures. Combust Flame 1994;96:201–11.
[257] Rienstra-Kiracofe JC, Allen WD, Schaefer III HF. The $C_2H_5 + O_2$ reaction mechanism: high level ab initio characterization. J Phys Chem A 2000;104:9823–40.
[258] Naik CV, Dean AV. Detailed kinetic modeling of ethane oxidation. Combust Flame 2006;145:16–37.
[259] Arutyunov VS, Rudakov VM, Savchenko VI, Sheverdenkin EV. Relative conversion of lover alkanes at their simultaneous partial gas-phase oxidation. Theor Found Chem Eng 2005;39(5):487–92.
[260] Meyer RE. LPG to chemicals by oxidation. Oil Gas J 1955;(7):82–6.
[261] Weiss JM. After the present emergence what? Chem Ind Sept 1941;49(3):348–52.
[262] Asinger F. Chemie und technologie der paraffin-kohlenwasserstoffe. Berlin: Akademie-Verlag; 1956 (Azinger F. Chemistry and technology of paraffin hydrocarbons. Moscow: Gostoptekhizdat; 1959 [in Russian]).
[263] Gudkov SF. Processing of hydrocarbon and associated gases. Moscow: Gostoptekhizdat; 1960 [in Russian].
[264] Powers ET. Production of chemical products by the direct oxidation of hydrocarbons. IV International petroleum congress. T.V. Chemical processing of oil and gas. Moscow: Gostoptekhizdat; 1956. p. 57–68. [in Russian].
[265] Direct oxidation of propane–butane – Celanese Corporation of America chemical plant. Pet Refin 1959; 38(11):234–5.
[266] Meyer RE. New avenue for capital – oxidation of LPG? Pet Refin 1956;35(12):172–8.

[267] Walker JF. Formaldehyde. 2nd ed. New York: Reinhold Publishing Corp.; 1953.
[268] Anisonyan AA, Gudkov SF, Enikolopyan NS, Kleimenov NA, Markevich AM, Nalbandyan AB, et al. Production of formaldehyde by the direct oxidation of natural gas by air oxygen. Gazov Prom 1957;(6):32–40 [in Russian].
[269] Ogorodnikov SK. Formaldehyde. Leningrad: Khimiya; 1984 [in Russian].
[270] Arutyunov VS. Prospective technologies based on the partial oxidation of hydrocarbon gases. Gazokhimiya 2009;2(6):50–61 [in Russian].
[271] Bukhgalter EB. Methanol and its use in the gas industry. Moscow: Nedra; 1986 [in Russian].
[272] Westerterp KR, Bodewes TN, Vrijland MSA, Kuczynski M. To new methanol converters. Hydrocarbon Process 1988;(11):69–73.
[273] Berty JM, Krishnan C, Elliot JR. Beat the equilibrium. Chemtech 1990;10:624–9.
[274] Patent RU 2200731; 20.03.2003.
[275] Basevich VYa, Vedeneev VI, Arutyunov VS, Romanovich. LB. Gas phase oxidation of methane to methanol in cyclic mode. Theor Found Chem Eng 1997;31:68–71.
[276] Patent RU2051202; 27.12.1995.
[277] A handbook of modern processes of gas processing 1992: oil, gas and petrochemistry abroad; 1992;9: 76–97.
[278] Purification of gases from sulfur compounds in gas fields. Moscow: Nedra; 1985 [in Russian].
[279] Semenova TA, Leites IL, editors. Purification of technological gases. Moscow: Khimiya; 1977 [in Russian].
[280] Benayoun D. Ifpecsol – simple method for gas processing. Gazov Prom 2001;(5):50–3 [in Russian].
[281] Patent RU 2283829; 24.02.2005.
[282] Mac Farlan A, Liu D. CANMET integrated acetic acid process: coproduction of chemicals and power from natural gas. In: Iglesia E, Fleish TH, editors. Proceedings of the 6th natural gas conversion symposium. Girdwood, Alaska, 2001. Studies in surface science and catalysis. Natural gas conversion VI, vol. 136. Amsterdam-London-New York-Oxford-Paris-Shannon-Tokyo: Elsevier Science B.V; 2001.
[283] Patent U.S. 5659077; 19.08.97.
[284] Chen L, Zhang J, Ning P, Chen Y, Wu W. Kinetics of methanol carbonylation to methyl formate catalyzed by sodium methoxide. J Nat Gas Chem 2004;13:225–30.
[285] Ma X, Li Z, Wang B, Xu G. In-situ IR studies on the oxidative carbonylation of methanol to dimethyl carbonate. ACS Fuel Chem Div Prepr 2003;48(1):426–7.
[286] Olah GA, Goeppert A, Prakash GKS. Beyond oil and gas: the methanol economy. Weinheim: Wiley-VCH Verlag GmbH & Co. KGaA; 2006.
[287] Cheung P, Bhan A, Sunley GJ, Iglesia E. Selective carbonylation of dimethyl ether to methyl acetate catalyzed by acidic zeolites. Angew Chem Int Ed 2006;45:1617–20.
[288] Liu D, Huang X, Hu L, Fang D, Ying W, Chen D. Synthesis of ethylidene diacetate from dimethyl ether, CO and H_2. J Nat Gas Chem 2010;19:165–8.
[289] Patent EP 2138093; 1982.
[290] Aricò F, Tundo P. Dimethyl carbonate as a modern green reagent and solvent. Russ Chem Rev 2010;79: 479–90.
[291] Peng W, Zhao N, Xiao F, Wei W, Sun Y. Direct vapor-phase carbonylation of methanol at atmospheric pressure on activated carbon-supported $NiCl_2$–$CuCl_2$ catalysts. Pure Appl Chem 2012;84:603–20.
[292] Li XL, Chen HY, Zhu ZY, Huang Z. Study of combustion and emission characteristics of a diesel engine operated with dimethyl carbonate. Energy Convers Manage 2006;47:1438–48.
[293] Tundo P, Selva M. The chemistry of dimethyl carbonate. Acc Chem Res 2002;35:706–16.
[294] He L, Liu H, Xiao C-X, Kou Y. Liquid-phase synthesis of methyl formate via heterogeneous carbonylation of methanol over a soluble copper nanocluster catalyst. Green Chem 2008;10:619–22.
[295] Savchenko VI, Makaryan IA, Fokin IG, Magomedov RN, Lipilin MG, Arutyunov VS. Low-tonnage GTL-processes based on the direct partial oxidation of hydrocarbon gases without syngas production. Neftepererab Neftekhim (in press) [in Russian].
[296] Druskin LI. Efficient use of natural gas at industrial installations. A handbook. Energoatomizdat; 1992 [in Russian].
[297] Hill SC, Smoot LD. Modeling of nitrogen oxides formation and destruction in combustion systems. Prog Energy Combust Sci 2000;26:417–58.

[298] Beer JM. Combustion technology developments in power generation in response to environmental challenges. Prog Energy Combust Sci 2000;26:301–27.
[299] Cocchi S, Giannini N, Provenzale M, Zucca A, Romano C, Ceccherini G. A simple model for NOx formation in diffusion gas turbine combustors: rig test validation with a wide range of fuel gases. GE Oil&Gas – Nuovo Pignone S.p.A., Firenze – Italy. 31st meeting on combustion, Italian Section of the Combustion Institute. VII-1. (http://143.225.115.225/proc/proc2008/data/papers/VII/VII-1.pdf).
[300] Su S, Beath A, Guo H, Mallet C. An assessment of mine methane mitigation and utilization technologies. Prog Energy Combust Sci 2005;31:123–70.
[301] Zamansky VM, Loc Ho, Maly PM, Seeker W.R. Gas phase reactions of hydrogen peroxide and hydrogen peroxide/methanol mixtures with air pollutants. In: 26th Symposium (international) on combustion/ Combustion Institute. 1996. p. 2125–32.
[302] Kotler VR, Pein R. Reduction of waste gases without flue gas purification at thermal power plants (USA experience). Fuel Energy Abstr 1995;36:356–62 [in Russian].
[303] Savchenko VI, Didenko LP, Sheverdenkin EV, Rudakov VM, Arutyunov VS. Equilibrium yield products in the steam and oxidative conversion of methane mixtures with C2–C3 hydrocarbons. Khim Fiz 2005;24:76–83 [in Russian].
[304] Karasevich AM, Zimakov BM, Storonskii NM, Khryukin VT. Prospects of development and resource base of coalbed methane in Russia. Gazov Prom 2004:30–5 [in Russian].
[305] Zaidenvarg VE, Troubetskoi KN, Murko VI, Nekhoroshii I Kh. Production and use of coal-water fuel. Moscow: Academy of Mining Sciences; 2001 [in Russian].
[306] Patent JP 55304/1978.
[307] Patent JP 45283/1983.
[308] Patent US 861439; 09.05.1986.
[309] Patent US 4802891; 21.03.1988.
[310] Arutyunov VS. Production of methanol-coal slurries based on the direct oxidation of coalbed methane. Gazov Prom 2012;(672) [special issue]:58–61 [in Russian].
[311] Zherebtsov SI. Interaction of low-metamorphic coals with methanol. Khim Tverd Topl 2007;(3):60–70 [in Russian].
[312] Adschiri T, Sato T, Shibuichi H, Okazaki S, Arai K. Extraction of Taiheiyo coal with supercritical water – HCOOH mixture. Fuel 2000;79:243–8.
[313] Vostrikov AA, Fedyaeva ON, Dubov DYu, Psarov SA, Sokol MYu. Dynamic and static conversion of brown coal in water in supercritical conditions. SKF-TP 2007;2:70–84 [in Russian].
[314] Arpentinier Ph, Cavani F, Trifiro F. The contribution of homogeneous reactions in catalytic oxidation process: safety and selectivity aspects. Catal Today 2005;99:15–22.
[315] Didenko LP, Linde VR, Savchenko VI. Partial catalytic oxidation and condensation of methane by oxygen and sulfur. Catal Today 1998;42:367–70.
[316] Mungen R, Kratzer MB. Partial combustion of methane with oxygen. Ind Eng Chem 1951;43:2782.
[317] Basevich VYa, Vedeneev VI, Arutyunov VS. Study of partial oxidation of hydrocarbons with reference to the production synthesis gas. Theor Found Chem Tech 1996;30:456–60.
[318] Zhu Q, Zhao X, Deng Y. Advances in the partial oxidation of methane to synthesis gas. J Nat Gas Chem 2004; 13:191–203.
[319] Abbott J. Gas heated reforming improves Fisher–Tropsch process. Oil Gas J 2002, Apr. 22:64–9.
[320] Sosna MH, Shovkoplays NYu. Gas heated reforming (GHR) reactors. Quality approved by decades of industrial experience. Gazokhimiya 2011;(3–4):38–40 [in Russian].
[321] Rostrup-Nielsen JR. Fuels and energy for future: the role of catalysis. Catal Rev 2004;46(3–4):247–70.
[322] Papavassiliou V, Pacouloute P, Wu KT, Drnevich R, Vlachos D, Hemmings J, et al. Catalytic partial oxidation pilot plant study. Ind Eng Chem Res 2010;49:94–103.
[323] Makhlin VA, Tsetseruk YaR. Modern technologies for syngas production from natural gas and associated gas. Khim Prom Segodnya 2010;(3):6–17 [in Russian].
[324] Dittrick P. GTL diesel accelerating. Oil Gas J July 24, 2006:17.
[325] Arutyunov VS, Lapidus AL. Gas chemistry as a key trend of development of power-chemical technologies of the 21st century. Ross Khim Zh 2003;(2):23–32 [in Russian].

[326] Chen CM, Bennett DL, Carolan MF, Foster EP, Schinski WL, Taylor DM. ITM syngas ceramic membrane technology for synthesis gas production. In: Bao X, Xu Y, editors. Proceedings of the 7th natural gas conversion symposium, June 6–10, Dalian, China. Studies in surface science and catalysis. Natural gas conversion VII, vol. 147. Amsterdam: Elsevier; 2004. p. 55.
[327] Mazanec T, Perry S, Tonkovich L, Wang Y. Microchannel gas-to-liquids conversion – thinking big by thinking small. In: Bao X, Xu Y, editors. Proceedings of the 7th natural gas conversion symposium, June 6–10, Dalian, China. Studies in surface science and catalysis. Natural gas conversion VII, vol. 147. Amsterdam: Elsevier; 2004. p. 169.
[328] Toledo M, Bubnovich V, Saveliev A, Kennedy L. Hydrogen production in ultrarich combustion of hydrocarbon fuels in porous media. Int J Hydrogen Energy 2009;34:1818–27.
[329] Kolbanovskii YuA, Plate NA. Power units in chemical engineering. Pet Chem 2000;40:289–98.
[330] Cornelissen R, Tober E, Kok J, van de Meer T. Generation of synthesis gas by partial oxidation of natural gas in a gas turbine. Energy 2006;31(15):3199–207.
[331] Arutyunov VS, Shmelev VM, Lobanov IN, Politenkova GG. A generator of synthesis gas and hydrogen based on a radiation burner. Theor Found Chem Tech 2010;44:20–9.
[332] Arutyunov VS, Shmelev VM, Sinev MYu, Shapovalova OV. Syngas and hydrogen production in a volumetric radiation burners. Chem Eng J 2011;176–177:291–4.
[333] Arutyunov VS, Shmelev VM, Shapovalova OV, Rakhmetov AN, Strekova LN. Partial conversion of hydrocarbons to syngas and hydrogen in volumetric radiation burners as a prospective way to enhance the performance characteristics of power engines. In: DeLuca L, Bonnal C, Haidn O, Frolov S, editors. Progress in propulsion physics. EUCASS advances in aerospace sciences book ser, vol. 4. EDP Sci – TORUS PRESS; 2013, ISBN 978-2-7598-0876-2; 2013. p. 489–500.
[334] Shmelev VM, Nikolaev VM, Arutyunov VS. Effective volumetric matrix burners. Gazokhimiya 2009;(4): 28–34 [in Russian].
[335] Sinev MYu, Fattakhova ZT, Lomonosov VI, Gordienko YuA. Kinetics of oxidative coupling of methane: bridging the gap between comprehension and description. J Nat Gas Chem 2009;18:273–87.
[336] Claridge JB, Green MLH, Tsang SC, York APE. Oxidative oligomerisation of methane to aromatics. Appl Catal A 1992;89:103–16.
[337] Holmen A. Direct conversion of methane to fuels and chemicals. Catal Today 2009;142:2–8.
[338] Ichikawa M. Recent progress on the methane-to-benzene catalytic technology and its industrial application. In: Plenary lecture, 8th natural gas conversion symposium, Natal, Brazil May 27–31, 2007.
[339] Dergachov AA, Lapidus AL. Catalytic aromatization of light alkanes. Russ J Gen Chem 2009;79:1244–51.
[340] Noskov AS, Parmon VN. Catalytic technologies to enhance fuel and raw material base of Russia at the expense of non-traditional sources of carbonaceous materials. Gazokhimiya 2008;(2):20–4 [in Russian].
[341] Arutyunov VS, Bykhovsky MYu, Sinev. MYu, Korchak VN. Oxidative conversion of hexane as a model of selective conversion of heavy components of hydrocarbon gases. Russ Chem Bull 2010;59:1528–32.
[342] Arutyunov VS, Magomedov RN. Gas-phase oxypyrolysis of light alkanes. Russ Chem Rev 2012;81(9):790–822.
[343] Magomedov RN, Proshina AYu, Arutyunov VS. Gas-Phase Oxidative Cracking of Ethane in a Nitrogen Atmosphere. Kinet Catal 2013;54(4):383–93.
[344] Magomedov RN, Proshina AYu, Peshnev BV, Arutyunov VS. Effects of the Gas Medium and Heterogeneous Factors on the Gas-Phase Oxidative Cracking of Ethane. Kinet Catal 2013;54(4):394–9.
[345] Arutyunov VS, Vedeneev VI. Methane pyrolysis in temperature range 1000–1700 K. Russ Chem Rev 1991;60: 1384–97.
[346] Keramiotis Ch, Vourliotakis G, Skevis G, Founti MA, Esarte C, Sanchez NE, et al. Energy 2012;43:103.
[347] McEnally CS, Pfefferle LD, Atakan B, Kohse-Hoinghaus K. Prog Energy Combust Sci 2006;32:247.
[348] Belov GP, Novikova EV. Polyketones as alternating copolymers of carbon monoxide. Russ Chem Rev 2004;73: 267–92.
[349] Asche F, Oglend A, Osmundsen P. Gas versus oil prices the impact of shale gas. Energy Policy 2012;47:117–24.
[350] Parker HW. Renewables not adaptable to large-scale installations. Oil Gas J Mar 16, 2009:24–30.
[351] Arutyunov V. Low-scale direct methane to methanol – modern status and future prospects. Catal Today 2013; 215:243–50.
[352] Sandrea R. New tool determines reserves of mature oil and gas fields. Oil Gas J Mar 23, 2009:33–6.

[353] Asadullah M, Taniguchi Y, Kitamura T, Fujiwara Y. One-step carboxylation reaction of saturated hydrocarbons with CO by $Co(OAc)_2$ catalyst under mild conditions. Appl Catal 2000;194:443–52.
[354] Lietti L, Tronconi E, Forzatti P. Synthesis of C_{2+} oxygenates from methanol at atmospheric pressure over alkali-promoted zinc-chromium oxide catalysts. Appl Catal 1991;70:73–86.
[355] Kaminsky EF, Havkin VA, Osipov LN, Kurganov VM. New technologies for production of motor fuels with advanced ecological parameters. Ross Khim Zh 1997;41(6):56–60.
[356] He H, Wang T, Zhu S. Continuous production of biodiesel fuel from vegetable oil using supercritical methanol process. Fuel 2007;86:442–7.
[357] Marschall K-J, Mleczko L. Short-contact time reactor for catalytic partial oxidation of methane. Ind Eng Chem Res 1999;38:1813–21.
[358] Patent SU 1145014; 23.12.81.
[359] Patent SU 1469788; 19.06.86.
[360] Patent SU 1336471; 14.05.85.
[361] Bistolfi M, Dente M, Ranzi E. In: IX International symposium on alcohol fuels. Firenze, 1; Nov 12–15, 1991. p. 105–12.

Appendix 1

A. 1 The Most Reliable Experimental Data on Gas-Phase Oxidative Conversion of Methane to Methanol

P, atm	*T*, °C	$[O_2]$, % *	Conversion of CH_4, %	Selectivity, %							Yield of CH_3OH, %	*t*, s	$\Delta CH_4/O_2$**	Notes	References
				S_{CH3OH}	S_{CH2O}	S_{HCOOH}	S_{CO}	S_{CO2}	S_{H2}	S_{C2+}					
106.4	341	11.0	7.9	22.3	0.75		47.0	31.1			1.76	720	0.72	Static reactor, material: steel	[16]
49	373	11.4	9.0	15.1	0.76		6.9	89.6			1.36	270	0.79		
50	430	3.2		51	4.1	0	22.4	22.4	0			5		Flow reactor, material: steel	[17]
50	410	5.3		29	1.4	0	34.8	34.8	0			5			
141	475	5.8	2.51	45	3.66	0.51					1.13	48	0.43	Flow reactor, material: copper	[87]
141	475	13.6	5.30	29.6	2.48	0.97					1.57	48	0.39		
50	475	3.0	2.19	21.3	0.05	2.8	45.7	29.7			0.47	1.8	0.73	Flow reactor, material: copper	[51]
100	475	3.0	2.51	32.3	0.01	0.5	38.6	26.7			0.81	1.8	0.84		
200	475	3.0	2.80	39.8	0.01	0.5	34.6	24.3			1.11	1.8	0.93		
300	475	3.0	4.00	36.6	0.01	0.2	42.5	7.3			1.46	1.8	1.33		
100	475	3.0	2.50	16.1	0.00	0.8	11.6	69.6			0.40	9	0.83		
200	475	3.0	2.60	26.8	0.01	0.4	41.2	29.2			0.70	9	0.87		
300	475	3.0	3.00	30.3	0.01	0.5	35.7	22.7			0.91	9	1.00		
300	375	3.0	2.93	47.4	0.01	0.7	43.3	9.9			1.39	9	0.98		
300	375	6.0	4.75	34.8	0.02	0.1	52.0	14.7			1.65	9	0.79		
45	800	9.7	~6	22.5	15.2	3.4	33.3	0.0	20.0		~1.0	~0.01	1.25	Adiabatic-compression reactor, conversion of $O_2 = 44\%$	[49]

(*Continued*)

A. 1 The Most Reliable Experimental Data on Gas-Phase Oxidative Conversion of Methane to Methanol—cont'd

P, atm	*T*, °C	$[O_2]$, % *	Conversion of CH_4, %	Selectivity, %							Yield of CH_3OH, %	*t*, s	$\Delta CH_4/O_2$**	Notes	References
				S_{CH3OH}	S_{CH2O}	S_{HCOOH}	S_{CO}	S_{CO2}	S_{H2}	S_{C2+}					
3515	247	7.9	4.06	26.14	0.57	1.01	26.80				*1.06*	5880	*0.51*	Static reactor, material: steel	[46]
3515	253	7.9	6.66	31.55	0.56	0.43	37.02	34.76			*2.10*	8280	*0.84*		
3515	262	7.9	5.21	10.47	0.15	0.93	9.47	78.98			*0.55*	7320	*0.66*		
3339	262	7.9	4.66	29.36	0.83	1.03	39.72	30.95			*1.37*	2460	*0.59*		
3515	264	7.9	6.29	40.10	0.64	0.45	58.13	39.28			*2.52*	420	*0.80*		
3515	265	7.9	6.42	33.29	0.73	0.45	42.31	26.79			*2.14*		*0.81*		
3515	269	7.9	5.32	35.93	0.77	0.89	39.93	26.60			*1.91*	27	*0.67*		
1700	310	8.0									*1.68*	1800		Static reactor, the yield was calibrated at. $[O_2]_0 = 8\%$	[48]
3400	310	8.0									*0.17*	1800			
1	456	10		≤20								~500		Static reactor, material: quartz	[25]
40	438	0.88	0.50	68.55	22.85						0.34	0.80	*0.57*	Flow reactor, material: stainless steel, quenching of products	[40]
40	446	1.29	1.04	67.54	18.76						0.70	0.80	*0.81*		
40	455	2.04	2.10	59.64	17.04						1.25	0.75	*1.03*		
40	461	3.23	2.90	54.40	17.00						1.58	0.75	*0.90*		
40	467	4.71	4.00	48.49	16.72						1.94	0.75	*0.85*		
40	483	7.25	6.10	34.58	13.83						2.11	0.75	*0.84*		

20	485	2.83	1.90	24.77	22.52	0.47	0.60	*0.67*		
40	476	4.19	4.00	52.89	18.89	2.12	0.75	*0.95*		
40	481	5.77	5.60	44.80	19.48	2.51	0.75	*0.97*		
40	486	7.75	6.10	27.91	13.29	1.70	0.71	*0.79*		
40	536	8.63	5.90	18.34	16.67	1.08	0.67	*0.68*		
40	541	15.90	9.80	15.35	17.05	1.50	0.67	*0.62*		
40	542	12.28	8.90	19.69	16.41	1.75	0.44	*0.72*		
80	480	11.47	9.10	28.79	13.71	2.62	0.92	*0.79*		
40	432	3.03	2.20	34.66	13.33	0.76	0.80	*0.73*		
40	455	2.04	2.10	59.64	17.04	1.25	0.75	*1.03*		
40	468	3.03	2.20	31.39	9.81	0.69	0.75	*0.73*		
40	473	5.38	4.00	25.56	10.65	1.02	0.75	*0.74*		
40	477	4.19	4.00	55.49	16.32	2.22	0.75	*0.95*		
40	479	4.59	4.20	52.30	18.05	2.20	0.75	*0.92*		
40	482	6.16	4.30	41.97	14.35	1.80	0.75	*0.70*		
40	488	5.19	4.20	18.32	10.18	0.77	0.75	*0.81*		
40	454	2.60	2.40	54.10	15.10	1.30	1.00	*0.92*		
40	455	2.04	2.10	59.64	17.04	1.25	0.75	*1.03*		
40	457	1.71	1.60	60.45	18.15	0.97	0.63	*0.94*		
40	464	1.79	2.45	49.58	13.12	1.21	0.11	*1.37*		
13	586	3.51	2.21	6.54	26.16	0.14	0.37	*0.63*		
100	430	3		~55		2.67	<1		Flow reactor, material: stainless steel	[358]
100	400	3		~55		2.8	<1			
100	440	3		~55		1.9	<1			

(Continued)

A. 1 The Most Reliable Experimental Data on Gas-Phase Oxidative Conversion of Methane to Methanol—cont'd

P, atm	*T*, °C	$[O_2]$, % *	Conversion of CH_4, %	Selectivity, %							Yield of CH_3OH, %	*t*, s	$\Delta CH_4/O_2$**	Notes	References
				S_{CH3OH}	S_{CH2O}	S_{HCOOH}	S_{CO}	S_{CO2}	S_{H2}	S_{C2+}					
100	340	2.8	4.81	55.71	0.11						2.68	0.22	*1.72*	Pilot plant, flow reactor, material: stainless steel, quenching of products	[359]
100	330	2.8	4.73	68.71	6.83						3.25	0.22	*1.69*		
100	370	2.8	4.77	51.81	6.52						2.47	0.20	*1.70*		
100	330	2.8	4.62	69.28	6.84						3.20	0.22	*1.65*		
50	380	7.5	9.97	25.37	10.13						2.53	~0.2	*1.33*	Pilot plant, flow reactor, material: stainless steel, separate supply of O_2 into 3 points along reactor (2.5% into each point)	[360]
100	380	7.5	9.97	44.83	9.63						4.47	~0.2	*1.33*		
100	380	13.5	14.26	40.39	13.39						5.76	~0.2	*1.06*		
30	350	5	13.36	92.07							12.5	10	*2.67*	Flow reactor, material: Pyrex	[44]
30	350	5	10.79	91.11							9.83	10	*2.16*	Flow reactor, material: Teflon	[44]
30	350	5	4.76	74.10							3.53	10	*0.95*	Flow reactor, material: stainless steel	[44]
25.3	415	8.7	9.9	37.6			54.5	7.9			3.72	99	*1.14*	Flow reactor, material: Pyrex	[42]
24.9	452	8.7	13.3	55.0			38.4	6.6			7.32	81	*1.53*		

26.5	476	8.7	8.9	26.2	61.6	12.2	2.33	87	*1.02*		
25.2	412	9.2	8.6	13.2	70.1	16.7	1.14	45	*0.93*		
25.2	445	9.2	8.7	43.2	56.8	0.0	3.76	44	*0.95*		
25.8	474	9.2	8.7	30.8	59.7	9.5	2.68	43	*0.95*		
33.7	410	9.4	7.7	16.4	74.3	9.3	1.26	111	*0.82*		
36.7	448	9.4	6.0	17.0	69.8	13.8	1.02	102	*0.64*		
35.5	411	8.0	1.6	55.1	36.3	8.6	0.88	281	*0.20*		
34.9	446	8.0	10.2	41.1	40.3	18.6	4.19	262	*1.28*		
33.9	413	5.2	4.7	48.8	51.2	0.0	2.29	107	*0.90*		
33.7	421	5.2	6.5	57.4	37.1	4.9	3.73	105	*1.25*		
36.1	449	5.2	4.1	42.9	57.1	0.0	1.76	108	*0.79*		
35.4	411	2.8	5.9	76.8	23.2	0.0	4.53	112	*2.11*		
35.5	431	2.8	5.7	76.8	23.2	0.0	4.38	109	*2.04*		
36.8	448	2.8	4.5	63.8	31.7	0.0	2.87	110	*1.61*		
50	451	6.7	9.5	76.0	12.6	11.4	7.22	120	*1.42*		
50	451	3.5	5.9	76.7	12.6	10.7	4.53	149	*1.69*		
50	453	8.0	7.5	56.2	23.9	19.3	4.22	93	*0.94*		
34.4	450	2.1	3.9	77.0	10.7	12.3	3.00	71	*1.86*		
34.7	453	6.0	8.1	64.6	20.7	14.7	5.23	112	*1.35*		
34	455	10.3	7.5	59.8	25.1	15.1	4.49	81	*0.73*		
65.6	468	2.6	5.3	81.5	9.6	8.9	4.32	120	*2.04*		
65.3	456	5.1	8.0	83.0	10.7	6.4	6.64	134	*1.57*		
65.4	456	7.4	11.0	66.5	22.6	10.9	7.32	147	*1.49*		
50	450	4.9	7.0	38.1			2.66	150	*1.43*	Flow reactor, material: Pyrex	[45]

(*Continued*)

A. 1 The Most Reliable Experimental Data on Gas-Phase Oxidative Conversion of Methane to Methanol—cont'd

P, atm	*T*, °C	$[O_2]$, % *	Conversion of CH_4, %	Selectivity, %							Yield of CH_3OH, %	*t*, s	$\Delta CH_4/O_2$**	Notes	References
				S_{CH3OH}	S_{CH2O}	S_{HCOOH}	S_{CO}	S_{CO2}	S_{H2}	S_{C2+}					
30	450	2.5	5.0	35.7							1.79	~200	2.00		
20	592	4.6	4.3	6.4	4.5		57.1	5.7	14.4	26.5	0.28	1.1	0.93	Flow reactor, material: Al_2O_3 (Alsint)	[80]
20	610	4.6	4.3	5.5	2.9		54.1	5.7	13.5	32.1	0.23	1.1	0.93		
20	630	4.6	4.3	4.3	2.1		52.8	5.2	12.8	35.1	0.18	1.1	0.93		
20	649	4.6	4.3	3.0	1.9		49.3	5.0	11.3	40.1	0.13	1.1	0.93		
20	670	4.6	4.3	2.2	1.1		47.5	5.2	10.7	43.8	0.09	1.1	0.93		
20	674	4.6	4.3	1.8	0.5		46.7	4.8	10.5	40.7	0.08	1.1	0.93		
20	690	4.6	4.3	1.3	0.4		43.2	4.6	10.3	45.3	0.05	1.1	0.93		
20	709	4.6	4.3	0.9	0.4		41.9	4.8	10.0	46.2	0.04	1.1	0.93		
20	729	4.6	4.3	0.7	0.2		39.0	5.2	10.6	48.0	0.03	1.1	0.93		
20	743	4.6	4.3	0.6	0.2		38.4	5.2	11.3	48.4	0.03	1.1	0.93		
50	350	5.7	10.0	89.5							9.0	264	1.75	Flow reactor, material: Pyrex	[140]
30	379	7.0	12.0	84.5							10.1	354	1.71	Flow reactor, material: stainless steel	[140]
70	448	3.3	3.9	65.7	1.8		27.4	5.1			2.56		1.18	Flow reactor, material: glass	[361]
50	450	5.0	6.84	27.4			60.4	8.8			1.87	9.5	1.37	Flow reactor, filed with quartz chips	[149]
31	630	14.0	9.3	1.8			68.5	21.5		8.2	0.17	0.8	0.56	Flow reactor, filed with quartz chips	[81]

41	585	14.0	9.8	4.0	65.2	11.5	18.4	0.39	0.6	*0.70*		
51	565	14.0	14.5	2.4	59.6	8.8	28.2	0.35	0.6	*1.04*		
62	550	14.0	12.6	1.5	56.7	9.6	31.2	0.19	0.6	*0.90*		
82	535	14.0	13.5	3.0	56.4	6.4	32.4	0.41	0.6	*0.96*		
102	515	14.0	14.8	2.2	54.1	6.1	35.6	0.33	0.7	*1.06*		
82	565	6.9	5.6	12.7	59.4	10.7	16.7	0.71	0.6	*0.81*		
82	565	10.8	8.2	7.5	56.8	7.1	26.3	0.62	0.6	*0.76*		
91	427	1.5	2.12	41.0	55.5	5.2		0.87	35	*1.4*	Flow reactor, material: quartz, averaged data	[98]
91	427	3.0	3.45	38.8	55.5	6.4		1.34	35	*1.2*		
91	427	4.3	4.81	36.4	55.5	7.5		1.75	35	*1.1*		
91	427	5.8	6.10	34.1	55.5	9.3		2.08	35	*1.1*		
91	427	9.0	8.72	28.9	55.5	15.0		2.52	35	*1.0*		
91	427	12.5	11.90	22.6	55.5	22.0		2.69	35	*1.0*		
30	401	2.5	3.1	47.0	44.4	8.4		1.5	1	*1.24*	Flow reactor, material: quartz	[88]
30	407	5.0	4.5	37.7	47.5	14.8		1.7	1	*0.90*		
30	421	7.5	6.4	28.1	55.3	16.6		1.8	1	*0.85*		
30	427	9.5	8.5	23.0	55.1	21.9		2.0	1	*0.89*		
15	438	5.0	4.1	27.7	60.3	12.0		1.1	2	*0.82*		
50	384	5.0	4.6	33.7	50.1	16.1		1.6	0.6	*0.92*		
30	435	5.0	5.7	36.3	46.6	17.2		2.1	0.2	*1.14*		
30	440	7.5	6.5	33.3	48.6	18.1		2.3	0.2	*0.87*		
30	447	9.5	9.0	23.5	57.5	19.0		2.1	0.2	*0.95*		
15	476	9.5	7.7	17.9	63.9	18.2		1.4	0.4	*0.81*		
50	435	9.5	6.7	23.6	49.9	26.6		1.8	0.12	*0.71*		

(*Continued*)

A. 1 The Most Reliable Experimental Data on Gas-Phase Oxidative Conversion of Methane to Methanol—cont'd

P, atm	*T*, °C	$[O_2]$, % *	Conversion of CH_4, %	Selectivity, %							Yield of CH_3OH, %	*t*, s	$\Delta CH_4/O_2$**	Notes	References
				S_{CH3OH}	S_{CH2O}	S_{HCOOH}	S_{CO}	S_{CO2}	S_{H2}	S_{C2+}					
50	437	4.0	4.0	37.4	1.6		50.7	6.6			1.5	5.8	1.00	Flow reactor, material: Pyrex	[150]
50	431	4.0	4.0	39.7	1.8		49.2	6.2			1.6	5.9	1.00		
41	460	3.3	2.2	23.0	0.8		42.6	32.5		1.1	0.5	1.0	0.67	Flow reactor, material: quartz	[53]
30	350	6.5	10.6	71	0		20	9.0			7.5	200	1.6	Flow reactor, material: sapphire, preliminary heating of gas to 250 °C	[43]
50	330	6.5	12.3	83	0		17	0			10.2	345	1.9		
70	330	6.5	11.0	78	0		16	8.6			8.6	483	1.7		
50	350	14.9	23	60	0		29	11			13.8	334	1.5		
100	400	2.8		38.5	6.1		1.40	0.23			1.1	~100		Flow reactor, material: stainless steel, selectivity is given relative to oxygen	[69]
75	400	2.8		39.9	8.0						1.1				
100	400	2.8		37.5	5.6		1.46	0.25			1.1	~100			
150	400	2.8		43.6	6.7		1.42	0.33			1.2				
90	400	2.8		34.9	7.1		1.34	0.32			1.0				
80	400	2.8		31.1	5.6		1.46	0.27			0.9				
70	400	2.8		34.6	5.3		1.57	0.20			1.0				
60	400	2.8		33.2	6.3		1.56	0.23			0.9				

50	400	2.8		29.7	6.2	1.33	0.26	0.8					
200	400	2.8		45.1	6.4	1.37	0.38	1.3					
150	400	2.8		37.6	4.2	1.38	0.36	1.1					
80	400	2.8		34.7	5.5	1.41	0.30	1.0					
75	400	2.8		34.3	6.2	1.52	0.20	1.0					
50	400	2.8		27.7	6.5			0.8					
95	400	2.8		35.0	5.2	1.48	0.38	1.0					
50	400	2.8		27.2	5.7			0.8					
30	400	2.8		17.5	4.9	1.19	0.22	0.5					
175	400	2.8		37.8	4.8	1.41		1.1					
30	400	2.8		16.3	5.2	1.64	0.32	0.5					
125	400	2.8		34.4	4.8	1.38	0.42	1.0					
40	502	2.6	2.3	37.1	5.5	46.0	4.5	0.9	1.1	*0.9*		Flow reactor, material: Al_2O_3 (Alsint), quenching of products	[67]
64	357	9.6	13.2	64.3		20.6	9.3	8.5	433	*1.4*		Flow reactor, material: Pyrex, preliminary heating of gas to 260 °C	[79]
64	358	11.0	15.2	62.2		22.3	9.9	9.5	433	*1.4*			
81	344	4.2	7.2	75.1		11.1	1.6	5.4	178	*1.7*			
81	347	5.7	8.6	69.3		15.2	3.8	5.9	178	*1.5*			
30	400	9.5	8.1	18.7				1.5	10	*0.9*		CSTR reactor, material: Pyrex	[101]
30	400	9.5	8.0	23.0				1.8	20	*0.8*			
30	400	9.5	7.7	23.1				1.8	30	*0.8*			

(*Continued*)

A. 1 The Most Reliable Experimental Data on Gas-Phase Oxidative Conversion of Methane to Methanol—cont'd

P, atm	*T*, °C	$[O_2]$, % *	Conversion of CH_4, %	Selectivity, %							Yield of CH_3OH, %	*t*, s	$\Delta CH_4/O_2$**	Notes	References
				S_{CH3OH}	S_{CH2O}	S_{HCOOH}	S_{CO}	S_{CO2}	S_{H2}	S_{C2+}					
30	400	7.5	5.8	28.0							1.6	20	*0.8*		
30	400	5.0	4.2	36.0							1.5	20	*0.8*		
30	400	2.5	2.2	45.0							1.0	20	*0.9*		
34	400	5.9	5.9	49.2	0.0		39.5	10.6	0.75		2.9	96	*1.1*	Flow reactor, material: quartz	[85]
50	400	5.9	7.1	52.2	0.6		35.5	10.8	0.91		3.7	60	*1.3*		
34	430	5.9	6.8	44.1	2.8		42.6	9.1	1.40		3.0	36	*1.2*		
34	430	5.9	7.0	53.6	1.9		34.7	8.9	0.93		3.8	60	*1.3*		
34	430	5.9	6.4	51.3	0.0		37.4	10.7	0.60		3.3	96	*1.2*		
50	430	5.9	7.3	53.8	1.3		33.8	10.1	0.89		4.0	60	*1.4*		
50	430	5.9	6.5	54.0	0.0		33.7	11.3	1.02		3.5	96	*1.2*		
3	475	17.5	12.2	13.8							1.7		*0.7*	Flow reactor, adsorption extraction of methanol	[105]
50	455	4.2	2.6	50.3							1.3		*0.6*		
50	450	9.1	13.2	62.3	0		32.6	4.8			8.2		*1.5*	Flow reactor, material: quartz, complete isolation of reaction from metallic surfaces.	[41]
50	450	9.1	9.2	31.6	0		50	18.8			2.9		*1.0*		
50	450	9.1	11.7	41.2	0		41.5	15.4			4.8		*1.3*		
50	460	9.1	9.1	40.5			59.4				3.7		*1.0*	Flow reactor, material: quartz	[86]

50	490	9.1	9.5	29.4		67.8			2.8		*1.0*		
4	575	7.0	5.5	20.0	15.5	58.9	4.8	0.5	1.1	2.5	*0.8*	Flow reactor, material: quartz	[68]
5	550	7.0	5.5	21.4	10.1	63.3	4.4	0.5	1.2	3.2	*0.8*		
6	525	7.0	5.6	22.9	7.5	59.9	5.6	1.0	1.3	4.0	*0.8*		
10	525	7.0	5.4	24.7	0.0	65.5	5.2	4.2	1.3	6.6	*0.8*		
20	450	7.0	5.3	24.3	0.0	59.9	8.0	3.8	1.3	14.5	*0.8*		
30	450	7.0	5.4	20.9	0.0	60.4	12.6	4.0	1.1	21.8	*0.8*		
100	410	1.73	2.2	74.6	0.0				1.6	17.6	*1.3*	Flow reactor, material: quartz	[113]
100	410	4.66	5.3	53.7	0.0				2.8	17.4	*1.1*		

* *relative to the sum* $CH_4 + O_2$.
** *estimations based on presented data (italic).*

Appendix 2

A.2 Data on the Yield of Products at DMTM Pilot Tests

				Yield, g/m^3 of Methane Passed Through Reactor													
P atm	*T*, °C	$[O_2]$, % *	Yield of Liquid	H_2O	CH_3OH	CH_2O	HCOOH	C_2H_5OH	C_3H_7OH	C_4H_9OH	$(CH_3)_2CO$	$CH_3C_2H_5CO$	$(CH_3)_2O$	$CH_3C_2H_5O$	*t*, s**	Notes	References
106.4	341	11.0			*25.1*	*0.7*									720	Static reactor, material: steel	[16]
49	373	11.4			*19.2*	*0.9*									270		
50	475	3.0	22.1	15.9	6.6	0.16	1.24								1.8	Flow reactor, material: copper. The yield of CH_3OH is given with the account of its incomplete extraction from outgoing gas.	[51]
100	475	3.0	30.5	19.5	11.5	0.25	0.27								1.8		
200	475	3.0	33.5	19.6	15.8	0.29	0.27								1.8		
300	475	3.0	41.9	22.6	20.8	0.27	0.20								1.8		
100	475	3.0	18.0	13.9	5.7	0.03	0.42								9		
200	475	3.0	19.2	12.1	9.9	0.18	0.19								9		
300	475	3.0	24.3	14.9	12.9	0.19	0.30								9		
300	375	3.0	40.3	22.7	19.7	0.5	0.4								9		
300	375	6.0	77.5	55.7	23.5	1.2	0.1								9		
100	370	2.8	35.6	17.35	14.45	1.85	0.17	0.57	0.061	0.050	0.32	0.08	0.39	0.31	*<1*	Flow reactor, material: stainless steel. Natural gas with C_2+ content 5%. The sum of aldehydes and the sum of acids are given.	[76]
100	390	2.8	37.6	14.36	16.34	3.65	0.19	0.73	0.060	0.056	0.63	0.08	0.67	0.54	*<1*		
100	410	2.8	39.6	12.62	19.30	4.16	0.21	0.83	0.063	0.059	0.72	0.09	0.85	0.69	*<1*		
100	430	2.8	36.1	12.68	16.25	4.62	0.22	0.66	0.054	0.051	0.46	0.07	0.58	0.47	*<1*		
100	450	2.8	29.0	11.89	12.18	3.87	0.21	0.22	0.038	0.032	0.18	0.04	0.19	0.15	*<1*		
50	410	2.8	31.4	14.77	11.04	4.29	0.16	0.27	0.022	0.019	0.18	0.16	0.27	0.22	*<1*		
60	410	2.8	33.3	14.78	12.67	4.25	0.16	0.35	0.027	0.023	0.24	0.17	0.35	0.28	*<1*		
70	410	2.8	34.4	14.20	14.13	4.16	0.17	0.43	0.031	0.028	0.30	0.15	0.43	0.36	*<1*		
80	410	2.8	37.3	14.55	16.37	4.12	0.18	0.51	0.037	0.034	0.38	0.14	0.54	0.44	*<1*		

(Continued)

A.2 Data on the Yield of Products at DMTM Pilot Tests—cont'd

				Yield, g/m^3 of Methane Passed Through Reactor													
P atm	*T*, °C	[O_2], % *	Yield of Liquid	H_2O	CH_3OH	CH_2O	HCOOH	C_2H_5OH	C_3H_7OH	C_4H_9OH	$(CH_3)_2CO$	$CH_3C_2H_5CO$	$(CH_3)_2O$	$CH_3C_2H_5O$	*t*, s**	Notes	References
90	410	2.8	39.0	13.66	18.66	4.21	0.20	0.63	0.043	0.039	0.45	0.12	0.66	0.53	<1		
100	410	2.8	39.5	12.59	19.25	4.15	0.21	0.83	0.063	0.059	0.71	0.09	0.86	0.69	<1		
100	410	2.0	31.7	12.14	14.64	2.20	0.14	0.60	0.051	0.041	0.34	0.07	0.69	0.79	<1		
100	410	2.4	38.0	13.57	18.30	2.77	0.17	0.75	0.065	0.053	0.42	0.09	0.86	0.96	<1		
100	410	2.8	39.5	11.79	20.06	3.98	0.21	0.68	0.059	0.040	0.55	0.07	0.92	1.15	<1		
100	410	3.2	45.0	13.62	21.86	4.97	0.23	0.56	0.054	0.036	0.66	0.05	1.41	1.54	<1		
100	410	3.6	49.2	14.35	22.84	6.42	0.24	0.35	0.039	0.010	0.88	0.04	1.95	2.09	<1		
100	410	4.0	52.6	15.15	24.08	7.24	0.22	0.34	0.026	0.005	0.97	0.02	2.14	2.40	<1		
100	400	2.8	34.6	17.13	14.29	2.11	0.25	0.48			0.40				~100	Flow reactor, material: stainless steel.	[69]
75	400	2.8	38	19.57	14.33	2.68	0.24	0.58			0.51						
100	400	2.8	35.6	18.23	14.17	1.97	0.28	0.49			0.42				~100		
150	400	2.8	39.4	19.46	16.31	2.36	0.30	0.66			0.41						
90	400	2.8	34.3	17.49	13.07	2.47	0.20	0.58			0.50						
80	400	2.8	31.6	17.10	11.63	1.96	0.25	0.37			0.36						
70	400	2.8	32.1	15.89	12.90	1.85	0.24	0.46			0.41						
60	400	2.8	31.6	17.19	11.57	2.07	0.15	0.27			0.22						
50	400	2.8	30.8	16.85	11.03	2.14	0.23	0.26			0.34						
200	400	2.8	41.4	21.57	16.60	2.20	0.36	0.52			0.27						
150	400	2.8	32.1	16.02	13.58	1.41	0.22	0.50			0.34						
80	400	2.8	33.2	16.67	13.21	1.96	0.27	0.51			0.51						
75	400	2.8	35.1	18.71	13.06	2.21	0.34	0.38			0.43						
50	400	2.8	27.7	14.14	10.55	2.31	0.14	0.35			0.25						

95	400	2.8	33.3	17.28	13.35	1.87	0.27	0.31	0.28			
50	400	2.8	29.5	16.56	10.37	2.05	0.17	0.20	0.24			
30	400	2.8	22.2	13.56	6.60	1.74	0.11	0.10	0.16			
175	400	2.8	35.1	17.11	14.53	1.72	0.15	0.69	0.43			
30	400	2.8	26.5	17.32	6.41	1.91	0.21	0.17	0.19			
125	400	2.8	34.1	17.26	13.44	1.77	0.14	0.54	0.48			
15	550	3.6	37.0	26.1	8.6	2.3				~2	Flow reactor, material: quartz.	[91]
20	550	3.6	37.0	25.8	9.1	2.2				~2		
25	550	3.6	36.4	24.0	9.8	2.6				~2		
30	550	3.6	35.7	23.1	9.8	2.8				~2		
40	550	3.6	36.7	23.5	10.9	2.3				~2		
40	550	3.6	42.0	25.3	13.7	3.0				~2		
50	550	3.6	48.3	30.1	15.7	2.5				~2		
50	550	3.6	51.4	31.9	17.0	2.5				~2		
60	550	3.6	45.3	25.8	17.4	2.2				~2		
60	550	3.6	50.9	31.1	17.4	2.5				~2		
70	550	3.6	50.4	27.9	20.0	2.5				~2		
80	550	1.4	16.3	10.5	4.7	1.2				~2		
80	550	2.6	41.2	26.0	10.9	4.3				~2		
80	550	4.0	61.4	42.0	15.5	3.9				~2		
80	550	4.5	67.2	47.0	15.9	4.3				~2		
80	550	5.1	74.2	51.3	18.7	4.3				~2		
80	550	6.2	78.5	51.3	21.4	5.8				~2		
80	550	6.7	96.7	73.0	20.6	3.1				~2		

(Continued)

A.2 Data on the Yield of Products at DMTM Pilot Tests—cont'd

			Yield, g/m^3 of Methane Passed Through Reactor														
P atm	*T*, °C	$[O_2]$, % *	Yield of Liquid	H_2O	CH_3OH	CH_2O	HCOOH	C_2H_5OH	C_3H_7OH	C_4H_9OH	$(CH_3)_2CO$	$CH_3C_2H_5CO$	$(CH_3)_2O$	$CH_3C_2H_5O$	*t*, s**	Notes	References
80	550	7.1	79.6	59.4	14.8	5.4									~2		
80	550	7.7	102.2	79.3	20.2	2.7									~2		
80	550	8.4	102.2	78.1	20.2	3.9									~2		
80	550	8.7	99.9	78.1	17.5	4.3									~3		
80	550	9.2	99.5	79.3	17.5	2.7									~2		
80	550	9.8	98.7	81.2	15.5	1.9									~2		
80	550	2.3	62.6	45.8	15.5	1.2									~2	Flow reactor, material: stainless steel.	[91]
80	550	2.4	75.8	56.3	17.9	1.6									~2		
80	550	3.6	53.6	38.5	14.4	0.8									~2		
80	550	3.6	66.4	48.2	17.1	1.2									~2		
80	550	3.6	72.6	50.5	20.6	1.6									~2		
80	550	3.9	105.3	77.3	26.8	1.2									~2		
80	550	4.8	88.2	72.7	14.0	1.6									~2		
80	550	5.9	93.3	78.9	13.2	1.2									~2		

* *relative to the sum* $CH_4 + O_2$.
** *estimations based on presented data (italic).*

Index

Note: Page numbers followed by "f" denote figures; "t" tables.